# CONTENTS

# To The Student

This is the second volume of Pure Mathematics for CAPE. I hope you read the introduction to the first volume as it contained lots of useful information about studying maths.

In case you didn't here are some of the important things you need to know about the Mathematics for CAPE series.

The books are written for students in the Caribbean preparing for the CAPE examination in Pure and Applied Mathematics. They are closely based on the CAPE syllabus. Volume 1 of Pure Maths was originally based on the 2004 syllabus (CXC A6/U2/04) while this volume is based on the 2007 syllabus (CXC A6/U2/07). There is therefore a slight overlap between the volumes, but essentially all topics are covered, and this volume follows the order of the 2007 syllabus for Unit 2, with a chapter on each of the topics, as well as an introductory chapter on modelling.

Each chapter contains teaching material, worked examples and exercises.

Definitions are marked with the symbol

Results and rules that you should learn are marked with the symbol

In worked examples, the method is usually set out step by step with each step introduced with a bullet mark ➢ so that you can follow the logic of the solution.

Proofs that you may need to reproduce are marked ✳

Hints are prefixed with the symbol

and similar, are warnings of common mistakes and pitfalls, marked $\boxed{!}$

Sometimes you are invited to discover results for yourselves, in which case there is an investigation $\boxed{?}$ set which is usually intended for group work.

There is great benefit in group work:

- ❖ it enables more to be found out in a shorter time

- ❖ discussing a problem among yourselves often enables you to find something out that you couldn't have done on your own

- ❖ you discover strengths in others in the group that you didn't realise they had, and you may well gain in confidence too, as you see the value of your own contribution

- ❖ you learn the value of working as a team

Sometimes an exercise concentrates on just one subtopic, and sometimes it covers 2 or 3 subtopics. Questions marked # are optional and are included for those who want a challenge.

The last exercise in each chapter consists entirely of questions taken from past CAPE questions, it is always labelled with an X – e.g. exercise 15X.

Multiple choice questions are not included in individual chapters, but Appendix 3 at the back of the book consists of multiple choice questions on the whole syllabus, subdivided into the 3 modules.

At the end of each chapter there is a summary of all the important facts and formulae that have been covered in the chapter.

Following this introduction you will find a list of symbols and abbreviations used in this book and in exam papers, and the definitions of some common mathematical terms. Also included are the pages of the official CAPE formula sheet that are relevant to the Pure Mathematics syllabus. These are the formulae that will be provided for you in the CAPE exams. Any other formulae that you need to use you will have to learn by heart.

At the end of each chapter there is a summary of the main points covered in that chapter; all the necessary formulae are included in this summary.

At the end of the volume, there are answers to the exercises. **Make sure you use these wisely.** When you are working an exercise, check the first few answers before completing the exercise. If you are getting the answers correct, great! but if you are getting them wrong it is important to identify your mistake before getting into bad habits.

Finally, if you are working through this book on your own, you may want to mix the chapters around a bit. For instance, chapters 18 – 21 are not dependent on previous chapters and can be used to break up some of the longer chapters. In particular, chapter 17 on integration is too long to complete without interruption. I would recommend taking a break before the section on differential equations (p 72). The first part (Counting) of chapter 23 (which is needed for chapter 20) could be tackled at this point.

Good luck!

*Elizabeth Campbell*
December 2007

# Symbols, Abbreviations and Definitions

## SYMBOLS

### Set Notation

| | |
|---|---|
| $\in$ | is an element of |
| $\notin$ | is not an element of |
| $\{x : \ldots\}$ | the set of all $x$ such that |
| n(A) | the number of elements in set A |
| $\varnothing$ | the empty or null set |
| U | the universal set |
| A′ | the complement of the set A |
| $\mathbb{W}$ | the set of whole numbers $\{0, 1, 2, 3, \ldots\}$ |
| $\mathbb{N}\,(=\mathbb{Z}^+)$ | the set of natural numbers (counting numbers or positive integers) $\{1, 2, 3, \ldots\}$ |
| $\mathbb{Z}$ | the set of integers |
| $\mathbb{Q}$ | the set of rational numbers |
| $\mathbb{R}$ | the set of real numbers |
| $\mathbb{C}$ | the set of complex numbers |
| $\subset$ | is a subset of |
| $\not\subset$ | is not a subset of |
| $\cup$ | union |
| $\cap$ | intersection |
| $[a, b]$ | the closed interval $\{x \in \mathbb{R} : a \leq x \leq b\}$ |
| $(a, b)$ | the open interval $\{x \in \mathbb{R} : a < x < b\}$ |
| $[a, b)$ | the interval $\{x \in \mathbb{R} : a \leq x < b\}$ |
| $(a, b]$ | the interval $\{x \in \mathbb{R} : a < x \leq b\}$ |

### Miscellaneous Symbols

| | |
|---|---|
| $\therefore$ | therefore |
| $\because$ | because |
| : | such that |
| $\neq$ | is not equal to |
| $\approx$ | is approximately equal to |
| $\equiv$ | is equivalent or identical to |

| | |
|---|---|
| $\perp$ | is perpendicular to |
| // or $\parallel$ | is parallel to |
| $\rightarrow$ | approaches, tends to |
| $\Rightarrow$ or $\rightarrow$ | implies |
| $A \Rightarrow B$ | A implies B |
| $A \Leftrightarrow B$ | A implies B and B implies A (equivalence) |
| $\exists$ | there exists |
| $\infty$ | infinity |
| $\Delta$ | the discriminant of a quadratic expression: $b^2 - 4ac$ for $ax^2 + bx + c$ |
| | the determinant of a matrix |

## Operations

| | |
|---|---|
| $\displaystyle\sum_{i=1}^{n} a_i$ | $a_1 + a_2 + a_3 + \ldots + a_n$ |
| $\sqrt{x}$ | the positive square root of the real number $x$ |
| $\lvert a \rvert$ | the modulus of $a$ — the value of $a$ with no sign attached |
| $n!$ | $n$ factorial   i.e. $n(n-1)(n-2)\ldots 2 \cdot 1$   $(0! = 1)$ |
| $\dbinom{n}{r}$ | the binomial coefficient $\dfrac{n!}{r!(n-r)!}$ for $n \in \mathbb{W}$ |
| | or   $\dfrac{n(n-1)(n-2)\ldots(n-r+1)}{r!}$ for $n \in \mathbb{Q}$ |
| $^{n}C_{r}$ | the number of combinations of $r$ from $n$,   $= \dbinom{n}{r}$ |
| $^{n}P_{r}$ | the number of permutations of $r$ from $n$,   $= \dfrac{n!}{(n-r)!}$ [also sometimes P $(n, r)$] |

## Functions

| | |
|---|---|
| f | the function f |
| $f(x)$ | the value of the function f at $x$ |
| $f : A \rightarrow B$ | the function f under which each element of the set A has an image in the set B |
| $f : x \mapsto y$ or $f : x \rightarrow y$ | the function f maps the element $x$ to the element $y$ |
| $f^{-1}$ | the inverse of the function f |
| fg | the composite function $f\{g(x)\}$ |
| $\displaystyle\lim_{x \to a} f(x)$ | the limit of $f(x)$ as $x$ tends to $a$ |
| $\delta x$ | an increment of (small increase in) $x$ |

$\dfrac{dy}{dx}$, $y'$ the derivative of $y$ with respect to $x$

$\dfrac{d^n y}{dx^n}$, $y^{(n)}$ the $n$th derivative of $y$ with respect to $x$

$f'(x)$, $f''(x)$, ..., $f^{(n)}(x)$ the first, second, ..., $n$th derivatives of $f(x)$ with respect to $x$.

e the exponential constant

$\ln x$ the natural logarithm of $x$ ($\log_e x$)

$\log x$, $\lg x$ the common logarithm of $x$ ($\log_{10} x$)

## Complex Numbers

$i$ $\sqrt{-1}$

$z$ a complex number, $z = x + iy$, where $x, y \in \mathbb{R}$

Re $z$ the real part of $z$ ($x$)

Im $z$ the imaginary part of $z$ ($y$)

$|z|$ the modulus of $z$ ($\sqrt{x^2 + y^2}$)

arg $z$ the argument of $z$ ($\tan^{-1}(y/x)$), where $-\pi < \arg z \le \pi$

$\bar{z}$ or $z^*$ the complex conjugate of $z$ ($x - iy$)

## Vectors

$\underline{a}$, $\mathbf{a}$, $\overrightarrow{AB}$ vectors

$\hat{\mathbf{a}}$ a unit vector in the direction of the vector $\mathbf{a}$

$|\mathbf{a}|$ the magnitude of the vector $\mathbf{a}$

$a \cdot b$ the scalar product of the vectors $\mathbf{a}$ and $\mathbf{b}$

$\mathbf{i}, \mathbf{j}, \mathbf{k}$ unit vectors in the directions of the positive Cartesian coordinate axes

$\begin{pmatrix} x \\ y \\ z \end{pmatrix}$ $x\mathbf{i} + y\mathbf{j} + z\mathbf{k}$

## Probability

S the sample space

A, B, ... the events A, B, ...

P(A) the probability of the event A

## Matrices

M a matrix M

$M^{-1}$ inverse of the non-singular matrix M

| | |
|---|---|
| $M^T$, $M_T$ | transpose of the matrix M |
| det M, \| M \| | determinant of the square matrix M  (sometimes $\Delta$) |

## COMMON ABBREVIATIONS

| | |
|---|---|
| LHS | left hand side |
| RHS | right hand side |
| tgt | tangent |
| eqn | equation |
| st | such that |
| wrt | with respect to |
| dp | decimal places |
| sf | significant figures |
| iff | if and only iff   ( $\Leftrightarrow$ ) |

## DEFINITIONS

| | |
|---|---|
| abscissa | the perpendicular (horizontal) line from a point P on a curve to the $y$-axis |
| | or   the $y$-coordinate of P  (c.f. ordinate) |
| absolute value | the positive value of any pair of opposite numbers |
| asymptote | a straight line that a curve continually approaches but never touches |
| axiom | a statement accepted as true without proof, an assumption, premise or postulate |
| binomial | a polynomial of two terms |
| coefficient | the constant which multiplies a power of a variable. |
| | e.g. in the expression $ax^2 + bx + c$,  $a$ and $b$ are the coefficients of $x^2$ and $x$ |
| collinear | 3 or more points are collinear if one straight line can be drawn through all of them |
| concurrent | 3 or more lines are concurrent if they all pass through the same point |
| constant | a fixed quantity, a variable with just one value |
| degree (1) | a measure of angle |
| degree (2) | *of a variable*: the power to which that variable is raised |
| | *of a term which is the product of more than 1 variable*: the sum of the degrees of the variables |
| | *of a polynomial*:  the greatest of the degrees of the terms |

| | |
|---|---|
| determinant | an expression associated with a square matrix |
| discriminant | an expression associated with a quadratic function ($b^2 - 4ac$ for $ax^2 + bx + c$) |
| explicit | *of a function*: a function where the dependent variable can be expressed in terms of the independent variable alone e.g. $y = f(x)$ (see implicit) |
| exponent | same as index |
| hypothesis | a given assumption |
| identity | an equation which is a true statement for every value of the variable(s) |
| implicit | *of a function*: a function where the dependent variable is not expressed as a function of the independent variable e.g. $x^2 + y^2 = 4$ (see explicit) |
| index | a symbol denoting a power |
| induction | a method of proof that shows the general validity of an assertion from the specific |
| modulus | the absolute value of a number, length of a vector, etc |
| non-singular | *of a matrix*: a matrix which has an inverse, i.e. whose determinant $\neq 0$ |
| numeral | a name or symbol for a number |
| ordinate | the perpendicular (vertical) line from a point P on a curve to the $x$-axis or the $x$-coordinate of P (c.f. abscissa) |
| power | the product of a number of equal factors a quantity subject to continual increase or decrease |
| quadrant | the coordinate axes divide the plane into 4 quadrants, and more particularly the four quarters of a circle. |
| reciprocal | the multiplicative inverse of a number: the reciprocal of $a$ is $\dfrac{1}{a}$ |
| root (1) | the inverse of a power, e.g. if $x^3 = 2$, then $x$ is the cube root of 2 |
| root (2) | the roots of an equation $f(x) = 0$ are the values of $x$ for which $f(x) = 0$ |
| surd | a square (or other) root that is an irrational number e.g. $\sqrt{2}$ (a radical) |
| tangent | a line that touches a curve (usually without intersecting it) |
| theorem | an assertion that is proved |
| variable | a quantity which may have an infinite number of values in the same expression |

# Mathematical Formulae

(These are the formulae contained in the formula booklet supplied for use during CAPE examinations. These and other formulae that you need to know are highlighted in the text as they are introduced, and are also included in the summaries at the end of each chapter.)

## Algebra

For the equation $ax^2 + bx + c = 0,$ $\quad x = \dfrac{-b \pm \sqrt{(b^2 - 4ac)}}{2a}$

For an arithmetic series:

$$u_n = a + (n-1)d, \qquad S_n = \frac{n}{2}\{2a + (n-1)d\}$$

For a geometric series:

$$u_n = ar^{n-1} \qquad\qquad S_n = \frac{a(r^n - 1)}{r - 1},\ r > 1, \quad S_n = \frac{a(1 - r^n)}{1 - r},\ r < 1.$$

Binomial Expansion:

$$(a+b)^n = a^n + \binom{n}{1}a^{n-1}b + \binom{n}{2}a^{n-2}b^2 + \ldots + b^n,\ \text{where } n \text{ is a positive integer}$$

Summations:

$$\sum_{r=1}^{n} r = \tfrac{1}{2}n(n+1), \qquad \sum_{r=1}^{n} r^2 = \tfrac{1}{6}n(n+1)(2n+1)$$

# Trigonometry

Sine Rule: $\dfrac{a}{\sin A} = \dfrac{b}{\sin B} = \dfrac{c}{\sin C}$.

Cosine rule: $\qquad a^2 = b^2 + c^2 - 2bc \cos A$.

Arc length of a circle: $s = r\theta$,   ($\theta$ in radians)

Area of a sector of a circle: Area $= \frac{1}{2}r^2\theta$,   ($\theta$ in radians)

Area of a triangle: Area $= \frac{1}{2}\,ab \sin C = \frac{1}{2}\,bc \sin A = \frac{1}{2}\,ca \sin B$.

## Trigonometric Identities:

$\cos^2 \theta + \sin^2 \theta = 1$,     $1 + \tan^2 \theta = \sec^2 \theta$,     $1 + \cot^2 \theta = \operatorname{cosec}^2 \theta$

$\sin (\alpha \pm \beta) = \sin A \cos \beta \pm \cos \alpha \sin \beta$

$\cos (\alpha \pm \beta) = \cos A \cos \beta \mp \sin \alpha \sin \beta$

$\tan (\alpha \pm \beta) = \dfrac{\tan \alpha \pm \tan \beta}{1 \mp \tan \alpha \tan \beta}$

$\cos 2\alpha = 2 \cos^2 \alpha - 1 = 1 - 2 \sin^2 \alpha$

$\sin 2\alpha = 2 \sin \alpha \cos \alpha$

$\sin \alpha + \sin \beta = 2\sin \dfrac{\alpha + \beta}{2} \cos \dfrac{\alpha - \beta}{2}$

$\sin \alpha - \sin \beta = 2\cos \dfrac{\alpha + \beta}{2} \sin \dfrac{\alpha - \beta}{2}$

$\cos \alpha + \cos \beta = 2\cos \dfrac{\alpha + \beta}{2} \cos \dfrac{\alpha - \beta}{2}$

$\cos \alpha - \cos \beta = 2\sin \dfrac{\alpha + \beta}{2} \sin \dfrac{\alpha - \beta}{2}$

If $\tan \dfrac{\alpha}{2} = t$, then $\sin \alpha = \dfrac{2t}{1+t^2}$ and $\cos \alpha = \dfrac{1-t^2}{1+t^2}$

# CHAPTER 14
# Mathematical Modelling

## WHAT IS MATHEMATICAL MODELLING?

Mathematical modelling is something you have done for most of your life and you do it almost every day, even though you are not usually aware of it.

When you tender the correct change to pay for something in the market, when you look at a graph or statistics in the newspaper, when you estimate how long it will take to walk to the doctor's... you are using mathematical models.

So, what do we mean by a mathematical model? The first thing to realise is that we are not using the word 'model' as we would when we talk of a model car or aeroplane, whether a toy or a mock-up by a designer. It means, rather, a representation, which can take any of many different forms.

A mathematical model translates a real life, or a hypothetical, situation into mathematical language – equations, formulae, graphs, diagrams, tables...

Here are some simple examples of mathematical models:

## Equation:

Hire purchase: A computer costing $700 can be purchased by paying a deposit of $100 and $n$ monthly instalments of $x$ each. This situation can be represented by the model $100 + nx = 700$. If the purchaser knows how long he wants to take to repay he can calculate the amount of each instalment from the model; or if she knows how much she can afford to pay each month she can calculate how long it will take to repay the balance.

## Table:

Same situation as above: To avoid having to do the arithmetic, the salesman could remodel the 'problem' as a table:

| no. of instalments | 1 | 2 | 3 | 4 | 5 | 6 | 8 | 9 | 10 | 12 |
|---|---|---|---|---|---|---|---|---|---|---|
| value of each instalment ($) | 600 | 300 | 200 | 150 | 120 | 100 | 75 | 66.67 | 60 | 50 |

## Rule of Thumb:

There are often situations where an approximate answer can be found using a well-known formula. For instance to roast a chicken most cooks would reckon to cook it for 20 minutes per pound plus an extra 15-20 minutes.

## Graph:

A doctor wants to know how his patient is progressing so he gets the nurse to take the patient's temperature once every 4 hours and record the results on a graph:

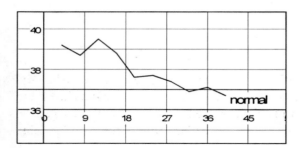

The above examples are very straightforward, but many real life problems are complicated and may need suitable strategies to find appropriate solutions to them so

*In problem solving*, mathematical modelling consists of analysing the situation or problem, identifying its essential features, deciding what mathematical model seems appropriate, performing any necessary mathematics and evaluating the results in the light of reality. Often the real world problem will need some simplification before a workable model can be found. If the mathematical results do not conform to the real life situation it may be possible to modify the model or it might be necessary to choose a different model.

The process of mathematical modelling can be shown in diagrammatic form

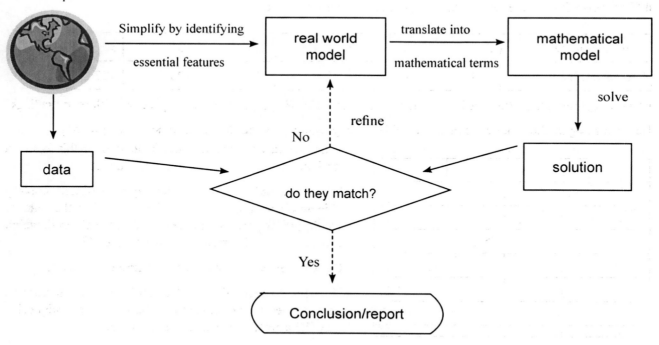

### An illustration of this process:

A javelin thrower wants to know at what angle to project his javelin to get the maximum range for his throw.

For the first model, the mathematician makes the following simplifications: treat the javelin as a particle, ignore air/wind resistance, regard the playing field as flat. He now has an imperfect real world model. Then he sets up equations from mechanics theory (using the data for this javelin thrower) to get his mathematical model.

He solves the equations. The thrower throws at different angles and a comparison is made between what really happens and what the mathematician expects to happen. If the results are close, the model is accepted and used for other throwers (different height, different velocity of projection etc), but if the results are widely different the mathematician must refine his model. He may need to model the javelin as a rod, or make allowance for resistance...

## APPLICATION OF MATHEMATICS TO PRACTICAL SITUATIONS

Throughout this course you will meet, or have already met, situations where an appropriate model must be used to solve a given problem. For instance in Unit 1 you used the principles of calculus to solve problems involving maxima and minima, you saw that problems involving tides could be solved using sine curves. Again, a scientist can often verify whether his experimental results conform to a particular form of equation by using logarithms to transform the equation to give a straight line.

Because you are working for an examination you will seldom be asked to do the modelling process from scratch for yourself at this stage, but you need to be aware that the mathematical tools that you are learning will in future be used in the modelling process. Since real life situations tend to be complex there almost always has to be simplification before the mathematical model is proposed. It is therefore important to be aware of what simplifications you are making, and why, so that the essential features of the problem are retained. Testing the theoretical results against the real life situation is also essential, so that the model can be revised if necessary before a solution is given.

## Example 14.1

In a manufacturing process the manufacturer wants to know how long it will take his product to cool under different conditions. He records the temperature at regular intervals.

| Time in minutes ($t$) | 5 | 10 | 15 | 20 | 25 | 30 |
|---|---|---|---|---|---|---|
| Temperature ($T$) | 508 | 416 | 340 | 279 | 228 | 187 |

where T represents the difference in temperature between the cooling object and the temperature of the surroundings.

He thinks maybe the product cools so that T is directly proportional to the time, so he plots a graph:

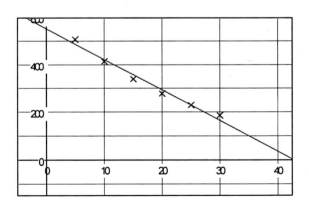

When he has plotted the graph he sees that it is not a straight line, so he has to change his model.

He understands that under the right conditions cooling objects obey Newton's law of cooling, that the **rate of cooling** is proportional to the time that has elapsed, which can be expressed $T = Ae^{-kt}$ . (See chapter 17)

He wants to check whether this is true in this case.

How can he check this? If his results can be transformed so that when the graph is plotted he will get a straight line he will know that he has the right model.

He therefore takes the equation $T = Ae^{-kt}$ and takes logarithms of both sides

$\ln T = \ln (Ae^{-kt}) = \ln A - kt \ln e.$     But $\ln e = 1$ so

$\ln T = \ln A - kt$    or   $\ln T = -kt + c$

If he plots $\ln T$ against $t$, instead of $T$ against $t$, he should get his straight line.

He adds an extra line to his table and then plots the graph:

| Time in minutes (t) | 5 | 10 | 15 | 20 | 25 | 30 |
|---|---|---|---|---|---|---|
| Temperature ($T$) | 508 | 416 | 340 | 279 | 228 | 187 |
| ln $T$ | 6.23 | 6.03 | 5.83 | 5.63 | 5.43 | 5.23 |

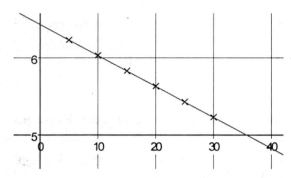

This time he has a straight line and he can interpret the graph to enable him to find the optimum conditions for his process.

In exercise 14A you will be asked to choose an appropriate model for the given situation and, where necessary, to state any assumptions or simplifications that you make in choosing the model, so that if this were a real life situation you would know where modifications would need to be made if the model proved unsatisfactory in the light of experience. You should be able to answer these questions using techniques you already know from CSEC and CAPE Unit 1.

The questions in exercise 14B involve techniques from the whole syllabus for Pure Mathematics Units 1 and 2, so you will probably want to leave the exercise until after you have finished the rest of the course.

## Choosing the Model

Often one of the hardest parts of solving a problem is choosing the most appropriate model. This is especially true in the real world. For this course, the only models that are likely to be needed are ones whose solution you know how to find, which considerably restricts the choice available!

Here are a few guidelines to help the choice:

❖ the question will often tell you what model to use

❖ if it doesn't:
   o rates of change usually lead to differential equations
   o fractional or percentage increases/decreases usually lead to series*
   o if a maximum or minimum value is required, differentiation to find stationary values could be the right approach

❖ differential equations and differentiation can only be applied to continuous functions

❖ answers in the question may display a recognisable pattern:
   o e to a power or ln usually arise from differential equations
   o other numbers to a power may come from a GP

 Read the question through very carefully, several times if necessary. Look for the trigger words mentioned above.

\* Although fractions and percentages will often suggest a series model, occasionally exponential growth is given in the form of a percentage increase. e.g. if a population, P, increases by $x$% per year, then $\dfrac{dP}{dt} = \dfrac{x}{100}P$. To decide which is the correct model, look for other indications in the question.

# EXERCISE 14A

Consider the following situations and decide what model you would use to solve them, and what assumptions or simplifications you have made when making that choice.

1.  A family wishes to enclose an area in which to keep their animals. They need to keep the goats, the dogs and the chickens separate, so will divide the enclosure into three sections:

    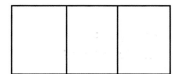 They have 200 m of straight fencing available and want to know what shape to make the enclosure to provide the largest possible area for the animals. What should the external dimensions of the enclosure be?

2.  The language of education in Curacao is Dutch, but most inhabitants speak one or more other languages as well.

    (a)  It is thought that 90% of the islanders speak Papiamento, 55% speak Spanish, and 49% speak both languages. What percentage of the population speak neither of these languages?

    (b)  It is also thought that 40% speak English and 12% speak Portuguese. 5% speak both English and Portuguese. What percentage of the population speak neither of these languages?

3.  After long study, tree scientists conclude that a eucalyptus tree will grow at a rate of $0.2 + 1.3t^{-4}$ m per year, where $t$ is the time in years. Find the number of metres that the tree will grow in the second year.

4.  $5000 is invested in a fund paying 4.8 % p.a. interest, paid twice-yearly. If the interest accumulates in the fund, what will the fund be worth after $t$ years?

The answers to exercise 14A can be found at the end of the chapter (page 9)

# EXERCISE 14B (To be worked on completion of the course)

1.  The rate at which oil is leaking from an engine sump at any time $t$ seconds is proportional to the volume of oil, V, in the sump at that instant. At time $t = 0$, V = A

    (a)  Show that $V = A\,e^{-kt}$ where $k$ is a positive constant.
    (b)  Sketch a graph to show the relation between V and $t$.
    (c)  Given further that $V = \frac{1}{2}A$ at $t = T$, show that $kT = \ln 2$.

2.  The population of a colony of insects increases in such a way that if it is N at the beginning of a week, then at the end of the week it is $a + bN$, where $a$ and $b$ are constants and $0 < b < 1$.

    Starting from the beginning of the week when the population is N, write down an expression for the population at the end of 2 weeks. Show that at the end of $c$ consecutive weeks the population is

    $$a\left(\frac{1-b^c}{1-b}\right) + b^c N.$$

    When $a = 2000$ and $b = 0.2$, it is known that the population takes about 4 weeks to increase from N to 2N. Estimate a value for N from this information.

3.   A man borrows a sum of money from a building society and agrees to pay the loan (plus interest) over a period of years.  If \$A is the sum borrowed and $r\%$ the yearly rate of interest charged show that the amount (\$ $P_n$) of each annual instalment which will extinguish the loan in $n$ years is given by the formula

$$P_n = \frac{A(R-1)R^n}{R^n - 1} \quad \text{where } R = 1 + \frac{r}{100}.$$

Assuming this formula calculate $P_n$ (correct to the nearest \$) if $A = 1000$, $n = 25$, $r = 6.5$.

Find, in its simplest form, the ratio $P_{2n} : P_n$.

4.   Oil flows into a cylindrical drum of radius 10 metre.  At time $t = 0$ the drum was empty, and at time $t$ the oil level in the drum is rising at a rate $10\left(\dfrac{1}{1+t} - \dfrac{1}{4}\right)$ metres per second.

(a)   What is the rate of increase in the volume $V$ of oil in the drum?

(b)   Sketch the graph of $f(t) = 10\left(\dfrac{1}{1+t} - \dfrac{1}{4}\right)$ for $0 \le t \le 3$

(c)   At what time is the rate of increase of oil level zero?

(d)   Write down an expression for $V$ at the time $t = 3$

(e)   Evaluate $V$ by ordinary integration

5.   In a model of mortgage repayment, the sum of money  owed to the Building Society is denoted by $x$ and time is denoted by $t$.  Both  $x$  and  $t$  are taken to be continuous variables.  The sum of money owed to the Building Society increases, due to interest, at a rate proportional to the sum of money owed.  Money is also repaid at a constant rate $r$.  When $x = a$, interest and repayment balance.

Show that, for $x > 0$, $\dfrac{dx}{dt} = \dfrac{r}{a}(x - a)$.  Given that when  $t = 0$, $x = A$, find  $x$  in terms of $t, r, a$ and A.

On a single clearly labelled sketch, show the graph of  $x$  against  $t$  in the two cases

(i)   $A > a$,

(ii)   $A < a$.

State the circumstances under which the loan is repaid in a finite time T (say) and show that, in this case

$$T = \frac{a}{r}\ln\left(\frac{a}{a - A}\right).$$

6.   The organiser of a sale, which lasted 3 hours and raised a total of \$1000, attempted to create a model to represent the relationship between $s$ and $t$, where  \$ $s$ is the amount which had been raised at time $t$ hours after the start of the sale.  In the model  $s$  and  $t$  were taken to be continuous variables.  The organiser assumed that the rate of raising money varied directly as the time remaining and inversely as the amount already raised.

Show that, for this model, $\dfrac{ds}{dt} = k\left(\dfrac{3-t}{s}\right)$ where  $k$  is a constant.

Solve this differential equation and show that the solution can be written in the form

$$\frac{s^2}{1000^2} + \frac{(3-t)^2}{3^2} = 1.$$

Hence

(a)   find the amount raised during the first hour of the sale.

(b)   find the rate of raising money 1 hour after the start of the sale.

7. The rate of destruction of a drug by the kidneys is proportional to the amount of the drug present in the body. The constant of proportionality is denoted by $k$. At time $t$ the quantity of drug in the body is $x$.

Write down a differential equation relating $x$ and $t$, and show that the general solution is $x = Ae^{-kt}$, where A is an arbitrary constant.

Before $t = 0$ there is no drug in the body, but at $t = 0$ a quantity Q of drug is administered. At $t = 1$ the amount of drug in the body is Qα where α is a constant such that $0 < \alpha < 1$.

Show that $x = Q\alpha^t$.

Sketch the graph of $x$ against $t$ for $0 < t < 1$.

When $t = 1$ and again when $t = 2$ another dose Q is administered. Show that the amount of drug in the body immediately after $t = 2$ is $Q(1 + \alpha + \alpha^2)$.

8. At time $t$ seconds the volume of a balloon is V m³. At any instant the rate of change of ln V w.r.t. $t$ is inversely proportional to the square root of $t$. Show that $V = ae^{b\sqrt{t}}$ where $a$ and $b$ are constants.

9. In places where very heavy rainfall occurs, 'storm drains' are used to collect rainwater and prevent flooding. A storm drain can be modelled as a rectangular tank with vertical sides, so that the volume of water in the tank is proportional to the depth of water. During a storm, rain water flows into the tank at a constant rate. Water drains away from the bottom of the tank at a rate proportional to the square root of the depth of water in the tank. At time $t$ hours the depth of water in the tank is $h$ metres. Explain how the information given above leads to the differential equation $\dfrac{dh}{dt} = A - B\sqrt{h}$, where A and B are positive constants.

When rainfall stops and no more water is flowing into the tank, the value of A is 0. For this case, find the value of B, given that it takes $\frac{1}{2}$ hour for the depth of water in the tank to fall from 1 m to 0.

In another case, $A = 2$ and $B = 1$, so that the differential equation is $\dfrac{dh}{dt} = 2 - \sqrt{h}$.

   (a) By means of the substitution $x = 2 - \sqrt{h}$, or otherwise, find $\displaystyle\int \dfrac{1}{2 - \sqrt{h}}\,dh$.

   (b) Given that the tank is initially empty, find the time taken for the depth of water in the tank to reach 1 m.

10. An employer offers the following schemes of salary payments over a five year period:
Scheme X: 60 monthly payments, starting with $1000 and increasing by $6 each month
[$1000, $1006, $1012, …]
Scheme Y: 5 annual payments, starting with $12 000 and increasing by $d each year
[$12 000, $(12 000 + d), …]

   (a) Over the complete five year period, find the total salary payable under scheme X.

   (b) Find the value of $d$ which gives the same salary for both schemes over the complete five year period.

11. According to UN data, the world population at the beginning of 1990 was approx 5.3 billion and growing at a rate of about 2% per year. Assuming an exponential growth model, estimate the world population at the beginning of the year 2015.

12. An investment of $2000 is made at the start of a year with a Finance Company. At the end of this year and at the end of each subsequent year the value of the investment is 11% greater than its value at the start of the year.

   (a) Find, to the nearest $, the value of the investment at the end of

      (i) the 5th year,

      (ii) the 10th year.

A client decided to invest $2000 at the start of EACH year.

(b)    Write down a series whose sum is the total value of this annual investment at the end of 12 years.

(c)    By finding the sum of your series, determine, to the nearest $, the value of the investment at the end of 12 years.

13.    In established forest fires, the proportion of the total area of forest which has been destroyed is denoted by $x$, and the rate of change of $x$ w.r.t. time, $t$ hours, is called the destruction rate. Investigations show that the destruction rate is directly proportional to the product of $x$ and $(1-x)$. Write down a differential equation in $x$ and $t$. A particular fire is initially noticed when one quarter of the forest is destroyed, and it is found that 12 hours later one half of the forest has been destroyed. Show that the constant of proportion is $\frac{1}{12}\ln 3$. Find how long it will take for 99% of the forest to be destroyed.

14.    Water is entering a tank at a constant rate of $p$ m³s⁻¹. At the same time the water is leaving the tank through a tap at the bottom at a rate proportional (with constant of proportion $k$) to the depth, $h$, of water in the tank. Given that the horizontal cross-section of the tank is 1 m², and that initially the depth of water in the tank is 2 m

(a)    form a differential equation for the rate of change in the depth of water in the tank..

(b)    solve the equation in (a)

(c)    Explain the physical significance of the situations when $k < \frac{1}{2}p$ and $k > \frac{1}{2}p$.

15.    The volume of liquid $V$ cm³ in a container when the depth is $x$ cm is given by

$$V = \frac{x^{\frac{1}{4}}}{(x+2)^{\frac{1}{2}}}, \ x > 0.$$

(a)    Find $\dfrac{dV}{dx}$ and determine the value of $x$ for which $\dfrac{dV}{dx} = 0$.

(b)    Calculate the rate of change of volume when the depth is 1 cm and increasing at a rate of 0.01 cm s⁻¹, giving you answer in cm³ s⁻¹ to 3 s.f.

16.    A plant grows in a pot which contains a volume $V$ of soil. At time $t$ the mass of the plant is $m$ and the volume of soil utilised by the roots is $\alpha m$, where $\alpha$ is a constant. The rate of increase of the mass of the plant is proportional to the mass of the plant times the volume of soil not yet utilised by the roots. Obtain a differential equation for $m$, and verify that it can be written in the form

$$V\beta \frac{dt}{dm} = \frac{1}{m} + \frac{\alpha}{V - \alpha m}, \ \text{where } \beta \text{ is a constant.}$$

The mass of the plant is initially $\dfrac{V}{4\alpha}$. Find, in terms of $V$ and $\beta$, the time taken for the plant to double its mass.

17.    During a spell of freezing weather in Canada, the ice on a pond has thickness $x$ mm at time $t$ hours after the start of freezing. At 3.00 p.m., one hour after the start of freezing weather, the ice is 2mm thick and it is desired to predict when it will be 4 mm thick.

(a)    In a simple model, the rate of increase of $x$ is assumed to be constant. For this model, express $x$ in terms of $t$ and hence determine when the ice will be 4 mm thick.

(b)    In a more refined model, the rate of increase of $x$ is taken to be proportional to $\frac{1}{x}$. Set up a differential equation for $x$ involving a constant of proportion $k$.

Solve the differential equation and hence show that the thickness of ice is proportional to the square root of the time elapsed from the start of freezing.

Determine the time at which the second model predicts that the ice will be 4 mm thick.

(c) What assumption about the weather underlies both models.

18. A patch of oil pollution in the sea is approximately circular in shape. When first seen its radius was 100 m and its radius was increasing at a rate of 0.5 m per minute. At a time $t$ minutes later, its radius is $r$ metres. An expert believes that, if the patch is untreated, its radius will increase at a rate which is proportional to $1/r^2$.

(a) Write down a differential equation for this situation, using a constant of proportion $k$.

(b) Using the initial conditions, find the value of $k$. Hence calculate the expert's prediction of the radius of the oil patch after 2 hours.

The expert thinks that if the oil patch is treated with chemicals then its radius will increase at a rate which is proportional to $\dfrac{1}{r^2(2+t)}$.

(c) Write down a differential equation for this new situation and, using the same initial conditions as before, find the value of the new constant of proportion.

(d) Calculate the expert's prediction of the radius of the treated oil patch after 2 hours.

19. The loss of heat from a hot water tank is proportional to the total surface area of the tank. It is therefore important to design a tank with the minimum possible surface area. For a closed cylindrical tank with radius $r$ cm and volume $50\,000$ cm³, show that an expression for the total surface area A cm² is

$$A = 2\pi r^2 + \frac{100\,000}{r}.$$

Use differentiation to find the value of $r$ corresponding to a stationary value of A and show that this stationary value is a minimum. Find the minimum value of A and determine whether this is greater than or less than the total surface area of a cubical tank with the same volume of $50\,000$ cm³.

20. A bottle is shaped so that when the depth of water is $x$ cm, the volume of water in the bottle is $(x^2 + 4x)$ cm², $x \geq 0$. Water is poured into the bottle so that at time $t$ s after pouring commences, the depth of water is $x$ cm and the rate of increase of the volume of the water is $(x^2 + 25)$ cm³ s⁻¹.

(a) Show that $\dfrac{dx}{dt} = \dfrac{x^2 + 25}{2x + 4}$.

(b) Given that the bottle was empty at $t = 0$, solve the differential equation to obtain $t$ in terms of $x$.

# ANSWERS TO EXERCISE 14A

You may have chosen a different model (there is often more than one model which will work in a given situation) and you may have thought of additional assumptions and limitations.

1. Let the width of the enclosure be $x$, making the length $100 - 2x$ and the area $100x - 2x^2$. The maximum area can be found by calculus: $x = 25$, length 50 m and area 1250 m².

We don't know what lengths the fencing comes in. If it comes in 2 m lengths which cannot easily be divided in half, the width will have to be either 24 or 26 m, leading to a slightly smaller area.

There could be trees on the property so that these dimensions could not be fitted in conveniently.

2.   Probably the easiest model to use here is a Venn diagram.

There is a problem with the wording – when the question speaks of the percentages who speak both languages, does it mean in addition to, or included in the numbers who speak the individual languages? In (a) there is no ambiguity because the sum of the percentages is more than 100%, so it must mean included in ( $P \cap S$ ), but if we only had part (b) it would be ambiguous.

3.   This is a rate of change question, so if the height of the tree at time $t$ is $h$ the model will be $\dfrac{dh}{dt} = 0.2 + 1.3t^{-4}$. This can be solved by integration.

A problem here could be the generality of the formula given. For a particular tree the soil conditions, the rainfall, the temperature, pests etc could all affect growth.

4.   This looks like a straightforward compound interest problem. But if the interest is paid twice-yearly does this mean that 2.4 % is paid every six months (which would lead to an annual interest rate of nearly 4.86 %) or is 4.8 % per annum achieved by paying approx 4.744 every six months?

Another problem (especially if the second alternative is used) is what happens if the investment is withdrawn at any time between the interest payment dates.

# CHAPTER 15

# 2.1a  Exponential and Logarithmic Functions

Think about the following questions – you don't have to solve them all

❖ You put $1 on the first square of a chess board, $2 on the next, $4 on the next, and so on for all 64 squares.  How much money have you placed on the board?  (First, guess the answer, and then calculate it if you can – how accurate were you?)

❖ You have won a TV quiz but there is one final question.  How do you want to receive the money?  There are two choices:

(a) you can have $100,000 the first year, 10% less the next year, 10% less the next year and so on for 10 years,  or

(b) you can have $10,000 the first year, 50% more the next year, 50% more the next year and so on for 10 years.

Which option would you choose, and why?

❖ The radioactive output of an imaginary element Zilium is 463 units, but it decreases by 50% every day.  How many days will it take for the radioactive output to be within the "safe level" of 4 units?

❖ A ball is dropped from a height of 1.5 m, at each bounce it rises to $\frac{3}{4}$ the height of the previous bounce.  How many times will it bounce before the height it achieves is less than 5 cm?

❖ A cup of tea is left to get cold.  What will its cooling curve look like?

What do all these questions have in common?

*They are all examples of what is known as exponential growth or decay.*

## DEFINITIONS

**Exponent** is another word for power or index, so an **exponential function** is a function where the variable is in the index.  For instance, in the first example, if $N$ is the number of dollars on the $n$th square, then $N = 2^{n-1}$.   $(n \in \mathbb{W})$

In general, an exponential function is in the form  $y = a^x$.  It occurs very frequently in real life situations, and, as in the examples above, at least one of the variables will often be a discrete variable.  However, in many situations, like a population of bacteria in a petri dish, the numbers are large enough to permit approximating the function to a continuous one.  The study of discrete functions has a branch of mathematics all to itself, but we will be dealing mainly with continuous exponential functions.  Before we discover some of the other interesting properties of exponential functions we will take a look at their inverses.

If the three numbers $n$, $b$ and $L$ are connected by the relation
$$n = b^L \quad \text{①}$$
then the index $L$ is said to be the **logarithm** of $n$ to **base** $b$ and we write
$$\log_b n = L \quad \text{②}$$
Alternatively we can say that the logarithm ($L$) of $n$ to the base $b$ is the power to which $b$ must be raised to give $n$. Statements ① and ② are equivalent, so

$$\boxed{b^L = n \iff \log_b n = L} \quad \text{③}$$

11

# THE RULES OF LOGARITHMS

These follow directly from the rules of indices and the equivalence ③ and they are

$$\log_a (xy) = \log_a x + \log_a y \qquad \text{(i)}$$

$$\log_a \left( \frac{x}{y} \right) = \log_a x - \log_a y \qquad \text{(ii)}$$

$$\log_a x^n = n \log_a x \qquad \text{(iii)}$$

$$\log_a a = 1 \qquad \text{(iv)}$$

$$\log_a 1 = 0 \qquad \text{(v)}$$

Here are the proofs:

If $p = \log_a x$ and $q = \log_a y$ then $a^p = x$ and $a^q = y$

So (i) $\qquad xy = a^p \times a^q = a^{p+q}$
giving (using ③) $\qquad \log_a (xy) = p + q$
$\qquad$ i.e. $\log_a (xy) = \log_a x + \log_a y$

also (ii) $\qquad \dfrac{x}{y} = \dfrac{a^p}{a^q} = a^{p-q}$

giving (using ③) $\qquad \log_a \left( \dfrac{x}{y} \right) = p - q$

$\qquad$ i.e. $\log_a \left( \dfrac{x}{y} \right) = \log_a x - \log_a y$

and (iii) $\qquad x^n = a^{pn}$
giving (using ③) $\qquad \log_a x^n = np$
$\qquad$ i.e. $\log_a x^n = n \log_a x$

also the two important corollaries:

(iv) If $r = \log_a a$ then $a^r = a$ i.e. $r = 1$ $\qquad$ so $\log_a a = 1$
(v) If $s = \log_a 1$ then $a^s = 1 = a^0$ i.e. $s = 0$ $\qquad$ so $\log_a 1 = 0$

# CONVENTIONS

Logarithms can have any positive number as base. The most common bases for logarithms are 10 and e, and these logarithms are to be found on scientific calculators.

Logarithms to base 10 are often called common logarithms and written as **lg** (or simply **log**, as on calculators), and base e logarithms are called natural or Napierian logarithms and written as **ln**. (We will discover about this important irrational number e later in this chapter) The rules derived above are true for any base, so long as the same base is used throughout.

## Example 15.1

Find $x$ if $\log_4 64 = x$
$\qquad \log_4 64 = x \quad \Leftrightarrow \quad 4^x = 64$
$\qquad\qquad\qquad\qquad\qquad 4^x = 4^3 \quad \therefore \ x = 3$

## Example 15.2

Express $\lg\left(\dfrac{10\sqrt{b}}{a}\right)$ in terms of $\lg a$ and $\lg b$

➤ Use rules (i) and (ii): $\qquad \lg\left(\dfrac{10\sqrt{b}}{a}\right) = \lg 10 + \lg\sqrt{b} - \lg a$

➤ Use rules (iii) and (iv): $\qquad\qquad\qquad = 1 + \tfrac{1}{2}\lg b - \lg a$

## Example 15.3

Express $\log 8 + 2\log 3 - \log 4$ as a single logarithm

$\log 8 + 2\log 3 - \log 4 = \log 8 + \log 9 - \log 4$

$$= \log\left(\frac{8\times 9}{4}\right)$$

$$= \log 18$$

# EXERCISE 15 A

1. Write each of the following statements in the form $a^x = y$

   (a) $\log_a c = d$ $\qquad$ (b) $\log_b s = t$ $\qquad$ (c) $\lg p = q$ $\qquad$ (d) $\ln x = y$

2. Find the value of $x$ in each of the following:

   (a) $\log_2 16 = x$ $\qquad$ (b) $\log_3 27 = x$ $\qquad$ (c) $\log_x 216 = 3$

   (d) $\log_x 16 = 0.5$ $\qquad$ (e) $\lg x = -2$ $\qquad$ (f) $\log_4 x = 3\tfrac{1}{2}$

   (g) $\log_8 32 = x$ $\qquad$ (h) $\log_9 27 = x$ $\qquad$ (i) $-\log_{16} x = 0.75$

3. Evaluate the following:

   (a) $\log_2 32$ $\quad$ (b) $\log_6 36$ $\qquad$ (c) $\log_5 125$ $\quad$ (d) $\log_2(\tfrac{1}{2})$ $\quad$ (e) $\log_2(\tfrac{1}{4})$ $\quad$ (f) $\log_a 1$

   (g) $\log_a a$ $\quad$ (h) $\log_4 2$ $\qquad$ (i) $\log_8(\tfrac{1}{2})$ $\quad$ (j) $\log_{27} 9$ $\quad$ (k) $\log_8 4$ $\quad$ (l) $\log_{27}(\tfrac{1}{3})$

4. Express the following in terms of $\log a$, $\log b$ and $\log c$

   (a) $\log(abc)$ $\qquad$ (b) $\log(a^2bc)$ $\qquad$ (c) $\log(ab^3c)$ $\qquad$ (d) $\log(b\sqrt{c})$ $\qquad$ (e) $\log\left(\dfrac{ab}{c}\right)$

   (f) $\log\left(\dfrac{b}{ac^3}\right)$ $\qquad$ (g) $\log\left(\dfrac{1}{ab}\right)$ $\qquad$ (h) $\log\left(a\sqrt{\dfrac{b}{c^3}}\right)$ $\qquad$ (i) $\lg(10a)$ $\qquad$ (j) $\lg\left(\dfrac{\sqrt{10a}}{b^2}\right)$

5. Express as a single logarithm

   (a) $2\log_a 5 + \log_a 4 - \log_a 10$ $\qquad$ ✱(b) $3\log_a 3 - \log_a 15 + 2\log_a 5$

   (c) $\log_a 12 - (\tfrac{1}{2}\log_a 9 + \tfrac{1}{3}\log_a 8)$ $\qquad$ (d) $2\log_a(\tfrac{1}{2}) + 3\log_a 4 - \log_a(\tfrac{4}{5})$

6. Express each of the following in the form $\log_a[f(x)]$ (i.e. as a single logarithm)

   (a) $3\log_a x$ $\qquad\qquad\qquad$ (b) $\log_a x + \log_a(x-1)$

   (c) $\log_a(x+2) - \log_a 2$ $\qquad$ (d) $\log_a(x^2-1) - \log_a(x-1)$

   (e) $3\log_a x - \log_a x(x-1)$ $\qquad$ (f) $2\log_a x + \log_a(x+1)$

   (g) $3\log_a x - \log_a(2x^2+x^3)$ $\qquad$ (h) $\tfrac{1}{2}\log_a x + \log_a(x+1) - \log_a\sqrt{x}$

## CHANGE OF BASE

It is sometimes useful to express the logarithm of $a$ to the base $b$ in terms of logarithms to some other base $c$.

Suppose $\qquad\qquad\qquad\qquad y = \log_b a$

Then $\qquad\qquad\qquad\qquad\qquad a = b^y$

So, by taking $\log_c$ of both sides, $\quad \log_c a = \log_c b^y$

$$= y \log_c b$$

$$\therefore \quad y = \frac{\log_c a}{\log_c b} \qquad\qquad \text{i.e. } \log_b a = \frac{\log_c a}{\log_c b}$$

also, putting $c = a$ gives $\qquad \log_b a = \frac{\log_a a}{\log_b a} \qquad\qquad \text{i.e. } \log_b a = \frac{1}{\log_a b}$

$$\log_b a = \frac{\log_c a}{\log_c b}$$

$$\log_b a = \frac{1}{\log_a b}$$

## Example 15.4

Evaluate $\log_3 8$

➢ change to base 10: $\quad \log_3 8 = \dfrac{\log_{10} 8}{\log_{10} 3}$

➢ evaluate: $\qquad\qquad = \dfrac{0.90309}{0.47712} = 1.89$

# USING LOGARITHMS TO SOLVE EXPONENTIAL EQUATIONS

The idea used in the previous section, of taking logarithms of both sides of an equation, is particularly useful when solving exponential equations.

## Example 15.5

Find $x$ if $2^x = 5$

We know that $2^2 = 4$, and $2^3 = 8$, so $x$ must lie somewhere between 2 and 3, but it would be tedious to find its exact value by trial and improvement.

So, take logarithms of both sides of the equation (any base can be used, but since you will be using a calculator, use either base 10 or base e)

➢ take logs of both sides $\qquad\qquad \lg(2^x) = \lg 5$

➢ use the third law of logarithms (iii) $\qquad x\lg 2 = \lg 5$

➢ solve for $x$ $\qquad\qquad\qquad\qquad x = \dfrac{\lg 5}{\lg 2} = 2.322$

## Example 15.6

Solve for $x$ the equation $3^{2x+1} = 5^{4-x}$

- ➢ Take logarithms of both sides:   $(2x + 1)\log 3 = (4 - x)\log 5$
- ➢ Multiply out:   $2x \log 3 + \log 3 = 4 \log 5 - x \log 5$
- ➢ Rearrange to make $x$ the subject:   $2x \log 3 + x \log 5 = 4 \log 5 - \log 3$

$$x = \frac{4\log 5 - \log 3}{2\log 3 + \log 5}$$

- ➢ Evaluate:   $x = 1.40$

## Example 15.7

Solve for $x$ the equation $3^{2x} - 8(3^x) + 15 = 0$

- ➢ Let $y = 3^x$ to give a quadratic in $y$:   $y^2 - 8y + 15 = 0$
- ➢ Factorise:   $(y - 3)(y - 5) = 0$
- ➢ Solve for $y$:   $y = 3$ or $y = 5$
- ➢ Resubstitute $x$:   $3^x = 3$ or $3^x = 5$
- ➢ Solve for $x$:   $x = 1$ or $x = \dfrac{\log 5}{\log 3} = 1.465$

# EXERCISE 15 B

1. Find the value of each of the following logarithms (to 3 s.f.) by changing them to base 10 logarithms, and then evaluating using your calculator

    (a) $\log_3 18$    (b) $\log_5 2$    (c) $\log_9 25$    (d) $\log_4 0.2$    (e) $\log_7 17$    (f) $\log_5 14$    (g) $\log_{12} 21$

2. Find $x$ from the equations

    (a) $4^x = 3$        (b) $7^x = 11$        (c) $11^x = 5$        (d) $10^x = 7$        (e) $12^x = 19$

3. Solve the following equations for $x$

    (a) $2^x = 9$        (b) $3^{x+1} = 15$        (c) $3^{x+1} = 2^x$        (d) $5^x = 8^{x-1}$        (e) $4^{x+1} = 3^{2-x}$

4. Solve the following equations for $x$

    (a) $2^{2x} - 5(2^x) + 6 = 0$        (b) $3^{2x} - 3^{x+2} + 8 = 0$        (c) $4^x - 13(2^x) + 40 = 0$

    (d) $5^{2x} - 7(5^x) + 10 = 0$        (e) $5^{2x} + 6 = 5^{x+1}$        (f) $2^{2x+1} - 5(2^x) + 2 = 0$

    (g) $3^x - 5 + (3^{-x}) = 0$        (h) $2^x + 3(2^{-x}) = 4$

## EXPONENTIAL FUNCTIONS

# EXERCISE 15 C

1. On the same axes, plot the graphs of the functions $y = 2^x$, $y = 3^x$ and $y = 4^x$ in the interval $[-3, 4]$.  What do you notice about these curves?  (A list of some of the things you might have noticed is given in the answers)

2. *Investigation* : What can you discover about the gradient of an exponential curve $y = a^x$ ?

    $\boxed{?}$   At the moment we can't differentiate $a^x$, and we don't have the algebra to work out the derivative from first principles.  So in this investigation we try to find the limit of the gradient of the curve at a number of points of the curve.

15

Divide the class into 3 groups, and each group take one of the curves $y = 2^x$, $y = 3^x$ and $y = 4^x$.

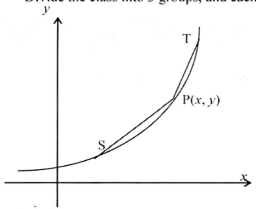

To find the gradient at P we need to find the limit of the gradient of one of the chords SP as S → P or PT as T → P. We cannot find this exactly, but we can as accurately as we need. Clearly the chord SP is slightly less steep than the tangent at P, and TP is slightly steeper, so the gradient at P will lie between the gradients of the chords.

The gradient of TP is $m = \dfrac{f(x+h) - f(x)}{h}$ where $h$ is a small increase in $x$.

Calculate $m$ for increasingly small values of $h$ (0.1, 0.01 …) and then find the corresponding values for $m$ for SP. You now have a fair approximation for the gradient at P. Do this for $x = 0, 1, 2, 3, 4, 5$, then tabulate your results including corresponding values for $x$, $y$ and $\dfrac{dy}{dx}$ by completing the table below.

| | | $y = 2^x$ | | $y = 3^x$ | | $y = 4^x$ | |
|---|---|---|---|---|---|---|---|
| $x$ | $y$ | $m = \dfrac{dy}{dx}$ | $y$ | $m = \dfrac{dy}{dx}$ | $y$ | $m = \dfrac{dy}{dx}$ |
| 0 | 1 | | 1 | | | |
| 1 | 2 | | 3 | | 4 | |
| 2 | 4 | | | | | |
| 3 | | | 27 | | | |
| 4 | | | | | 256 | |
| 5 | 32 | | 243 | | | |

See what patterns you can find from your table, then compare with the results found by the other two groups. (a complete table with values correct to 6 d.p. is given at the end of this chapter on p. 24).

# 'THE' EXPONENTIAL FUNCTION $y = e^x$

Did you notice that for each curve $y = a^x$, $\dfrac{dy}{dx} = ky$ for some constant $k$?

Looking at the tables of results we see that for $y = 2^x$, $k = 0.693147$, for $y = 3^x$, $k = 1.098612$, and for $y = 4^x$, $k = 1.386294$. This suggests that there should be a number between 2 and 3 for which $k = 1$. This number is given the letter e and its value is actually 2.718281828459…

Like π, e is an irrational number which keeps cropping up in mathematics. We found e by putting $k = 1$; in other words we have defined the exponential function $y = e^x$ as the function for which the gradient at any point is equal to the value of the function (the $y$-coordinate).

  Any function of the form $y = a^x$ is known as **an** exponential function

the function $y = e^x$ is known as **the exponential function**, and

$$\text{if } y = e^x, \text{ then } \frac{dy}{dx} = e^x$$

Because the exponential function is so important, its inverse function, $y = \log_e x$ is given special notation $y = \ln x$, as we saw earlier in the chapter.

### #*The Exponential Function as a Series*

We have defined the exponential function as that function whose derivative at any point is equal to the value of the function itself: $f'(x) = f(x)$. How can we be sure that such a function exists, and, if it does, find the value of e to any required degree of accuracy?

Let us suppose that there exists a polynomial function $f(x)$ for which the derivative $f'(x) \equiv f(x)$.

Then if $f(x) = a_0 + a_1 x + a_2 x^2 + a_3 x^3 + a_4 x^4 + \ldots + a_r x^r + \ldots$        ①

$\qquad f'(x) = a_1 + 2a_2 x + 3a_3 x^2 + 4a_4 x^3 + \ldots + ra_r x^{r-1} + \ldots$

Since we require that $f'(x) \equiv f(x)$, we get

$\qquad a_1 + 2a_2 x + 3a_3 x^2 + 4a_4 x^3 + \ldots \equiv a_0 + a_1 x + a_2 x^2 + a_3 x^3 + a_4 x^4 + \ldots$

Equating coefficients gives: $a_1 = a_0$ , $a_2 = \frac{1}{2}a_1$ , $a_3 = \frac{1}{3}a_2$ , $a_4 = \frac{1}{4}a_3$ , etc

or $\quad a_1 = a_0$ , $a_2 = \frac{1}{2}a_0$ , $a_3 = \frac{1}{2}\cdot\frac{1}{3}a_0$ , $a_4 = \frac{1}{2}\cdot\frac{1}{3}\cdot\frac{1}{4}a_0$ , $\ldots.. a_r = \dfrac{1}{r!}a_0$ (where $r! = 1\cdot2\cdot3\cdot4\cdot5\ldots(r-1)r$)

Now $f(x)$ is the exponential function $f(x) = e^x$.        ②

When $x = 0$, ① becomes $f(0) = a_0$ and ② becomes $f(0) = e^0 = 1$ $\therefore$ $a_0 = 1$
the coefficients are therefore $a_0 = 1$, $a_1 = 1$, $a_2 = \frac{1}{2}$ , $a_3 = \frac{1}{2}\cdot\frac{1}{3}$ , $a_4 = \frac{1}{2}\cdot\frac{1}{3}\cdot\frac{1}{4}$ etc.

generalising and using factorial notation (see chapter 23) $a_r = \dfrac{1}{r!}$ (where $r! = 1\cdot2\cdot3\cdot4\cdot5\ldots(r-1)r$ )

and $\qquad \boxed{e^x = 1 + \frac{1}{1!}x + \frac{1}{2!}x^2 + \frac{1}{3!}x^3 + \cdots + \frac{1}{r!}x^r + \cdots}$

To find the value of e we need to put $x = 1$. This gives $e = 1 + \frac{1}{1!} + \frac{1}{2!} + \frac{1}{3!} + \cdots + \frac{1}{r!} + \cdots$

It is beyond the scope of this course to show that this series converges to a limit, but if you put the first few terms into your calculator you will find that $e \approx 2.7183$ to 4 d.p.

## Graphs of the Exponential and Logarithmic Functions

| ? |

Copy and complete the tables below for values of $e^x$ and $\ln x$, using a calculator where necessary.

Approximate all values to 1 decimal place.

| $x$ | −3 | −2 | −1 | 0 | 0.5 | 1 | 1.5 | 2 | 2.5 | 3 |
|-----|-----|-----|-----|-----|-----|-----|-----|-----|-----|-----|
| $e^x$ | (0.049) | 0.1 | | | 1.6 | | | 7.4 | | |

| $x$ | −1 | 0 | 0.1 | 0.3 | 0.4 | 0.6 | 0.8 | 1 | 2 | 3 | 4 | 5 | 7 | 9 | 12 |
|-----|-----|-----|-----|-----|-----|-----|-----|-----|-----|-----|-----|-----|-----|-----|-----|
| $\ln x$ | (err) | | −2.3 | | | −0.5 | | | 0.7 | | | 1.6 | | 2.2 | 2.5 |

On the same pair of axes and using a scale of 1 cm for 1 unit on both axes, draw the graphs of the two curves $y = e^x$ and $y = \ln x$, $-5 \le x \le 12$, $-5 \le y \le 12$.

Your graphs should look something like this:

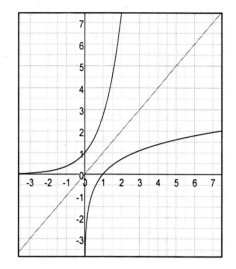

Now study both graphs carefully and see how many properties of the two functions you can find from the graphs. After you have done this compare your findings with the ones below:

## *Properties of the exponential and logarithmic functions*

$y = e^x$                                $y = \ln x$

➤ passes through the point $(0, 1)$       passes through the point $(1, 0)$

➤ is a continuous function            is a continuous function

➤ is an increasing function          is an increasing function

➤ the rate of increase increases      the rate of increase decreases

➤ $x$-axis is an asymptote to the curve    $y$-axis is an asymptote to the curve

➤ range is $y > 0$                domain is $x > 0$

                                 so only positive numbers have logarithms

➤ since $y = e^x$ and $y = \ln x$ are inverse functions, their graphs are reflections in the line $y = x$.

Also, since e and ln are inverses     $\boxed{\ln(e^x) = e^{\ln x} = x}$

## *Further problems involving exponential and logarithmic functions*

You now have all the tools you need for solving problems involving exponential and logarithmic functions.

## Example 15.8

Given that $u = \log_9 x$, find, in terms of $u$,

(i) $x$,    (ii) $\log_3 x$,    (iii) $\log_9(27x)$,    (iv) $\log_x 81$

 since $u$ is given as a base 9 logarithm, numbers within the logarithms in the question should be given as powers of 9: $3 = 9^{\frac{1}{2}}$, $27 = 9^{\frac{3}{2}}$ etc.

    (i) ➤ Turn log equation into index equation: $x = 9^u$

(ii) ➤ Change base:

$$\log_3 x = \frac{\log_9 x}{\log_9 3}$$

$$= \frac{u}{\log_9 9^{\frac{1}{2}}}$$

$$= 2u$$

(iii) ➤ Use first law:

$$\log_9(27x) = \log_9 27 + \log_9 x$$

$$= \log_9 9^{\frac{3}{2}} + u$$

➤ Use third law:

$$= \tfrac{3}{2} + u$$

(iv) ➤ Use third law:

$$\log_x 81 = 2\log_x 9$$

➤ Use corollary:

$$= \frac{2}{\log_9 x} = \frac{2}{u}$$

## Example 15.9

Solve the simultaneous equations

$$\log_2 x + 2\log_4 y = 4 \qquad ①$$

$$x + 12y = 52 \qquad ②$$

➤ Rewrite ① in terms of base 4 logarithms: $\dfrac{\log_4 x}{\log_4 2} + 2\log_4 y = 4\log_4 4 \qquad ③$
  (remember, $\log_a a = 1$)

➤ Simplify ($\log_4 2 = \log_4 4^{\frac{1}{2}} = \frac{1}{2}$ so ③ becomes): $2\log_4 x + 2\log_4 y = 4\log_4 4$

➤ divide by 2, then apply rules (i) and (iii): $\log_4 xy = \log_4 4^2$

➤ Antilog: $xy = 16$

➤ Now solve simultaneous equations by substitution: $(52 - 12y)y = 16$

  leading to solutions $y = 4, x = 4$ and $y = \frac{1}{3}, x = 48$

 If you are sketching exponential or logarithmic functions, remember that you can sketch more complicated functions using techniques involving transformations.

## Example 15.10

Sketch the functions (a) $y = 2e^x + 1$, (b) $y = \ln(x + 2)$, giving the coordinates of any points where the graph cuts the coordinate axes and the equations of any asymptotes.

(a) $y = 2e^x + 1$ basic curve $y = e^x$ (intercept $(0, 1)$, asymptote $y = 0$)

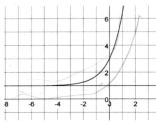

Transformations: first vertical stretch, then translation 1 up
Stretch moves intersection with $y$-axis up to $(0, 2)$
Translation moves it up further to $(0, 3)$
Stretch does not affect asymptote (why not?)
Translation moves asymptote up to $y = 1$

(b)    Basic curve $y = \ln x$ (intercept $(1, 0)$, asymptote $x = 0$)
        Transformation: translation 2 left
        Translation moves intersection with $x$-axis to $(-1, 0)$
        and asymptote to $x = -2$

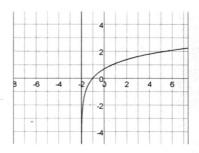

# EXERCISE 15 D (miscellaneous)

1.    Find $t$ if $\log_9 t = -\tfrac{3}{2}$

2.    Solve $3e^{\ln y} + 2\ln(e^y) = 10$

3.    Solve $5^z = 4$  to 2 decimal places

4.    Find the exact value of
      (a)    $\log_4 64$                    (b)    $\log_a \sqrt{a}$

5.    Solve $2^z = 5$  to 2 decimal places

6.    Use the substitution $y = 3^x$ to solve the equation $3^{2x} - 3^x = 2$ to 3 significant figures.

7.    Solve the equation $2(3^{2x}) + 5(3^x) - 3 = 0$  to 3 significant figures.

8.    If $\log_5 x = 9$, find (without using a calculator) $\log_x 5$

9.    If $\log_9 x = a$ and $\log_3 y = b$, express $xy$ and $\dfrac{x}{y}$ as powers of 3.

      If $xy = 243$ and $\dfrac{x}{y} = 3$, calculate $a$ and $b$.

#10. If $\lg 4 + 2 \lg p = 2$, find $p$.

11.   Find the values of $x$ which satisfy the equation
            $4\log_3 x = 9 \log_x 3$

12.   Given that $y = \log_b 45 + \log_b 25 - 2 \log_b 75$, express $y$ as a single logarithm in base $b$.
      In the case when $b = 5$, state the value of $y$.

13.   Solve the equation $\lg(1 - 2x) - 2 \lg x = 1 - \lg(2 - 5x)$

14.   Solve the equation $3^{y+1} = 4^y$

15.   Solve the equation $2\lg(x + 2) + \lg 4 = \lg x + 4\lg 3$

16.   Solve the equation $2^{2x} - 2^{x+3} + 7 = 0$

17.   Given that $x = 3^y$, $y > 0$, express in terms of $y$
      (a)    $\log_3 x$                    (b)    $\log_x 3$

18.   Solve, for $x$, the equation $e^{2x} - 3e^x + 2 = 0$

20

19. Sketch the curve $y = e^{2x} - 2$ and state the coordinates of the point where the curve meets the x-axis .

20. Sketch, on separate diagrams, the functions (a) $y = 2e^x + 1$, (b) $y = \ln(x + 2)$, giving the coordinates of any points where the graph cuts the coordinate axes and the equations of any asymptotes.

21. (a) On the same axes sketch the exponential functions $y = e^{2x}$ and $y = e^x + 2$.
    Label and name the coordinates of the points where the curves intersect the axes.
    (b) Find the coordinates of the point(s) of intersection of the curves in (a)

22. The functions f and g are defined as follows
    $$f : x \mapsto \ln(x + 4) \quad x > -4$$
    $$g : x \mapsto e^x$$
    (a) Sketch $y = f(x)$, giving the coordinates of any points where the graph cuts the coordinate axes and the equations of any asymptotes.
    (b) Find an expression for $f^{-1}(x)$ .
    (c) State the range of $f^{-1}$.
    (d) Find the value of $gf(0)$.

23. Solve the simultaneous equations

    $$2 \log_y x = 1$$
    $$xy = 64$$

24. Solve the simultaneous equations
    $$\lg x + 2\lg y = 3$$
    $$x^2 y = 125$$

25. Given that $2^x 4^y = 128$ and that $\ln(4x - y) = \ln 2 + \ln 5$, calculate the value of x and y.

26. Solve, giving your answers as exact fractions, the simultaneous equations
    $$8^y = 4^{2x+3}$$
    $$\log_2 y = \log_2 x + 4.$$

27. Solve the simultaneous equations
    $$3^p = 9(27)^q,$$
    $$\log_2 7 - \log_2(11q - 2p) = 1.$$

## EXERCISE 15 X

1. Given $e^x - e^{-x} = 4$, show that $x = \ln(2 + \sqrt{5})$   [5]

2. Given that $x = 2^y$, $y > 0$, express in terms of $y$
   (a) $\log_2 x$        (b) $\log_x 2$   [4]

   Hence, or otherwise, solve the equation
   $$\log_2 x = 8 \log_x 2 + 2$$   [6]

3.

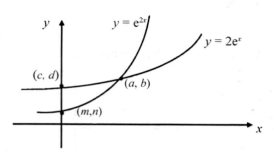

In the diagram, *not drawn to scale*, find the coordinates $(c, d)$, $(a, b)$ and $(m, n)$. [7]

4. Solve, for $x$, the equation $e^{2x} - 3e^x + 2 = 0$ [6]

5. Solve for $a$, the equation $\log_2 a + \dfrac{2}{\log_2 a} = 3, \quad a > 0$ [6]

6. Solve the equation $3^x = 2^{x+1}$ [4]

7. (i) Sketch on the same rectangular coordinate system the exponential and logarithmic functions f and g given by $f(x) = e^x$ and $g(x) = \ln x$

Label and name the coordinates of the points where the curves intersect the axes. [4]

(ii) State the relationship between f and g. [2]

8. The variation of the solubility, S, of a chemical substance with the temperature $\theta$ can be modelled by the equation $S = 60.2(1.01)^{\theta}$, where $\theta$ is an integer. Calculate the value of $\theta$ when S = 100. [2]

9. Solve the equation $e^{2x} - 2e^x = 3$, for real $x$. [6]

10. Solve the simultaneous equations for $x > 0, y > 0$

$$xy = 4$$
$$2 \ln x = \ln 2 + \ln y$$

[6]

11. Show that $\log_x y = \dfrac{1}{\log_y x}$ for $x, y$ real and positive.

Hence, if $\log_x y = 2 + 3 \log_y x$, find $y$ as functions of $x$. [8]

12. In the diagram below (not drawn to scale), the line $y = 3$ cuts the curve $y = e^{2x}$ at the point $(a, b)$.

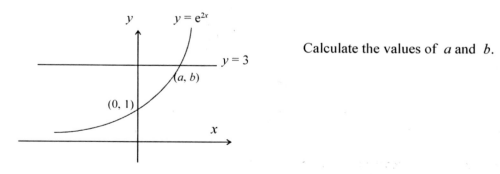

Calculate the values of $a$ and $b$. [4]

13. Solve, for $x$, the equation $e^{2x} - 3e^x - 4 = 0$ [3]

14. (a) If $2 \log_a 2 + \log_a 10 - 3 \log_a 3 = 3 + \log_a 5$, $a > 0$, find the value of $a$. [5]

(b) Find the value(s) of $x \in \mathbb{R}$ which satisfy $2 \log_3 x = \log_3 (x + 6)$. [5]

(c) Complete the table below for values of $2^x$ and $e^x$ using a calculator, where necessary.

| $x$ | $-2$ | $-1$ | $0$ | $0.5$ | $1$ | $1.5$ | $2$ | $2.5$ | $3$ |
|---|---|---|---|---|---|---|---|---|---|
| $2^x$ | 0.3 | 0.5 | 1.0 | 1.4 | 2.0 | 2.8 | 4.0 | 5.7 | 8.0 |
| $e^x$ | 0.1 | 0.4 | 1.0 | 1.6 | 2.7 | 4.5 | 7.4 | 12.2 | 20.1 |

Approximate all values to 1 decimal place. [4]

(d) On the same pair of axes and using a scale of 2 cm for 1 unit on the $x$-axis, 1 cm for 1 unit on the $y$-axis, draw the graphs of the two curves $y = 2^x$ and $y = e^x$ for $-2 \le x \le 3$. [7]

(e) Use your graphs to find

$x \ln 2 = x \ln e$

(i) the value of $x$ satisfying $2^x = e^x$ [2]

(ii) the SMALLEST INTEGER $x$ for which $e^x - 2^x > 3$. $x =$ [2]

15. Solve the equations

(a) $5^{x-2} = 20$ [4]

(b) $\log_2 x + \dfrac{2}{\log_2 x} = 3$, $x > 0$ [5]

$y^2 - 3y + 2 = 0$
$y^2 - 2y - y + 2 = 0$
$y(y-2) - 1(y-2) = 0$
$y = 1 \wedge 2$

16. The function $f(x)$ is defined by $f(x) = a_0 + a_1 x + a_2 x^2 + a_3 x^3 + \dots$ where $a_1, a_2, a_3 \in \mathbb{R}$.

(a) Obtain $f'(x)$. [1]

(b) Given that $f'(x) = \lambda\, f(x)$, $\lambda \ne 0$, $\lambda \in \mathbb{R}$, express the coefficients $a_1, a_2, a_3$, in terms of $\lambda$ and $a_0$. [4]

$a_1 + 2a_2 x + 3a_3 x^2 + \dots = \lambda a_0 + \lambda a_1 x + \lambda a_2 x^2 + \dots$

17. Given $\log_{10} 3 = m$, $\log_{10} e = n$, express in terms of $m$ and/or $n$

(a) $\log_{10}\left(\frac{3}{10}\right)$ [3]

(b) $\log_e 9$ [4]

$\dfrac{\log_{10} 9^m}{\log_{10} e} = \dfrac{2 \log_{10} 3}{\log_{10} e} = \dfrac{2m}{n}$

18. Find the value of $x \in \mathbb{R}$ for which $3^x = 7$. [3]

19. The diagram below, **not drawn to scale**, shows two points, P $(p, 0.368)$ and R $(3.5, r)$ on $f(x) = e^x$ for $x \in \mathbb{R}$.

(a) Copy the diagram above and on the same axes, sketch the graph of g $(x) = \ln x$. [3]

(b) Describe clearly the relationship between f $(x) = e^x$ and g $(x) = \ln x$. [3]

(c) Using a calculator, find the value of      (i) $r$, [1]
     (ii) $p$. [2]

20. Given that $\log_a(bc) = x$, $\log_b(ca) = y$, $\log_c(ab) = z$ and $a \neq b \neq c$, show that $a^x b^y c^z = (abc)^2$.　　　[3]

21. Find the values of $x \in \mathbb{R}$ for which $e^x + 3e^{-x} = 4$.　　　[8]

22. Solve, for $x$, the equations

　(a)　$\log_2 x + \dfrac{2}{\log_2 x} = 3$, $x > 0$　　$y^2 - 3y + 2 = 0$　　　[5]

　(b)　$3^x = 5^{x-1}$　　　[3]

*Answers to opening questions:*
$1.8 \times 10^{19}$;　　(b) almost \$1 million (rather than \$0.65 million);　7 days;　12 bounces

*Results of the investigation in exercise 15C*

Most calculators owned by students calculate with up to 12 significant figures. For the calculations needed in this investigation this does not lead to very accurate results. The table below has been produced using a computer calculator, with results given to 6 d.p. In every case the limit from the right and the limit from the left are the same at this degree of accuracy, and therefore give the gradient correct to 6 d.p.

| | | $y = 2^x$ | | $y = 3^x$ | | $y = 4^x$ |
|---|---|---|---|---|---|---|
| $x$ | $y$ | $\dfrac{dy}{dx}$ | $y$ | $\dfrac{dy}{dx}$ | $y$ | $\dfrac{dy}{dx}$ |
| 0 | 1 | 0.693147 | 1 | 1.098612 | 1 | 1.386294 |
| 1 | 2 | 1.386294 | 3 | 3.295837 | 4 | 5.545177 |
| 2 | 4 | 2.772589 | 9 | 9.887511 | 16 | 22.181017 |
| 3 | 8 | 5.545177 | 27 | 29.662532 | 64 | 88.722839 |
| 4 | 16 | 11.090355 | 81 | 88.987595 | 256 | 354.891356 |
| 5 | 32 | 22.180710 | 243 | 266.962786 | 1024 | 1419.565426 |

# What you have learnt in Chapter 15

### *The Definition of a Logarithm:*

the logarithm ($L$) of $n$ to the base $b$ is the power to which $b$ must be raised to give $n$.

i.e. $\quad b^L = n \iff \log_b n = L$

### *The Rules of Logarithms:*

$$\log_a (xy) = \log_a x + \log_a y \qquad \text{(i)}$$

$$\log_a \left( \frac{x}{y} \right) = \log_a x - \log_a y \qquad \text{(ii)}$$

$$\log_a x^n = n \log_a x \qquad \text{(iii)}$$

### *Other Facts about Logarithms*

$$\log_a a = 1$$

$$\log_a 1 = 0$$

### *Change of Base*

$$\log_b a = \frac{\log_c a}{\log_c b}$$

$$\log_b a = \frac{1}{\log_a b}$$

### *To Solve the Equation $a^x = b$ Take Logarithms of Both Sides*

$$x \log a = \log b$$

### *The Exponential Function is Defined as the Function for which*

$$\text{if } y = e^x, \text{ then } \frac{dy}{dx} = e^x$$

### *The Logarithmic Function is the inverse of the Exponential Function*

$$y = e^x \iff x = \ln y$$

$$\ln (e^x) = e^{\ln x} = x$$

# CHAPTER 16

# 2.1b  Differentiation II

## RECAPITULATION

Before starting new work for this unit, here is a reminder of what we learnt in Unit 1.

### *Differentiation from First Principles*

$$f'(x) = \lim_{h \to 0} \left\{ \frac{f(x+h) - f(x)}{h} \right\}$$

### *Differentiation of standard functions of  x*

For all rational values of  $n$

If   $f(x) = x^n$   then   $f'(x) = nx^{n-1}$

If   $f(x) = \sin ax$  then  $f'(x) = a \cos ax$

If   $f(x) = \cos ax$  then  $f'(x) = -a \sin ax$

If   $f(x) = \tan ax$  then  $f'(x) = a \sec^2 ax$

where $x$ is measured in radians

### *Rules for combining Derivatives*

If $f(x)$ and $g(x)$  are differentiable functions and  $a$  is a constant, then

$$\frac{d}{dx}\{f(x) \pm g(x)\} = f'(x) \pm g'(x)$$

and   $\dfrac{d}{dx}\{a f(x)\} = a \dfrac{d}{dx}\{f(x)\}$

### *Product and Quotient Rules*

If  $u$  and  $v$  are differentiable functions of $x$  and

if  $y = uv$,  then  $\dfrac{dy}{dx} = v\dfrac{du}{dx} + u\dfrac{dv}{dx}$

if  $y = \dfrac{u}{v}$,  then  $\dfrac{dy}{dx} = \dfrac{v\dfrac{du}{dx} - u\dfrac{dv}{dx}}{v^2}$

### *Chain Rule*

If $y$ is a function of $u$, and $u$ is a function of $x$, then

$$\frac{dy}{dx} = \frac{dy}{du} \times \frac{du}{dx}$$   and corollary   $\dfrac{dx}{dy} = \dfrac{1}{\dfrac{dy}{dx}}$

## Stationary Points

---

Critical/stationary/turning points occur when $f'(x) = \dfrac{dy}{dx} = 0$.

At a turning point:

$y$ is maximum if $f'(x)$ is changing from positive to negative
  or $f''(x)$ is negative,

$y$ is minimum if $f'(x)$ is changing from negative to positive
  or $f''(x)$ is positive.

---

# DIFFERENTIATION OF EXPONENTIAL FUNCTIONS

We saw in chapter 15 that by definition, $\dfrac{d}{dx}(e^x) = e^x$.

Other powers of $e$ can then be differentiated using the chain rule

## Example 16.1

Differentiate $y = e^{5x}$ w.r.t. $x$

➤ Make a substitution:  Let $u = 5x$, then $y = e^u$

➤ Differentiate both functions:  $\dfrac{du}{dx} = 5$ and $\dfrac{dy}{du} = e^u$

➤ Apply the chain rule:  $\dfrac{dy}{dx} = \dfrac{dy}{du} \times \dfrac{du}{dx} = e^u.5$

➤ Give the answer in terms of $x$:  $\dfrac{dy}{dx} = 5\,e^{5x}$

## Example 16.2

Differentiate $y = e^{x^2+3}$ w.r.t. $x$

➤ Make a substitution:  Let $u = x^2 + 3$, then $y = e^u$

➤ Differentiate both functions:  $\dfrac{du}{dx} = 2x$ and $\dfrac{dy}{du} = e^u$

➤ Apply the chain rule:  $\dfrac{dy}{dx} = \dfrac{dy}{du} \times \dfrac{du}{dx} = e^u.2x$

➤ Give the answer in terms of $x$:  $\dfrac{dy}{dx} = 2x\,e^{x^2+3}$

# EXERCISE 16 A

Questions 1 – 4 revise differentiation of powers of $x$, trig. functions, the chain rule, and products and quotients, as well as differentiation of exponential functions.

1. Differentiate w.r.t. $x$ (remember to rewrite roots and reciprocals as indices **before** differentiating)

    (a) $5x^2$
    (b) $x^{-3}$
    (c) $x^{\frac{1}{3}}$
    (d) $\sqrt{x}$
    (e) $5x^{-2}$
    (f) $\cos x$

    (g) $\dfrac{3}{x^3}$
    (h) $2\sqrt[3]{x}$
    (i) $\dfrac{4}{\sqrt{x}}$
    (j) $10\sqrt{x^5}$
    (k) $\dfrac{1}{\sqrt[5]{x}}$
    (l) $\tan x$

2. Differentiate the following functions w.r.t. $x$

    (a) $e^{3x}$
    (b) $e^{-5x}$
    (c) $e^{8x+5}$
    (d) $e^{x^2}$

    (e) $e^{ax+b}$
    (f) $e^{\cos x}$
    (g) $e^{2x^3-x}$
    (h) $e^{-\tan x}$

3. Differentiate the following functions w.r.t. $x$

    (a) $(3x-1)^4$
    (b) $(x^3-2)^7$
    (c) $(2x^2-4x)^5$
    (d) $(3-5x)^8$
    (e) $\sqrt{(x^2-1)}$

    (f) $\dfrac{1}{x^2+3}$
    (g) $\dfrac{1}{\sqrt{3-2x}}$
    (h) $(px+q)^{n+1}$
    (i) $4x+(2x-7)^5$
    (j) $2-(1+x^2)^{-2}$

    (k) $\sin 5x$
    (l) $\cos 7x$
    (m) $\tan 2x$
    (n) $\sin 4x$

    (o) $\sin x^2$
    (p) $\sin^2 x$
    (q) $\cos (x+\frac{1}{4}\pi)$
    (r) $\tan (\frac{1}{3}\pi - x)$

    (s) $\sin (px+q)$
    (t) $\cos (px+q)$
    (u) $\tan (px+q)$

4. Differentiate w.r.t. $x$

    (a) $x^2(5-\sqrt{x})$
    (b) $x \sin x$
    (c) $(2x-3)\cos x$
    (d) $\sin x \tan x$

    (e) $\dfrac{x^2-1}{x^2+1}$
    (f) $\dfrac{2-\sqrt{x}}{\sqrt{x}+5}$
    (g) $\dfrac{\sin x}{2+x}$
    (h) $\dfrac{x}{\cos x}$

    (i) $xe^x$
    (j) $e^{2x}\sin x$
    (k) $\dfrac{e^{x^2}}{\sin x}$
    (l) $\dfrac{x^2+1}{e^{2x}+1}$

5. ✳ Differentiate $e^{-x}$ w.r.t. $x$ (a) using the chain rule, (b) using the quotient rule

6. ✳ Differentiate $e^{f(x)}$ w.r.t. $x$

You should have found by now that you can do many of the chain rule questions without writing out the substitution in full. This is not only an acceptable thing to do, it is important that you learn to do it as it saves a lot of time. Writing the process as a rule looks rather complicated, but it is quite simple in practice.

If you have a composite function (sometimes known as a function of a function) it can be written as $y = f\{g(x)\}$. Then the derivative is $\dfrac{dy}{dx} = f'\{g(x)\}\, g'(x)$.

An example will show how this works in practice (you don't write the first two steps down on their own)

## Example 16.3

Differentiate $(x^2+3x)^5$

  ➢ Differentiate the bracket as if it were a single letter:     $5(x^2+3x)^4$

  ➢ Differentiate the contents of the bracket:     $2x+3$

  ➢ Multiply together:     $5(2x+3)(x^2+3x)^4$

Now try these questions without writing out a substitution whenever possible

# EXERCISE 16 B

1. Differentiate the following functions w.r.t. $x$

   (a) $(x-2)^5$   (b) $(2-3x)^7$   (c) $(x^2-5x)^3$   (d) $\sqrt{x^2+1}$   (e) $(ax+b)^n$

2. Differentiate the following functions w.r.t. $x$

   (a) $\sin 3x$   (b) $\cos 4x$   (c) $\tan(2x + \frac{1}{4}\pi)$   (d) $\cos x^2$   (e) $\sin(ax+b)$

3. Differentiate the following functions w.r.t. $x$

   (a) $e^{7x}$   (b) $e^{2x+3}$   (c) $e^{x^2-1}$   (d) $e^{\sin x}$   (e) $e^{px+q}$

4. Differentiate the following functions w.r.t. $x$

   (a) $(5x^2-3x)^3$   (b) $e^{-2x}$   (c) $\cos 7x$   (d) $(2x+4)^{-2}$   (e) $e^{\tan x}$

   (f) $(x^3-1)^5$   (g) $\sin(x^2)$   (h) $\cos^2 x$   (i) $e^{3x-1}$   (j) $4e^{-x^2}$

In the above exercises you have derived a number of results which are worth remembering

| $f(x)$ | $f'(x)$ |
|---|---|
| $(ax+b)^n$ | $an(ax+b)^{n-1}$ |
| $\sin(ax+b)$ | $a\cos(ax+b)$ |
| $\cos(ax+b)$ | $-a\sin(ax+b)$ |
| $\tan(ax+b)$ | $a\sec^2(ax+b)$ |
| $e^{(ax+b)}$ | $ae^{(ax+b)}$ |
| $e^{f(x)}$ | $f'(x)\,e^{f(x)}$ |

# DIFFERENTIATION OF LOGARITHMIC FUNCTIONS

✳ **Differentiation of ln $x$**

If $y = \ln x$ then $x = e^y$.

Differentiating **w.r.t.** $y$:  $\dfrac{dx}{dy} = e^y$

but $e^y = x$,   so   $\dfrac{dx}{dy} = x$

By the corollary of the chain rule   $\dfrac{dy}{dx} = \dfrac{1}{\dfrac{dx}{dy}} = \dfrac{1}{x}$

so   $\dfrac{d}{dx}(\ln x) = \dfrac{1}{x}$

**Differentiation of ln f$(x)$**

Let $y = \ln f(x)$

then $y$ is a function of a function and the chain rule must be used to find its derivative

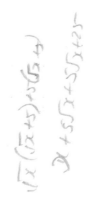

Let $f(x) = u$, so $y = \ln u$

Differentiating: $\dfrac{du}{dx} = f'(x), \ \dfrac{dy}{du} = \dfrac{1}{u}$

Use chain rule: $\dfrac{dy}{dx} = \dfrac{dy}{du} \times \dfrac{du}{dx}$

$$= \left(\frac{1}{u}\right) \times f'(x)$$

$$= \frac{f'(x)}{f(x)}$$

so $\dfrac{d}{dx}\{\ln f(x)\} = \dfrac{f'(x)}{f(x)}$

In particular, if $f(x) = ax$, then $f'(x) = a$

so $\dfrac{d}{dx}(\ln ax) = \dfrac{a}{ax} = \dfrac{1}{x}$

Summarising:

$$\frac{d}{dx}(\ln x) = \frac{1}{x}$$

$$\frac{d}{dx}(\ln ax) = \frac{1}{x}$$

$$\frac{d}{dx}\{\ln f(x)\} = \frac{f'(x)}{f(x)}$$

# Example 16.4

Find the derivative of $\ln(4x^2 + 7)$

➤     If $f(x) = 4x^2 + 7$, then $f'(x) = 8x$

➤     $\dfrac{d}{dx}\{\ln(4x^2 + 7)\} = \dfrac{f'(x)}{f(x)} = \dfrac{8x}{4x^2 + 7}$

# Example 16.5

Find the derivative of $\ln(2x + 4)^5$

 Although this can be solved by taking $f(x) = (2x + 4)^5$, it is much easier first to rewrite the problem, using the third law of logarithms:

➤     If $y = \ln(2x + 4)^5$, then $y = 5\ln(2x + 4)$

➤     Let $f(x) = 2x + 4$, so $f'(x) = 2$

➤     $\dfrac{dy}{dx} = 5\dfrac{f'(x)}{f(x)}$

$$= 5\frac{2}{2x + 4}$$

$$= \frac{10}{2x + 4}$$

## DIFFERENTIATION OF INVERSE TRIG FUNCTIONS

If you need to find the angle $\theta$ for which $\sin\theta = a$ you use the inverse sine key on your calculator. We normally write this either as $\theta = \sin^{-1} a$ or $\theta = \arcsin a$. We are therefore defining a new function – inverse sine. Now the sine function $y = \sin x$ is a many-to-one function, so it can only have an inverse if its domain is restricted so that it becomes one-to-one; this is normally done by restricting the domain to $-\frac{1}{2}\pi \leq x \leq \frac{1}{2}\pi$, and we know that its range is $-1 \leq \sin x \leq 1$. The inverse sine function is therefore $y = \sin^{-1}x$ on the domain $-1 \leq x \leq 1$ with a range $-\frac{1}{2}\pi \leq y \leq \frac{1}{2}\pi$.

Figure 16.1a shows the function $y = \sin^{-1}x$ and figure 16.1b shows the two functions $y = \sin x$ and $y = \sin^{-1}x$ on the same axes so that you can clearly see that they are inverses.

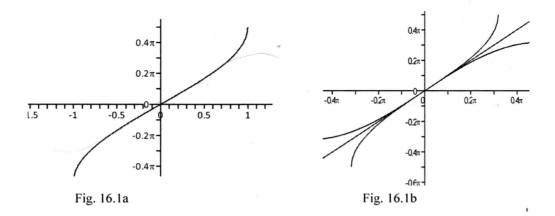

Fig. 16.1a                                           Fig. 16.1b

For $y = \tan x$ to be a one-to-one function its domain must also be restricted to $-\frac{1}{2}\pi \leq x \leq \frac{1}{2}\pi$ while its range is infinite. The inverse function $y = \tan^{-1}x$ therefore has the domain $x \in \mathbb{R}$ and range $-\frac{1}{2}\pi \leq y \leq \frac{1}{2}\pi$ and is shown in fig. 16.2

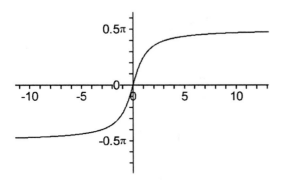

Fig. 16.2

Inverse trig functions can be differentiated using implicit differentiation, see page 36, or, as we did for the log function (being the inverse of the exponential function), differentiating w.r.t. $y$.

| ! | Remember that the rules for differentiating/integrating any trig function only apply when angles are measured in **radians**. |

Notice also that although $\sin^2 x$ stands for $\sin x$ squared, $\sin^{-1}x$ represents the inverse of $\sin x$, not, as you might suppose, its reciprocal. For the reciprocal $\left(\dfrac{1}{\sin x}\right)$ we write $(\sin x)^{-1}$.

## Example 16.6

Find the derivative of the function f when $f(x) = \sin^{-1}x$, $-1 \le x \le 1$

> If $y = \sin^{-1}x$           then $x = \sin y$

> Differentiating w.r.t. $y$:        $\dfrac{dx}{dy} = \cos y$   i.e.   $\dfrac{dy}{dx} = \dfrac{1}{\cos y}$

> We must give $\dfrac{dy}{dx}$ as a function not of $y$ but of $x$

> Since $x = \sin y$ and $\cos^2 y + \sin^2 y = 1$:     $\cos y = \sqrt{1 - \sin^2 y} = \sqrt{1 - x^2}$

> so $\dfrac{dy}{dx} = \dfrac{1}{\sqrt{1 - x^2}}$   or   $f'(x) = \dfrac{1}{\sqrt{1 - x^2}}$

*[handwritten:]* $y = \cos^{-1} x$   $\cos y = x$   $-\sin y \, dy = dx$   $\dfrac{dy}{dx} = \dfrac{-1}{\sqrt{1 - x^2}}$

## Example 16.7

Find the derivative of the function f when $f(x) = \tan^{-1}\left(\dfrac{x}{a}\right)$, $x \in \mathbb{R}$

> If $y = \tan^{-1}\left(\dfrac{x}{a}\right)$          then $x = a \tan y$

> Differentiating w.r.t. $y$:        $\dfrac{dx}{dy} = a \sec^2 y$   i.e.   $\dfrac{dy}{dx} = \dfrac{1}{a \sec^2 y}$

> Using $\sec^2 y = 1 + \tan^2 y$ and $a \tan y = x$:    $\dfrac{dy}{dx} = \dfrac{a}{a^2 + x^2}$   or   $f'(x) = \dfrac{a}{a^2 + x^2}$

Summarising:

$$\frac{d}{dx}(\sin^{-1} x) = \frac{1}{\sqrt{1 - x^2}}, \ |x| < 1 \qquad \frac{d}{dx}(\tan^{-1} x) = \frac{1}{1 + x^2}, \ x \in \mathbb{R}$$

$$\frac{d}{dx}\left(\sin^{-1}\left(\frac{x}{a}\right)\right) = \frac{1}{\sqrt{a^2 - x^2}}, \ |x| < a \qquad \frac{d}{dx}\left(\tan^{-1}\left(\frac{x}{a}\right)\right) = \frac{a}{a^2 + x^2}, \ x \in \mathbb{R}$$

It is important to learn these formulae, as you will need them for integration.

## EXERCISE 16 C

1.   Differentiate the following w.r.t. $x$

    (a) $\ln 5x$          (b) $3 \ln 2x$          (c) $e^x + \ln x$          (d) $5 \ln x - 3$

2.   Differentiate the following w.r.t. $x$

    (a) $\ln (x^2 - 5)$       (b) $3 \ln (2x + 7)$       (c) $\ln (\sin x)$       (d) $\ln (\cos x)$

    (e) $\ln (2x - x^3)$      (f) $\ln (4x^2 - 1)^7$      (g) $\ln \sqrt{8x - 5}$      (h) $\ln \dfrac{2}{x - 1}$

3. Use the method of examples 16.6 and 16.7 to show that

(a) if $y = \tan^{-1} x,$ then $\dfrac{dy}{dx} = \dfrac{1}{1+x^2}$ 　　　　(b) if $y = \cos^{-1} x,$ then $\dfrac{dy}{dx} = \dfrac{-1}{\sqrt{1-x^2}}$

(c) if $y = \sin^{-1}\left(\dfrac{x}{a}\right),$ then $\dfrac{dy}{dx} = \dfrac{1}{\sqrt{a^2 - x^2}}$ $\quad (\,|x| < a)$

4. Differentiate the following w.r.t. $x$, stating the values of $x$ for which your result is valid:

(a) $\sin^{-1}(\tfrac{1}{2}x)$ 　　　(b) $\tan^{-1}(\tfrac{1}{3}x)$ 　　　(c) $\cos^{-1}(\tfrac{1}{4}x)$

(d) $\sin^{-1}(4x)$ 　　　(e) $\tan^{-1}(2x)$ 　　　(f) $\cos^{-1}(3x)$

## DIFFERENTIATION OF MORE COMPLICATED FUNCTIONS

You now know how to differentiate most basic functions and have several rules for differentiating more complicated functions – the product rule, the quotient rule and the chain rule. Many problems can be solved by applying a combination of these rules.

### Example 16.8

Find the derivative of the function f when $f(x) = e^{\cos x} \sin x$

Here we have a product of two functions one of which is composite.

First let the two functions be $u$ and $v$, then differentiate $u$ and $v$ using the chain rule, then use the product rule.

➢ Make substitutions: 　　　Let $u = e^{\cos x}$ 　　and 　　$v = \sin x$

➢ Differentiate: 　　　then $\dfrac{du}{dx} = -e^{\cos x}\sin x$ and $\dfrac{dv}{dx} = \cos x$

➢ Apply the product rule: 　　$\dfrac{dy}{dx} = v\dfrac{du}{dx} + u\dfrac{dv}{dx}$

$$= \sin x(-e^{\cos x}\sin x) + e^{\cos x}\cos x$$

$$= (\cos x - \sin^2 x)e^{\cos x}$$

### Example 16.9

Find the derivative of the function f when $f(x) = \dfrac{e^{x^2}}{\ln(2x)}, x > 0, x \neq \tfrac{1}{2}.$

➢ Make substitutions: 　　　Let $u = e^{x^2}$ 　　and 　　$v = \ln(2x)$

➢ Differentiate: 　　　then $\dfrac{du}{dx} = 2xe^{x^2}$ and $\dfrac{dv}{dx} = \dfrac{2}{2x} = \dfrac{1}{x}$

➢ Apply the quotient rule: 　$\dfrac{dy}{dx} = \dfrac{v\dfrac{du}{dx} - u\dfrac{dv}{dx}}{v^2}$

$$= \dfrac{2xe^{x^2}\ln(2x) - \dfrac{e^{x^2}}{x}}{\{\ln(2x)\}^2}$$

$$= \dfrac{e^{x^2}}{\{\ln(2x)\}^2}\left(2x\ln(2x) - \dfrac{1}{x}\right)$$

# EXERCISE 16D

Find the derivatives of the following functions:

1. $f(x) = e^x \tan x$

2. $f(x) = x^2 \cos 3x$

3. $f(x) = e^x \ln x$

4. $f(x) = x^3 \sin 2x$

5. $f(x) = (2x + 1)^3 \sin x$

6. $f(x) = \ln(e^x + 3)^3$

7. $f(x) = \dfrac{\cos 2x}{1 + \tan 2x}$

8. $f(x) = \dfrac{e^{x^2}}{1 - x^2}$

9. $f(x) = e^{5x} \ln 3x$

#10. $y = \sin(\cos x)$

11. $y = \cos(x + \ln x)$

12. $y = e^{(x \sin x)}$

13. $y = x \sin^{-1} x$

14. $y = x^2 \tan^{-1} x$

15. $y = e^x \tan^{-1} x$

## SECOND DERIVATIVES

As we found when studying maxima and minima, we sometimes need to differentiate a function twice.

If $y = f(x)$, then the second derivative is $f''(x) = \dfrac{d^2 y}{dx^2}$.

$\dfrac{d^2 y}{dx^2}$ can be found by differentiating $\dfrac{dy}{dx}$.

 When the second derivative is needed, it is a good idea to simplify the first derivative as much as possible before proceeding.

## Example 16.10

If $x = e^{2t} \sin t$, show that $\dfrac{d^2 x}{dt^2} - 4\dfrac{dx}{dt} + 5x = 0$

➤ Find $\dfrac{dx}{dt}$:  $\dfrac{dx}{dt} = 2e^{2t} \sin t + e^{2t} \cos t$

➤ Simplify:  $\dfrac{dx}{dt} = e^{2t}(2 \sin t + \cos t)$

➤ Differentiate again:  $\dfrac{d^2 x}{dt^2} = 2e^{2t}(2 \sin t + \cos t) + e^{2t}(2 \cos t - \sin t)$

$= e^{2t}(3 \sin t + 4 \cos t)$

➤ Substitute in formula:  $\dfrac{d^2 x}{dt^2} - 4\dfrac{dx}{dt} + 5x = e^{2t}(3 \sin t + 4 \cos t) - 4e^{2t}(2 \sin t + \cos t) + 5e^{2t} \sin t$

$= e^{2t}\{(3 - 8 + 5)\sin t\} + (4 - 4)\cos t$

$= 0$

The next exercise contains a mixture of questions.

 Look carefully at each question and decide which rules you are going to need to use – product, quotient and/or chain rules. Then take the question methodically, step by step. If you do that and write it out clearly, you will find it is not as frightening as it may at first appear.

# EXERCISE 16 E

1.  Find the derivative of the function f at the point $x$, if

    (a)  $f(x) = \cos^2(3x - 1)$
    (b)  $f(x) = e^{2x} \ln 3x$
    (c)  $f(x) = \tan(5x^2 - 3)$

    (d)  $f(x) = x \ln x - x, \ x > 0$
    (e)  $f(x) = x^2 \cos x^2$
    (f)  $f(x) = \ln(2x^2 - 4x), \ x > 2$

    (g)  $f(x) = \tan^2 2x$
    (h)  $f(x) = e^{5x} \sin x$
    (i)  $f(x) = x^4(x^2 + 3)^5$

    (j)  $f(x) = e^{x^3} \ln x, \ x > 0.$
    (k)  $f(x) = e^{\sin x} \tan x$
    (l)  $f(x) = \sin^7 x$

    (m)  $f(x) = e^{-x} \sin x$
    (n)  $f(x) = \ln(\cos x)$
    (o)  $f(x) = \ln\left(\dfrac{1 + \cos x}{1 + \sin x}\right)$

2.  Find the gradient of the curve $y = f(x)$ at the point P on the curve and the equation of the tangent at P

    (a)  $f(x) = x \sin(2x + \frac{1}{2}\pi)$ and P is $(0, 0)$
    (b)  $f(x) = \dfrac{x^2 - 3x}{2 + 3x}$ and P is $(-1, -4)$
    (c)  $f(x) = \tan^{-1} x$ and P is $(2, 1.1)$

3.  Find the gradient of the curve $y = f(x)$ at the point P on the curve and the equation of the normal at P

    (a)  $f(x) = \sqrt{4 + 3x^2}$ and P is $(2, 4)$
    (b)  $f(x) = x\sqrt{x + 3}$ and P is $(1, 2)$
    (c)  $f(x) = \sin^{-1} x$ and P is $(0, 0)$

4.  Find the gradient of the curve $y = \ln \sqrt{(1 + \sin 2x)}$ at the point where $x = \frac{1}{2}\pi$.
    Hence, find the equation of the tangent at that point.

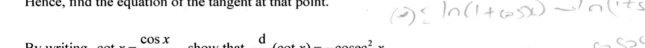

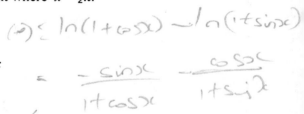

5.  By writing $\cot x = \dfrac{\cos x}{\sin x}$ show that $\dfrac{d}{dx}(\cot x) = -\operatorname{cosec}^2 x$

6.  Find $\dfrac{d}{dx}(\sec x)$

7.  Find $\dfrac{dx}{dt}$ if $x = \dfrac{\sin t}{1 + \cos t}$

8.  Find $\dfrac{dy}{dx}$ if $y = \dfrac{e^x - e^{-x}}{e^x + e^{-x}}$

9.  Find $\dfrac{dy}{dx}$ if $y = \ln\left(\dfrac{1 + x}{1 - x}\right)$

#10. If $y = x(2x + 3)^5$, find $\dfrac{d^2 y}{dx^2}$, simplifying your answer as far as possible.

11.  If $y = \dfrac{1 + x}{1 - x}$ find $\dfrac{d^2 y}{dx^2}$, simplifying your answer as far as possible.

12.  Find the value of $\dfrac{d^2 y}{dx^2}$ when $x = 0$ if $y = e^x \sin 2x$

13.  If $y = e^x \sin x$ show that $\dfrac{d^2 y}{dx^2} - 2\dfrac{dy}{dx} + 2y = 0$.

14.  If $y = \dfrac{\sin x}{x}$ prove that $x\dfrac{d^2 y}{dx^2} + 2\dfrac{dy}{dx} + xy = 0$

#15. Given that $y = ae^{-2x} \sin (x + b)$, where $a$ and $b$ are constants, verify that
$$\dfrac{d^2 y}{dx^2} + 4\dfrac{dy}{dx} + 5y = 0$$

16.  Given that the derivative at $x$ of the exponential function $e^x$ is $e^x$,

(i)  Find the derivative at $x$ of the logarithmic function $\ln x$.
(ii) hence, find the derivative at $x$ of the function $e^x \ln x$.

17.  Use the fact that $e^{-x} = \dfrac{1}{e^x}$ to show that $\dfrac{d}{dx}(e^{-x}) = -e^{-x}$. (i.e. use quotient rule)

# IMPLICIT AND PARAMETRIC EQUATIONS OF CURVES

The equations of some curves cannot easily (if at all) be expressed in the form $y = f(x)$. We have already seen in module 1.2 that the equation of a circle, centre $(a, b)$ and radius $r$ is

$$(x - a)^2 + (y - b)^2 = r^2 \qquad \text{(or } x^2 + y^2 - 2ax - 2by + a^2 + b^2 - r^2 = 0). \qquad ①$$

Although $y$ ***can*** be made the subject of this equation, the result is unwieldy (see below), and anyway there are many other curves where $y$ ***cannot*** be made the subject of the equation. An equation like ①, which connects $x$ and $y$ but is not in the form $y = f(x)$ is known as an **implicit** equation for the curve (as opposed to $y = f(x)$, which is an explicit equation for the curve). Another way, which we also saw in module 1.2, in which the equation of a curve can be expressed is by use of a **parameter**, or third variable.

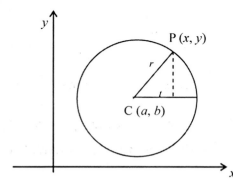

If we look again at the circle, centre $C(a, b)$ and radius $r$ we see that the coordinates of any point $P(x, y)$ on the circle are given by $x = a + r \cos t$ and $y = b + r \sin t$, where $t$ is the angle that $CP$ makes with the direction of the $x$-axis.

$x$ and $y$ have been expressed in terms of a third variable, the parameter $t$, and the equations
$$x = a + r \cos t$$
$$y = b + r \sin t \qquad\qquad ②$$
are the **parametric equations** of the curve.

Eliminating $t$ from these equations leads to equation ①.

The explicit equation for the circle given in implicit equation ① and parametric equations ② is
$y = b + \sqrt{r^2 - (x - a)^2}$ for the top half of the circle and $y = b - \sqrt{r^2 - (x - a)^2}$ for the bottom half of the circle. You can see that this is complicated and would be tricky to differentiate, so if the gradient function is required for the circle it would be easier to differentiate either the implicit equation ① or the parametric equations ② if this can be done.

## *Differentiation of Parametric Equations*

### The First Derivative

Parametric equations of a curve are in the form $x = f(t)$ and $y = g(t)$

Differentiating w.r.t. $t$ gives $\dfrac{dx}{dt} = f'(t)$ and $\dfrac{dy}{dt} = g'(t)$

Applying the chain rule gives $\dfrac{dy}{dx} = \dfrac{dy}{dt} \times \dfrac{dt}{dx} = g'(t) \div f'(t)$

## Example 16.11

A curve is defined by the parametric equations $x = \dfrac{1}{1+t}$, $y = t^2 + 4$, $t \neq -1$

(a)    Find $\dfrac{dy}{dx}$ in terms of the parameter $t$, and hence find the coordinates of the turning point on the curve.

(b)    Write down the equation of the tangent at the turning point

    (a)  ➤  Differentiate $x$ and $y$ w.r.t. $t$:      $\dfrac{dx}{dt} = \dfrac{-1}{(1+t)^2}$,   $\dfrac{dy}{dt} = 2t$

       ➤  Find $\dfrac{dy}{dx}$:          $\therefore \dfrac{dy}{dx} = \dfrac{dy}{dt} \div \dfrac{dx}{dt} = -2t(1+t)^2$

       ➤  Find turning point:     $\dfrac{dy}{dx} = 0$ when $t = 0$, $\therefore$ turning point occurs at the point where $t = 0$, i.e at the point $(1, 4)$

    (b)  ➤  Find equation of tangent:        Since the tangent at the turning point is parallel to the $x$-axis, its equation is $y = 4$.

### The Second Derivative

## Example 16.12

Consider the curve defined by the parametric equations $x = \dfrac{1}{t}$, $y = 5t^2 - 1$.

$$\dfrac{dx}{dt} = \dfrac{-1}{t^2}, \quad \dfrac{dy}{dt} = 10t \qquad \therefore \dfrac{dy}{dx} = \dfrac{dy}{dt} \times \dfrac{dt}{dx} = -10t^3$$

We have found $\dfrac{dy}{dx}$ as a function of $t$. To find $\dfrac{d^2 y}{dx^2}$ we have to differentiate $\dfrac{dy}{dx}$ *w.r.t. $x$*, but this cannot be done

directly since $\dfrac{dy}{dx}$ is a function of $t$, not of $x$. When the parametric equations of a curve can be easily converted into a Cartesian equation, so that $y = f(x)$, then finding the first and second derivatives is straightforward, but in this case it is not so, so we must find another method.

$$\dfrac{d^2 y}{dx^2} = \dfrac{d}{dx}\left(\dfrac{dy}{dx}\right) = \dfrac{d}{dx}(-10t^3)$$

but by the chain rule $\dfrac{d}{dx}\{f(t)\} = \dfrac{d}{dt}\{(f(t)\} \times \dfrac{dt}{dx}$

so $\dfrac{d^2 y}{dx^2} = \dfrac{d}{dt}(-10t^3) \div \dfrac{dx}{dt} = (-30t^2) \div \left(\dfrac{-1}{t^2}\right) = 30t^4$ or $30x^{-4}$

## Differentiation of Implicit Equations

### The First Derivative

An implicit equation can usually be differentiated by differentiating each term of the equation in turn **with respect to x**, whether it contains $x$'s, $y$'s or both, remembering that $\dfrac{d}{dx}(y) = \dfrac{dy}{dx}$ and, by the chain rule, $\dfrac{d}{dx}\{f(y)\} = f'(y)\dfrac{dy}{dx}$ and that any products like $xy$ must be differentiated using the product rule.

## Example 16.13

(a)  Find $\dfrac{dy}{dx}$ if $4x^2 + 3xy - 5y^2 = 2$.

(b)  Hence find the gradient of the tangent at the point $(1, 1)$ on the curve.

(a)  ➤  Differentiate every term w.r.t. $x$:  $\dfrac{d}{dx}(4x^2) + 3\dfrac{d}{dx}(xy) - 5\dfrac{d}{dx}(y^2) = \dfrac{d}{dx}(2)$  ①

➤  $\left[\text{Notice that the second term is a product so: } \dfrac{d}{dx}(xy) = y + x\dfrac{dy}{dx}\right]$

➤  ① becomes:  $8x + 3\left(y + x\dfrac{dy}{dx}\right) - 5\left(2y\dfrac{dy}{dx}\right) = 0$

➤  Collect terms containing $\dfrac{dy}{dx}$ on one side:  $(3x - 10y)\dfrac{dy}{dx} = -(8x + 3y)$

➤  Give the answer:  $\dfrac{dy}{dx} = \dfrac{8x + 3y}{10y - 3x}$

(b)  ➤  Substitute $(1, 1)$ into this formula:  at $(1, 1)$  $\dfrac{dy}{dx} = \dfrac{8 + 3}{10 - 3} = \dfrac{11}{7}$

### The Second Derivative

## Example 16.14

Find $\dfrac{d^2y}{dx^2}$ for the circle $x^2 + y^2 = 4$

➤  Find $\dfrac{dy}{dx}$:  $2x + 2y\dfrac{dy}{dx} = 0$  ①

$\therefore \dfrac{dy}{dx} = -\dfrac{x}{y}$  ②

To find $\dfrac{d^2y}{dx^2}$ we must differentiate again. It is possible to differentiate either ① or ②.

For this example there is not much difference in difficulty between the two possibilities, but although differentiating ② involves differentiating a quotient implicitly, differentiating ① would introduce terms involving $\dfrac{d^2y}{dx^2}, \dfrac{dy}{dx}$ and $\left(\dfrac{dy}{dx}\right)^2$ as well as $x$ and $y$, which becomes very complicated if you have to give your answer in terms of $x$ and $y$ alone. Usually it is best to differentiate the explicit equation ② for $\dfrac{dy}{dx}$.

➤ Differentiate ②: $\dfrac{d^2y}{dx^2} = -\dfrac{y(1) - x\dfrac{dy}{dx}}{y^2}$

➤ Substitute for $\dfrac{dy}{dx}$: $\dfrac{d^2y}{dx^2} = -\dfrac{y - x\left(-\dfrac{x}{y}\right)}{y^2}$

➤ Simplify: $\dfrac{d^2y}{dx^2} = -\dfrac{(y^2 + x^2)}{y^3}$

➤ but $x^2 + y^2 = 4$ so $\dfrac{d^2y}{dx^2} = -\dfrac{4}{y^3}$

## Example 16.15

Find the coordinates of the maximum point on the curve $x^2 + 3y^2 - xy = 11$  ①

➤ At a maximum point $\dfrac{dy}{dx} = 0$ and $\dfrac{d^2y}{dx^2}$ is negative

➤ Find $\dfrac{dy}{dx}$: $2x + 6y\dfrac{dy}{dx} - y - x\dfrac{dy}{dx} = 0$ $\therefore$ $\dfrac{dy}{dx} = \dfrac{2x - y}{x - 6y}$  ②

➤ $\therefore$ $\dfrac{dy}{dx} = 0$ when $2x = y$

➤ substitute for $y$ in ① to find coordinates of critical points:

$$x^2 + 3(2x)^2 - x(2x) = 11 \therefore x^2 = 1 \text{ i.e. } x = \pm 1$$

critical points are $(1, 2)$ and $(-1, -2)$

➤ find $\dfrac{d^2y}{dx^2}$: $\dfrac{d^2y}{dx^2} = \dfrac{(x - 6y)\left(2 - \dfrac{dy}{dx}\right) - (2x - y)\left(1 - 6\dfrac{dy}{dx}\right)}{(x - 6y)^2}$

➤ since we only need to know whether $\dfrac{d^2y}{dx^2}$ is positive or negative at the turning points we do not need to simplify this expression, and we can immediately put $\dfrac{dy}{dx} = 0$, also the denominator is squared so must be positive and can be ignored

➤ at a turning point $\dfrac{d^2y}{dx^2} = \dfrac{2(x - 6y) - (2x - y)}{(x - 6y)^2}$

➤ at $(1, 2)$, $\dfrac{d^2y}{dx^2} < 0$ and at $(-1, -2)$, $\dfrac{d^2y}{dx^2} > 0$

➤ maximum point is at $(1, 2)$

## Example 16.16

If $e^x y = \sin x$, show that $\dfrac{d^2y}{dx^2} + 2\dfrac{dy}{dx} + 2y = 0$

Although this problem could be solved by writing $y = e^{-x}\sin x$ (you can try it if you like), it would involve differentiating a product twice and it is in fact much simpler to differentiate implicitly.

➤ Differentiate every term w.r.t. $x$:      $\dfrac{d}{dx}(e^x\,y) = \dfrac{d}{dx}(\sin x)$    ①

➤ $\left[\text{Notice that the first term is a product so: } \dfrac{d}{dx}(e^x y) = e^x y + e^x \dfrac{dy}{dx}\right]$

➤ ① becomes:      $e^x y + e^x \dfrac{dy}{dx} = \cos x$

➤ Differentiate again:      $e^x y + e^x \dfrac{dy}{dx} + e^x \dfrac{dy}{dx} + e^x \dfrac{d^2 y}{dx^2} = -\sin x$

➤ But $\sin x = e^x y$, so :      $e^x y + 2e^x \dfrac{dy}{dx} + e^x \dfrac{d^2 y}{dx^2} = -e^x y,$

         giving $2e^x y + 2e^x \dfrac{dy}{dx} + e^x \dfrac{d^2 y}{dx^2} = 0$

➤ Since $e^x \neq 0$, divide by $e^x$    $\therefore$   $\dfrac{d^2 y}{dx^2} + 2\dfrac{dy}{dx} + 2y = 0$

# EXERCISE 16 F

1. For the following curves defined parametrically find $\dfrac{dy}{dx}$ in terms of $t$ and the gradient of the curve for the given value of $t$ : ( 2 mks each)

   (a) $x = 4t^2, \; y = 8t$      $(t = 2)$        (b) $x = t^2 + 4, \; y = \sqrt{t}$      $(t = 1)$

   (c) $x = 2t, \; y = \dfrac{1}{t}$      $(t = 2)$        (d) $x = 3 + 2t, \; y = 4t^2 - 9$      $(t = -2)$

   (e) $x = 2\cos t, \; y = 5\sin t$    $(t = \tfrac{1}{4}\pi)$      (f) $x = 2\sin t, \; y = \cos^2 t$    $(t = \tfrac{1}{6}\pi)$

   (g) $x = \cos t, \; y = \sin 2t$    $(t = \tfrac{1}{6}\pi)$      (h) $x = \ln t, \; y = 1 - t^2$    $(t = \tfrac{1}{2})$

   (i) $x = e^t \sin t, \; y = e^t \cos t$   $(t = 0)$      (j) $x = e^{\sin t}, \; y = e^{\cos t}$    $(t = \tfrac{1}{4}\pi)$

   (k) $x = 1 + t^2, \; y = t + \sqrt{t}$   $(t = 1)$      (l) $x = \sin t, \; y = \ln 2t$    $(t = \tfrac{1}{3}\pi)$

2. Find $\dfrac{dy}{dx}$ in terms of $x$ and $y$ for the curves whose implicit equations are: ( 2 mks each)

   (a) $3y^2 - 4x^2 - 3 = 0$      (b) $x^2 + y - y^2 = 5$      (c) $y^2 + 3y - 2x = 7$

   (d) $y^2 - xy + x - y = 2$      (e) $y^2 = 4e^x + 2$      (f) $e^y = 5x^2 - 4$

   (g) $y + y^2 = \sin x$      (h) $\sin y = x^2 + 5$      (i) $xe^y = x + 3$

   (j) $x^3 + 3x^2 y = 2y^2$      (k) $x \cos y + y \sin x = 0$      (l) $y^2 + x \ln y = 0$

3. Find the equation of the tangent to the curve whose parametric equations are $x = t^2, \; y = t - \dfrac{1}{t}$ at the point on the curve where $t = 2$. ②

4. Find the gradient of the normal to the curve $x^3 + 2xy = 32 + y^2$ at the point $(3, 5)$

5. Find $\dfrac{dy}{dx}$ for the circle $x^2 + 16x + y^2 - 4y - 101 = 0$. Hence    4 mks

   (a) find the equation of the tangent to the circle at the point $(4, -3)$.

   (b) find the points on the circle where the gradient is undefined (i.e. the tangent is vertical)

✳ 6.  A curve has parametric equations $x = \theta - \sin\theta$, $y = \cos\theta$. Find the coordinates of a point at which the gradient of this curve is 0.　　3 mks

7.  Given that $y^2 - 5xy + 8x^2 = 2$, prove that $\dfrac{dy}{dx} = \dfrac{5y - 16x}{2y - 5x}$.　　2mks

8.  Find the equations of the tangents to the curve $y^2 + 3xy + 4x^2 = 14$ at the points where $x = 1$.

9.  A curve is given parametrically by $x = t^2$, $y = t^3$.
    Find an equation for the normal to the curve at the point $(4, 8)$.

10.  The parametric equations of a curve are $x = t - e^{2t}$, $y = e^{2t}$, where $t$ takes all real values.

  (a)  Express $\dfrac{dy}{dx}$ in terms of $t$.

  (b)  Hence find the value of $t$ for which the gradient of the curve is 3, giving your answer in logarithmic form.

11.  A curve C is given by the parametric equations $x = t + \dfrac{1}{t}$, $y = t - \dfrac{1}{t}$, $t > 0$.

  (a)  Find $\dfrac{dy}{dx}$ in terms of $t$ and deduce that the tangent to C at the point A $(2, 0)$ has equation $x - 2 = 0$.
  (b)  For every point $(x, y)$ on C, show that $x^2 - y^2 = 4$.

#12.  A curve C is given by the equations
    $x = 2\cos t + \sin 2t$, $y = \cos t - 2\sin 2t$, $0 < t < \pi$, where $t$ is a parameter.

  (a)  Find $\dfrac{dx}{dt}$ and $\dfrac{dy}{dt}$ in terms of $t$.

  (b)  Find the value of $\dfrac{dy}{dx}$ at the point P on C where $t = \frac{1}{4}\pi$.

  (c)  Find an equation for the normal to the curve at P.

13.  A curve C has equation $x^2 + 4y^2 - 4x - 12y - 12 = 0$.

  (a)  By differentiation find the gradient of C at the point P $(6, 3)$.

  (b)  Find an equation of the tangent to the curve at P.

14.  By rewriting $y = 2^x$ as a logarithm equation and using implicit differentiation, show that if $y = 2^x$, $\dfrac{dy}{dx} = 2^x \ln 2$.

15.  Find $\dfrac{d^2 y}{dx^2}$ for each of the following curves:

  (a)  $x = t + 2$, $y = t^2 - 3$

  (b)  $x = 3 - 2t^2$, $y = \dfrac{1}{t}$

  (c)  $x = \sin t$, $y = \cos 2t$ (remember to simplify $\dfrac{dy}{dx}$ before continuing, and simplify the final answer)

16. The parametric equations of a curve are $x = e^t$, $y = \sin t$.

    Find $\dfrac{dy}{dx}$ and $\dfrac{d^2y}{dx^2}$ as functions of $t$.

    Hence show that $x^2\dfrac{d^2y}{dx^2} + x\dfrac{dy}{dx} + y = 0$.

17. The parametric equations of a curve are $x = \sin t$, $y = \cos t + \sin t$.

    By finding $\dfrac{dy}{dx}$ and $\dfrac{d^2y}{dx^2}$ find the turning points on the curve, and state whether they are maximum or minimum.

18. If $e^x y = \cos x$, show that $\dfrac{d^2y}{dx^2} + 2\dfrac{dy}{dx} + 2y = 0$

19. If $x^2 y = 3\cos 2x$, show that $x^2\dfrac{d^2y}{dx^2} + 4x\dfrac{dy}{dx} + (4x^2 + 2)y = 0$

20. If $x^2 + y^2 = 2y$, find $\dfrac{dy}{dx}$ in terms of $x$ and $y$. Prove that $\dfrac{d^2y}{dx^2} = \dfrac{1}{(1-y)^3}$.

21. If $e^t x = \sin 2t$, show that $\dfrac{d^2x}{dt^2} + 2\dfrac{dx}{dt} + 5x = 0$

22. Find the gradient of the curve $xy^3 - x^2 y = 6$ at the point $(1, 2)$

23. Find the critical point/s on the curves
    (a) $3x^2 + x^2 y + y = 2$        (b) $y + xy - x^2 = 3$      and determine their nature

# EXERCISE 16 X

1. Find the derivative of the function f at the point $x$, if
   (a) $f(x) = \sin^2(2x + 3)$        (b) $f(x) = e^{3x}\ln 2x$      [2]    [2]

2. A function f is defined on the set $\mathbb{R}$ of real numbers by

   $f : x \mapsto \sqrt{5 + 4x^2}$.

   Calculate the derivative of f at $x = 1$.      [2]

3. Find the derivative with respect to $x$ for EACH of the following functions:

   (a) $f(x) = e^{x^2}\ln\left(\dfrac{x}{2}\right)$, $x > 0$.      (b) $f(x) = \dfrac{x^2 - 5x}{2 + 5x}$, for $x \neq -\tfrac{2}{5}$.      [3]    [3]

4. Differentiate, with respect to $x$,
   (a) $f(x) = e^{\sin x}\cos x$        (b) $f(x) = \tan(3x^2 + 5)$.      [4]    [4]

5. Find the derivative of the function f when
   (a) $f(x) = x^2\tan 3x$        (b) $f(x) = x\ln x - x$, $x > 0$      [3]    [3]

6. (a) Find the derivative of $x^2 \sin x^2$ with respect to $x$. [3]

(b) Find the gradient of the curve $y^2 + xy = 4x^2$ at the point $(1, -2)$. [4]

7. Let $\sin u = x$ and $\cos v = x$, where $u$ and $v$ are acute angles.

(i) Prove that $u + v = \tfrac{1}{2}\pi$ [3]

(ii) Show that $\dfrac{du}{dx} = \dfrac{1}{\sqrt{1 - x^2}}$. [4]

(iii) Hence, find $\dfrac{dv}{dx}$. [2]

8. Given that the derivative at $x$ of the exponential function $e^x$ is $e^x$,

(i) Find the derivative at $x$ of the logarithmic function $\ln x$. [3]

(ii) hence, find the derivative at $x$ of the function $e^x \ln x$. [2]

9. The parametric equations of a curve are given by $x = t^3\, e^t$ and $y = t\, e^t$.

(i) Show that $\dfrac{dy}{dx} = \dfrac{t+1}{t^2(t+3)}$. [4]

(ii) Hence, state the values of $t$ for which the derivative of the curve is not defined. [2]

(iii) Determine the equation of the tangent at the stationary point on the curve. [3]

10. Using the derivatives of $\sin x$ and $\cos x$ with respect to $x$, show that the derivative of $\tan x$ with respect to $x$ is $\sec^2 x$. [5]

11. A curve has equation $y = \tan(x + \tfrac{1}{6}\pi)$, where $0 < x < \tfrac{1}{3}\pi$.

(i) Show that $\dfrac{dy}{dx} = 1 + y^2$. [3]

(ii) Hence, calculate the exact coordinates of the point on the curve at which the gradient is equal to 2. [6]

(iii) Find an expression for $\dfrac{d^2 y}{dx^2}$ in terms of $y$. [2]

12. Use the fact that $e^{-x} = \dfrac{1}{e^x}$ to show that $\dfrac{d}{dx}(e^{-x}) = -e^{-x}$. [3]

13. If $x = e^{-3t}\sin 3t$, show that $\dfrac{d^2 x}{dt^2} + 6\dfrac{dx}{dt} + 18x = 0$. [8]

14. Given that the derivative at $x$ of the exponential function $e^x$ is $e^x$, show that

(i) by using $e^{-x} = \dfrac{1}{e^x}$, the derivative of $e^{-x}$ is $-e^{-x}$. [4]

(ii) if $u = e^x + e^{-x}$, $v = e^x - e^{-x}$, then $\dfrac{du}{dx} = v$ and $\dfrac{dv}{dx} = u$ [6]

(iii) $\dfrac{d^2 u}{dx^2} = u$ [3]

15. Differentiate with respect to $x$ from first principles $(1 + x^2)^{-1}$ [6]

16. Differentiate with respect to $x$,
    (a) $e^{2x}\sin 3x$ [3]
    (b) $(\ln x)^2$ [3]

17. Find $\dfrac{dy}{dx}$ when
    (i) $y = x^2 \tan 3x$ [4]
    (ii) $y = \dfrac{t}{1 + t^2}$, $x = 3t^2$ [7]

18. Differentiate with respect to $x$,
    (a) $y = \ln(3x^2)$ $x \neq 0$ [2]
    (b) $y = \sin^2 x \cos x$ [3]

19. Find the gradient at the point $(1, 1)$ on the curve $2xy + y^2 - 3 = 0$ [5]

20. Differentiate, with respect to $x$, EACH function below. Simplify your answers as far as possible
    (a) $\dfrac{e^x}{x + 1}$ [5]
    (b) $\tan^2(x^3)$ [5]

21. The parametric equations of a curve are given by $x = 3 - 2t$, $y = t(1 - t)$.
    (a) Find $\dfrac{dy}{dx}$ in terms of $t$. [4]
    (b) A normal to the curve is parallel to the line $x + y = 2$. Find the equation of this normal. [6]

22. Differentiate, with respect to $x$,
    (a) $f(x) = 2\sin^2(4x + 1)$  (b) $f(x) = -e^{2x+3}\cos 3x$ [4] [5]

23. Given that $x = 2\theta - \sin 2\theta$ and $y = 1 - \cos 2\theta$, show that
    (a) $\dfrac{dy}{dx} = \cot\theta$ [4]
    (b) $\dfrac{d^2 y}{dx^2} = \dfrac{-1}{4\sin^4}$ [4]

24. Differentiate with respect to $x$ the following:
    (a) $y = e^{\cos x}$ [2]
    (b) $y = x^2\sqrt{(1 + x)}$ [3]

25. Given that $u = \tan\theta$, use the quotient rule to show that $\dfrac{du}{d\theta} = \sec^2\theta$ [4]

26. Show that if $y = \sin x \cos x$ then

    (a) $\dfrac{dy}{dx} = 2\cos^2 x - 1$        [3]

    (b) $\dfrac{d^2 y}{dx^2} + 4y = 0$        [3]

27. A curve is given parametrically by $x = (3 - 2t)^2$, $y = t^3 - 2t$. Find

    (a) $\dfrac{dy}{dx}$ in terms of $t$.        [4]

    (b) the gradient of the normal to the curve at the point $t = 2$.        [2]

28. Differentiate with respect to $x$ the following:

    (a) $y = e^{2x + \sin x}$        [3]

    (b) $y = \tan 3x + \ln(x^2 + 4)$        [4]

29. (a) Find the gradient of the curve $x^2 + xy = 2y^2$ at the point P $(-2, 1)$.        [5]

    (b) Hence, find the equation of the normal to the curve at P.        [3]

30. If $y = \sin 2x + \cos 2x$

    (a) find $\dfrac{dy}{dx}$        [3]

    (b) show that $\dfrac{d^2 y}{dx^2} + 4y = 0$        [4]

31. If $f(x) = x^3 \ln^2 x$, show that

    (a) $f'(x) = x^2 \ln x (3\ln x + 2)$        [5]

    (b) $f''(x) = 6x \ln^2 x + 10x \ln x + 2x$        [5]

# What you have learnt in Chapter 16

## Derivatives of Some Common Functions

| $f(x)$ | $f'(x)$ |
|---|---|
| $ax^n$ | $anx^{n-1}$ |
| $\sin(ax+b)$ | $a\cos(ax+b)$ |
| $\cos(ax+b)$ | $-a\sin(ax+b)$ |
| $\tan(ax+b)$ | $a\sec^2(ax+b)$ |
| $e^{(ax+b)}$ | $a\,e^{(ax+b)}$ |
| $\ln ax$ | $\dfrac{1}{x}$ |
| $\sin^{-1}\left(\dfrac{x}{a}\right) \quad |x|<a$ | $\dfrac{1}{\sqrt{a^2-x^2}}$ |
| $\tan^{-1}\left(\dfrac{x}{a}\right)$ | $\dfrac{a}{a^2+x^2}$ |

### The Chain Rule

if $y$ is a function of $u$ and $u$ is a function of $x$, then $\dfrac{dy}{dx}=\dfrac{dy}{du}\times\dfrac{du}{dx}$

**or** if $y=f\{g(x)\}$, then $\dfrac{dy}{dx}=f'\{g(x)\}\,g'(x)$

in particular: $\dfrac{d}{dx}\{\ln f(x)\}=\dfrac{f'(x)}{f(x)}$

### The Product Rule

$$\frac{dy}{dx}=v\frac{du}{dx}+u\frac{dv}{dx}$$

### The Quotient Rule

$$\frac{dy}{dx}=\frac{v\dfrac{du}{dx}-u\dfrac{dv}{dx}}{v^2}$$

### Parametric Form: $x=f(t)$ and $y=g(t)$

$$\frac{dy}{dx}=\frac{dy}{dt}\times\frac{dt}{dx}=\frac{dy}{dt}\div\frac{dx}{dt}$$

or $\dfrac{dy}{dx}=g'(t)\div f'(t)$

### Implicit Differentiation

$$\frac{d}{dx}\{f(y)\}=f'(y)\frac{dy}{dx}$$

# CHAPTER 17
# 2.1.c Integration II

## INTEGRATION OF STANDARD FUNCTIONS

Since integration is the inverse of differentiation, you should now be able to integrate many simple functions:

| $f(x)$ | $\int f(x)\,dx$ |
|---|---|
| $e^x$ | $e^x + c$ |
| $e^{(ax+b)}$ | $\frac{1}{a}e^{(ax+b)} + c$ |
| $\sin x$ | $-\cos x + c$ |
| $\sin(ax+b)$ | $-\frac{1}{a}\cos(ax+b) + c$ |
| $\cos x$ | $\sin x + c$ |
| $\cos(ax+b)$ | $\frac{1}{a}\sin(ax+b) + c$ |
| $\sec^2 x$ | $\tan x + c$ |
| $\sec^2(ax+b)$ | $\frac{1}{a}\tan(ax+b) + c$ |
| $\dfrac{1}{\sqrt{1-x^2}},\ |x|<1$ | $\sin^{-1} x + c$ |
| $\dfrac{1}{\sqrt{a^2-x^2}},\ |x|<a$ | $\sin^{-1}\left(\dfrac{x}{a}\right) + c$ |
| $\dfrac{1}{1+x^2}$ | $\tan^{-1} x + c$ |
| $\dfrac{a}{a^2+x^2}$ | $\tan^{-1}\left(\dfrac{x}{a}\right) + c$ |
| $(ax+b)^n\ \ n\neq -1$ | $\dfrac{(ax+b)^{n+1}}{a(n+1)} + c$ |
| $\dfrac{1}{x} = x^{-1}\ (x\neq 0)$ | $\ln|x| + c$ or $\ln A|x|$ |
| $\dfrac{f'(x)}{f(x)}$ | $\ln|f(x)| + c$ or $\ln A|f(x)|$ |

$\frac{d(\tan x)}{dx} = \sec^2 x$
$\frac{d(\sin x)}{dx} = \cos x$
$\frac{d(\cos x)}{dx} = -\sin x$

47

# Integration Leading to Logarithmic Functions

Notice the last two results in the table. Where does the modulus sign come from?

Since $\dfrac{d}{dx}(\ln x) = \dfrac{1}{x}$ you would expect that $\displaystyle\int \dfrac{1}{x}\,dx = \ln x + c$ ①. But $\ln x$ only exists if $x > 0$ so ① is only true for $x > 0$. However the function $f(x) = \dfrac{1}{x}$ is defined for $x \in \mathbb{R}$, $x \neq 0$, so what happens to $\displaystyle\int \dfrac{1}{x}\,dx$ when $x < 0$?

Since $\dfrac{1}{x} = \dfrac{-1}{-x}$, we can write, when $x < 0$, $\displaystyle\int \dfrac{1}{x}\,dx = \int \dfrac{-1}{-x}\,dx$ which equals $\ln(-x)$ ② and this is valid since $-x$ is a positive number.

Combining ① and ② gives $\displaystyle\int \dfrac{1}{x}\,dx = \ln|x| + c$, $x \neq 0$, and similarly $\displaystyle\int \dfrac{f'(x)}{f(x)}\,dx = \ln|f(x)| + c$

**A note on the arbitrary constant when an integration results in a logarithmic function**

Since $\ln A$ is a constant, when an integration results in a logarithmic function $\ln A$ can be used instead of $c$ as the arbitrary constant. This can simplify the working and answers in certain problems, such as when partial fractions are used and especially when solving differential equations leading to exponential functions.

## Example 17.1

Integrate the following functions w.r.t. $x$

(a) $\dfrac{3}{x}$ (b) $\dfrac{5 + 15e^{3x}}{x + e^{3x}}$ (c) $\dfrac{5}{25 + x^2}$ (d) $\dfrac{1}{\sqrt{9 - x^2}}$

(in these examples every step in the working is shown to explain the method, you do not need to show so much working)

(a) ➤ $\dfrac{3}{x} = 3\left(\dfrac{1}{x}\right)$ so $\displaystyle\int \dfrac{3}{x}\,dx = 3\int \dfrac{1}{x}\,dx$

    ➤ integrate:             $= 3\ln|x| + c$

    ➤ write constant as logarithm: $= 3\ln|x| + \ln A$

    ➤ apply log laws:        $= \ln|Ax^3|$

(b) ➤ $1 + 3e^{3x}$ is the derivative of $x + e^{3x}$, so the integral is a logarithm

    ➤ take out factor 5:     $\displaystyle\int \dfrac{5 + 15e^{3x}}{x + e^{3x}}\,dx = 5\int \dfrac{1 + 3e^{3x}}{x + e^{3x}}$

    ➤ integrate:               $= 5\ln|x + e^{3x}| + c$

    ➤ replace $c$ with $\ln A$:      $= 5\ln|x + e^{3x}| + \ln A$

    ➤ apply log laws:        $= 5\ln A|x + e^{3x}|$ or $\ln|A(x + e^{3x})^5|$

(c)   ➤   you need to be able to recognise this as the derivative of an inverse trig function

    ➤   $a = 5$, so $\displaystyle\int \frac{5}{25 + x^2}\, dx = \tan^{-1}\left(\frac{x}{5}\right)$

(d)   ➤   $\displaystyle\int \frac{1}{\sqrt{9 - x^2}}\, dx = \sin^{-1}\left(\frac{x}{3}\right)$

$y = \sin^{-1}\frac{x}{3}$

$\sin y = \frac{x}{3}$

$\dfrac{-\cos\left(\frac{1}{2}x\right)}{\frac{1}{2}}$

# EXERCISE 17 A

1. Integrate the following functions w.r.t. $x$ :

(a) $e^{5x}$        (b) $12e^{-3x}$        (c) $e^{4x - 3}$        (d) $2e^{5 - x}$

(e) $3 - e^{-x}$        (f) $e^x + e^{-x}$        (g) $e^{4x - 1} + x^2$        (h) $\frac{1}{2}e^{2x + 1}$

(i) $\sin 3x$        (j) $\cos 5x$        (k) $\cos \frac{1}{2}x$        (l) $\sin \pi x$

(m) $\sec^2 4x$        (n) $\cos \frac{1}{3}x + \sin 2x$        (o) $\sec^2 (3x + \frac{1}{2}\pi)$        (p) $\cos (4x - \frac{1}{3}\pi)$

(q) $(5x - 3)^4$        (r) $\sqrt{(3x + 4)}$        (s) $\dfrac{12}{(3x + 5)^2}$        (t) $\dfrac{1}{\sqrt{5x + 2}}$

(u) $\dfrac{5}{x}$ (be careful)        (v) $\dfrac{2x}{x^2 - 1}$        (w) $\dfrac{2x + 3}{2x^2 + 6x - 1}$        (x) $\dfrac{e^{2x}}{5 + e^{2x}}$

2. Integrate the following functions w.r.t. $x$ :

(a) $\sin (2x + \frac{1}{4}\pi)$        (b) $e^x + x^{-1}$        (c) $\dfrac{8x}{x^2 - 1}$        (d) $\dfrac{2}{(x - 1)^2}$

(e) $\dfrac{7}{x} - \dfrac{6}{2x + 3}$        (f) $20(2x - 5)^{\frac{3}{2}}$        (g) $4 + e^{-(x + 5)}$        (h) $2x^3 - 3x^2 + 2x^{-1}$

(i) $\dfrac{\cos x}{1 + \sin x}$        (j) $(4 - 3x)^5$        (k) $\sqrt{(2x - 7)}$        (l) $\dfrac{x^2}{2 - x^3}$

(m) $\dfrac{1}{\sqrt{1 - x^2}}$        (n) $\dfrac{4}{x^2 + 1}$        (o) $\dfrac{3}{\sqrt{9 - x^2}}$        (p) $\dfrac{75}{x^2 + 25}$

3. Evaluate the following definite integrals:

(a) $\displaystyle\int_0^{\frac{\pi}{2}} \cos \tfrac{1}{3}x\, dx$        (b) $\displaystyle\int_0^1 e^{2x}\, dx$        (c) $\displaystyle\int_0^1 (3 - 2x)^3\, dx$        (d) $\displaystyle\int_0^{\frac{\pi}{3}} \sin \tfrac{1}{2}x\, dx$

(e) $\displaystyle\int_1^3 \frac{dx}{x}$        (f) $\displaystyle\int_0^{\ln 3} 2e^x\, dx$        (g) $\displaystyle\int_0^3 \frac{4}{\sqrt{1 + x}}\, dx$        (h) $\displaystyle\int_2^4 \frac{2x - 3}{x^2 - 3x + 5}\, dx$

(i) $\displaystyle\int_0^2 \frac{8}{x^2 + 4}\, dx$        (j) $\displaystyle\int_0^{\frac{1}{2}} \frac{3}{\sqrt{1 - x^2}}\, dx$

# PARTIAL FRACTIONS

You know how to add algebraic fractions.

For instance if $\qquad f(x) = \dfrac{5}{2x+1} + \dfrac{3}{x-2}$ ① 

then $\qquad f(x) = \dfrac{5(x-2)+3(2x+1)}{(2x+1)(x-2)}$

so $\qquad f(x) = \dfrac{11x-7}{(2x+1)(x-2)}$ ②

As it stands, ② cannot be integrated, but you should be able to recognise that the two terms of ① can both be integrated to give ln functions. It would therefore be useful, if given a function like ②, to convert it into a function like ①.

So, can we write $f(x) \equiv \dfrac{5x+4}{(x-1)(x+2)}$, for instance, as the sum of two simpler fractions? We can, and when

we do so we say that we are expressing $f(x)$ in partial fractions (or decomposing $f(x)$ into partial fractions).

## Example 17.2

Express $f(x) \equiv \dfrac{5x+4}{(x-1)(x+2)}$ in partial fractions.

➤ Choose the factors of the denominator of $f(x)$ as denominators of separate fractions and let

$$\frac{5x+4}{(x-1)(x+2)} \equiv \frac{A}{(x-1)} + \frac{B}{(x+2)}$$

➤ Putting RHS onto a common denominator: $\dfrac{5x+4}{(x-1)(x+2)} \equiv \dfrac{A(x+2)+B(x-1)}{(x-1)(x+2)}$

➤ Since denominators are the same, so must the numerators be, so

$$5x+4 \equiv A(x+2)+B(x-1)$$ ③

➤ This is an identity so the coefficient of $x$ and the constant must be equal on both sides, giving

$$A=3 \text{ and } B=2, \text{ so } f(x) \equiv \frac{5x+4}{(x-1)(x+2)}$$

$$\equiv \frac{3}{x-1} + \frac{2}{x+2}$$

There are two ways in which you could have found that $A=3$ and $B=2$

➤ Method I: multiply out the brackets and equate coefficients

$$5x+4 \equiv Ax+2A+Bx-B \implies A+B=5 \text{ and } 2A-B=4 \implies A=3 \text{ and } B=2$$

➤ Method II: eliminate $B$ by putting $x=1$, and $A$ by putting $x=2$

when $x=1$, ③ becomes $5+4 = A\times 3 + B\times 0$ ∴ $A=3$

when $x=-2$, ③ becomes $-10+4 = A\times 0 + B\times -3$ ∴ $B=2$

Method II is usually simpler, but as you will see, there are some situations where it cannot completely solve the problem of finding partial fractions, in which case method I or a combination of the two methods can be used.

Example 17.2 is an example of the simplest type of decomposition into partial fractions – the denominator is the product of two linear factors. This is sometimes known as a *type I* partial fraction.

## Type I Partial Fractions

The denominator is the product of (any number of) linear factors. e.g. $\dfrac{px^2 + qx + r}{(ax + b)(cx + d)(ex + f)}$

To solve, assume the fraction can be written as the sum of terms consisting of a constant divided by one of the linear factors, and follow the steps given in example 17.1.

The method is the same however many factors there are, though we are only concerned with 2 factors.

## Example 17.3

Express $f(x) \equiv \dfrac{x - 3}{(x + 3)(x - 1)}$ in partial fractions.

➢ Choose the factors of the denominator of $f(x)$ as denominators of separate fractions and let

$$f(x) \equiv \frac{x - 3}{(x + 3)(x - 1)} \equiv \frac{A}{(x + 3)} + \frac{B}{(x - 1)}$$

➢ Putting RHS onto a common denominator: $\dfrac{x - 3}{(x + 3)(x - 1)} \equiv \dfrac{A(x - 1) + B(x + 3)}{(x + 3)(x - 1)}$

➢ Equating numerators: $\qquad x - 3 \equiv A(x - 1) + B(x + 3)$

➢ Eliminate $B$ by putting $x = -3$: $\quad -4A = -6 \implies A = \frac{3}{2}$

➢ Eliminate $A$ by putting $x = 1$: $\quad 4B = -2 \implies B = -\frac{1}{2}$

➢ Give the answer: $\quad f(x) \equiv \dfrac{x - 3}{(x + 3)(x - 1)} \equiv \dfrac{3}{2(x + 3)} - \dfrac{1}{2(x - 1)}$

## Type II Partial Fractions

The denominator contains an irreducible quadratic factor. e.g. $\dfrac{px^2 + qx + r}{(ax^2 + bx + c)(dx + e)}$

Again the partial fractions will have the factors of the original expression as denominators, but this time, it is not enough to assume that the quadratic denominator has a constant numerator, it must have a linear numerator

## Example 17.4

Express $f(x) \equiv \dfrac{x^2 + 2x - 3}{(x^2 + 1)(x - 2)}$ in partial fractions

➢ Choose the factors of the denominator of $f(x)$ as denominators of separate fractions and let

$$f(x) \equiv \frac{x^2 + 2x - 3}{(x^2 + 1)(x - 2)} \equiv \frac{Ax + B}{(x^2 + 1)} + \frac{C}{(x - 2)}$$

➢ Putting RHS onto a common denominator: $\dfrac{x^2 + 2x - 3}{(x^2 + 1)(x - 2)} \equiv \dfrac{(Ax + B)(x - 2) + C(x^2 + 1)}{(x^2 + 1)(x - 2)}$

➢ Equating numerators: $\quad x^2 + 2x - 3 \equiv (Ax + B)(x - 2) + C(x^2 + 1)$ ①

➢ Find $C$ by putting $x = 2$: $\quad 4 + 4 - 3 = 5C \implies C = 1$

➢ $A$ and $B$ cannot be found by this method, so

> Expand ①: $\qquad x^2 + 2x - 3 \equiv Ax^2 - 2Ax + Bx - 2B + Cx^2 + C$
> Equate coefficients: of $x^2$: $\quad A + C = 1 \Rightarrow A = 0$
> $\qquad\qquad\qquad$ of constant: $\quad C - 2B = -3 \Rightarrow B = 2$
> Give the answer: $\quad f(x) \equiv \dfrac{x^2 + 2x - 3}{(x^2 + 1)(x - 2)} \equiv \dfrac{2}{(x^2 + 1)} + \dfrac{1}{(x - 2)}$

## Example 17.5

Express $f(x) \equiv \dfrac{x+1}{(x^2 - 3)(x + 2)}$ in partial fractions

> Choose the factors of the denominator of f(x) as denominators of separate fractions and let
> $$f(x) \equiv \dfrac{x+1}{(x^2 - 3)(x + 2)} \equiv \dfrac{Ax + B}{x^2 - 3} + \dfrac{C}{x + 2}$$
> Putting RHS onto a common denominator: $\dfrac{x+1}{(x^2 - 3)(x + 2)} \equiv \dfrac{(Ax + B)(x + 2) + C(x^2 - 3)}{(x^2 - 3)(x + 2)}$
> Equating numerators: $\quad x + 1 \equiv (Ax + B)(x + 2) + C(x^2 - 3)$ $\qquad$ ①
> Find C by putting $x = -2$: $\quad -1 = C \Rightarrow C = -1$
> Expand ①: $\qquad x + 1 \equiv Ax^2 + 2Ax + Bx + 2B + Cx^2 - 3C$
> Equate coefficients: of $x^2$: $\quad A + C = 0 \Rightarrow A = 1$
> $\qquad\qquad\qquad$ constant: $\quad 2B - 3C = 1 \Rightarrow B = -1$
> Give the answer: $\quad f(x) \equiv \dfrac{x+1}{(x^2 - 3)(x + 2)} \equiv \dfrac{x-1}{x^2 - 3} - \dfrac{1}{x + 2}$

### Type III Partial Fractions

The denominator contains a repeated linear factor. e.g. $\dfrac{px^2 + qx + r}{(ax + b)^2 (cx + d)}$

This time, since you have **3** factors in the denominator, you need 3 partial fractions, and their denominators will be $(ax + b)^2$, $ax + b$ and $cx + d$.

## Example 17.6

Express $f(x) \equiv \dfrac{7x + 4}{(x + 1)^2 (x - 2)}$ in partial fractions

> Choose $(x + 1)^2$, $(x + 1)$ and $(x - 2)$ as denominators of separate fractions and let
> $$f(x) \equiv \dfrac{7x + 4}{(x + 1)^2 (x - 2)} \equiv \dfrac{A}{(x + 1)^2} + \dfrac{B}{x + 1} + \dfrac{C}{x - 2}$$
> Putting RHS onto a common denominator:
> $$\dfrac{7x + 4}{(x + 1)^2 (x - 2)} \equiv \dfrac{A(x - 2) + B(x + 1)(x - 2) + C(x + 1)^2}{(x + 1)^2 (x - 2)}$$

> Equating numerators: $\quad 7x + 4 \equiv A(x-2) + B(x+1)(x-2) + C(x+1)^2$ ①

> Find $A$ by putting $x = -1$: $\qquad -3 = -3A \Rightarrow A = 1$

> Find $C$ by putting $x = 2$: $\qquad 18 = 9C \Rightarrow C = 2$

> Find $B$ by equating coefficient of $x^2$ in ①: $\quad 0 = B + C \Rightarrow B = -C = -2$

> Give the answer: $\quad f(x) \equiv \dfrac{7x+4}{(x+1)^2(x-2)} \equiv \dfrac{1}{(x+1)^2} - \dfrac{2}{x+1} + \dfrac{2}{x-2}$

(Type III partial fractions can also be decomposed using a method similar to that used for type II, giving two fractions but still 3 constants. Using this method example 17.6 would be written as

$$f(x) \equiv \frac{7x+4}{(x+1)^2(x-2)} \equiv \frac{Px+Q}{(x+1)^2} + \frac{R}{(x-2)}$$

leading to the result $\quad f(x) \equiv \dfrac{7x+4}{(x+1)^2(x-2)} \equiv \dfrac{-2x-1}{(x+1)^2} + \dfrac{2}{x-2}$

Although this result is true, it is not helpful if the partial fractions are being used as a tool for integration.)

## EXERCISE 17 B

Express the following functions in partial fractions:

1. (a) $\dfrac{1}{x(x+1)}$    (b) $\dfrac{1}{(x+2)(x+3)}$    (c) $\dfrac{3-8x}{x-x^2}$    (d) $\dfrac{4x-9}{(x-2)(x-3)}$

   (e) $\dfrac{x+24}{x^2-x-12}$    (f) $\dfrac{7}{(x+2)(2x-3)}$    (g) $\dfrac{2x}{x^2+5x+6}$    (h) $\dfrac{5x-1}{(x-3)(x+4)}$

2. (a) $\dfrac{4}{x(x^2+1)}$    (b) $\dfrac{5}{(x+1)(x^2+4)}$    (c) $\dfrac{8x-1}{(x-2)(x^2+1)}$    (d) $\dfrac{x^2+3}{x(x^2+2)}$

   (e) $\dfrac{2}{(x-1)(x^2+1)}$    (f) $\dfrac{x+10}{(x^2+2)(x-2)}$    (g) $\dfrac{3x^2-4}{(x^2-2)(x+1)}$    (h) $\dfrac{10x}{(x-1)(x^2+9)}$

3. (a) $\dfrac{4}{(x+1)(x-1)^2}$    (b) $\dfrac{2}{x^2(x-1)}$    (c) $\dfrac{x+4}{(x+1)(x-2)^2}$    (d) $\dfrac{8+x}{(x-1)(x+2)^2}$

   (e) $\dfrac{5-x}{(1-2x)(x+1)^2}$    (f) $\dfrac{x}{(x-1)(x-2)^2}$    (g) $\dfrac{3}{x(3x-1)^2}$    (h) $\dfrac{x^2-1}{x^2(2x+1)}$

### *Improper Partial Fractions*

All the partial fractions we have met so far have involved proper fractions only – the highest power in the numerator has been lower than the highest power in the denominator.

When an expression which is an improper fraction needs to be expressed as partial fractions it must first be reduced to a polynomial and a proper fraction, and then the proper fraction expressed as partial fractions.

Where the degree of the numerator is greater than that of the denominator this is done by long division, but if the degree of numerator and denominator are the same, the method in example 17.7 is usually simpler. In either case, the first step is to express numerator and denominator as polynomials rather than factors – multiply out the brackets.

## Example 17.7

Express $f(x) \equiv \dfrac{x(x+1)}{(x-1)(x-2)}$ in partial fractions.

➤ multiply out numerator and denominator: $f(x) \equiv \dfrac{x^2+x}{x^2-3x+2}$

➤ rewrite the numerator so that it contains the denominator: $f(x) \equiv \dfrac{(x^2-3x+2)+4x-2}{x^2-3x+2}$

➤ write as a constant and proper fraction: $f(x) \equiv 1 + \dfrac{4x-2}{x^2-3x+2} \equiv 1 + g(x)$

➤ express the proper fraction in partial fractions: $g(x) \equiv \dfrac{4x-2}{(x-1)(x-2)} \equiv \dfrac{-2}{x-1} + \dfrac{6}{x-2}$

➤ give the answer: $f(x) \equiv \dfrac{x(x+1)}{(x-1)(x-2)} \equiv 1 - \dfrac{2}{x-1} + \dfrac{6}{x-2}$

## Example 17.8

Express $f(x) \equiv \dfrac{x^3}{(x-1)(x+3)}$ in partial fractions.

Here the degree of the numerator is greater than that of the denominator, so long division is needed.

➤ Multiply out the denominator: $f(x) \equiv \dfrac{x^3}{x^2+2x-3}$

➤ Divide the denominator into the numerator as far as the constant term in the quotient:

$$
\begin{array}{r}
x \;\; - \;\; 2 \phantom{)} \\
x^2+2x-3 \overline{)\, x^3 + 0x^2 + 0x + 0} \\
\underline{x^3 + 2x^2 - 3x} \phantom{00000} \\
-2x^2 + 3x \phantom{000} \\
\underline{-2x^2 - 4x + 6} \phantom{00} \\
7x - 6 \phantom{00}
\end{array}
$$

➤ Write $f(x)$ as a polynomial and proper fraction: $f(x) = x - 2 + \dfrac{7x-6}{(x-1)(x+3)}$

➤ Turn the proper fraction into partial fractions: $\dfrac{7x-6}{(x-1)(x+3)} \equiv \dfrac{1}{4(x-1)} + \dfrac{27}{4(x+3)}$

➤ Give the answer: $f(x) \equiv x - 2 + \dfrac{1}{4(x-1)} + \dfrac{27}{4(x+3)}$

# EXERCISE 17 C

1.  Express the following functions in partial fractions:

(a) $\dfrac{x^2 + 1}{x^2 - 1}$

(b) $\dfrac{x^2 + 1}{(x + 2)(x + 4)}$

(c) $\dfrac{x^2 + 3x - 13}{x^2 + x - 2}$

(d) $\dfrac{3x^2 - 5x + 3}{x(x - 3)}$

(e) $\dfrac{2x^3 + x^2 + x}{x^2 - 1}$

(f) $\dfrac{x^3 + 3x + 2}{(x^2 + 1)(x + 1)}$

(g) $\dfrac{2x^3 - x^2 - 3x + 4}{(x + 1)(x - 1)^2}$

(h) $\dfrac{x^4 - 3x^3 - 3}{x^3 - x^2}$

2.  Express the following functions in partial fractions:

(a) $\dfrac{6}{(2x - 1)(2x + 1)}$

(b) $\dfrac{10x}{(x - 1)(x^2 + 9)}$

(c) $\dfrac{5}{(x - 2)(x + 3)^2}$

(d) $\dfrac{1}{(2 - x)(1 + 2x)}$

(e) $\dfrac{2x^2 + 2x + 3}{x^2 - 1}$

(f) $\dfrac{14}{(x^2 + 3)(x + 2)}$

(g) $\dfrac{4}{(x + 3)(x - 1)^2}$

(h) $\dfrac{3x + 1}{(2x - 1)(x + 2)}$

(i) $\dfrac{x^3 + 3x^2 + 10}{(x + 1)(x + 4)}$

(j) $\dfrac{2x - 3}{x^2 - 1}$

(k) $\dfrac{2x^2 - 7}{(x - 3)(2x + 5)}$

(l) $\dfrac{2x^3 + 11}{(x^2 + 4)(x - 3)}$

## *Integration using Partial Fractions*

As mentioned earlier, one of the uses of partial fractions is to facilitate the integration of expressions involving algebraic fractions.

If we look at some of the examples used earlier in the chapter we will see how this works.

Suppose you had to find $I = \displaystyle\int \dfrac{5x + 4}{(x - 1)(x + 2)}\,dx$.

In this form you cannot do it, but we found in example 17.1 that

$$\frac{5x + 4}{(x - 1)(x + 2)} \equiv \frac{3}{x - 1} + \frac{2}{x + 2} \quad \text{and therefore}$$

$$\int \frac{5x + 4}{(x - 1)(x + 2)}\,dx \equiv \int \left\{ \frac{3}{x - 1} + \frac{2}{x + 2} \right\}dx$$

Both terms of this new integral can be integrated to give a logarithm function, and

$I = 3 \ln |x - 1| + 2 \ln |x + 2| + c,$

which can be simplified using the laws of logarithms to

$I = \ln |(x - 1)^3 (x + 2)^2| + c$ or $\ln A |(x - 1)^3 (x + 2)^2|.$

All proper *type I* partial fractions can be integrated in this way to give a log function answer.

*Type II* is more tricky – the second fraction integrates straightforwardly as a log function but the fraction with $x^2$ in the denominator might integrate in a number of different ways; we will look particularly at two that you should be able to do easily with the rules you already know, using inverse tan and log functions or a combination

(a) If there is no $x$ in the numerator, and there is a plus sign in the denominator use inverse tan

(b) If there is no constant in the numerator use a log function

(c) if the first fraction is of the form $\dfrac{px + q}{ax^2 + c}$ , where $c$ is positive and $p, q \neq 0$  it can be written as two separate fractions to which (a) and (b) above apply

*Type III* fractions can be integrated, but care must be taken.

Example 17.6 gave us:  $\dfrac{7x + 4}{(x+1)^2(x-2)} \equiv \dfrac{1}{(x+1)^2} - \dfrac{2}{x+1} + \dfrac{2}{x-2}$ .

Here the second and third terms integrate to give log functions, but the first term is of the form $(ax + b)^n$, where $n = -2$.

> **!** Don't forget when decomposing into partial fractions that improper fractions must first be written as a polynomial and a fraction

# Example 17.9

Find the following integrals:

(a) $\displaystyle\int \dfrac{x - 3}{(x+3)(x-1)}\,dx$  (b) $\displaystyle\int \dfrac{x^2 + 2x - 3}{(x^2 + 1)(x - 2)}\,dx$  (c) $\displaystyle\int \dfrac{x + 3}{(x-1)(x^2 + 3)}\,dx$  (d) $\displaystyle\int \dfrac{25}{(x-2)^2(x+3)}\,dx$

(a)  (Type I)

> ➤ Decompose the integrand into partial fractions:  $\dfrac{x - 3}{(x+3)(x-1)} \equiv \dfrac{3}{2(x+3)} - \dfrac{1}{2(x-1)}$

> ➤ Write as an integral:  $I = \displaystyle\int \dfrac{x-3}{(x+3)(x-1)}\,dx = \int \left\{ \dfrac{\frac{3}{2}}{(x+3)} - \dfrac{\frac{1}{2}}{(x-1)} \right\} dx$

> ➤ Integrate:  $I = \frac{3}{2}\ln|x+3| - \frac{1}{2}\ln|x-1| + c$

> ➤ Simplify the result:  $I = \ln\left( A\sqrt{\left|\dfrac{(x+3)^3}{(x-1)}\right|} \right)$

> **!** Notice that in the second line the fraction constants of the partial fractions were written in the numerator to facilitate the integration.

(b)  (Type II)

> ➤ Decompose the integrand into partial fractions:  $\dfrac{x^2 + 2x - 3}{(x^2 + 1)(x - 2)} \equiv \dfrac{2}{(x^2 + 1)} + \dfrac{1}{(x - 2)}$

> ➤ Write as an integral:  $I = \displaystyle\int \dfrac{x^2 + 2x - 3}{(x^2 + 1)(x - 2)}\,dx \equiv \int \dfrac{2}{(x^2 + 1)}\,dx + \int \dfrac{1}{(x - 2)}\,dx$

> ➤ Integrate:  $I = 2\tan^{-1}x + \ln|x - 2| + c$

(c)  (Type II)

➤ Decompose the integrand into partial fractions:  $\dfrac{x+3}{(x-1)(x^2+3)} \equiv \dfrac{1}{(x-1)} - \dfrac{x}{(x^2+3)}$

➤ Write as an integral:  $I = \displaystyle\int \dfrac{x+3}{(x-1)(x^2+3)}\,dx \equiv \int \left\{ \dfrac{1}{(x-1)} - \dfrac{x}{(x^2+3)} \right\}dx$

➤ Integrate:  $I = \ln|x-1| - \frac{1}{2}\ln|x^2+3| + c$

➤ Simplify the result:  $I = \ln A \left| \dfrac{(x-1)}{\sqrt{x^2+3}} \right|$

(d)  (Type III)

➤ Decompose the integrand into partial fractions:  $\dfrac{25}{(x-2)^2(x+3)} = \dfrac{5}{(x-2)^2} - \dfrac{1}{x-2} + \dfrac{1}{x+3}$

➤ Write as an integral:  $I = \displaystyle\int \dfrac{25}{(x-2)^2(x+3)}\,dx = \int \left\{ \dfrac{5}{(x-2)^2} - \dfrac{1}{x-2} + \dfrac{1}{x+3} \right\}dx$

➤ Integrate:  $I = \dfrac{5}{-1}(x-2)^{-1} - \ln|x-2| + \ln|x+3| + c$

➤ Simplify the result:  $I = \dfrac{-5}{(x+2)} + \ln\left|\dfrac{x+3}{x-2}\right| + c$

# EXERCISE 17 D

1. Find the following integrals:

(a) $\displaystyle\int \dfrac{4x-1}{(x+2)(x-1)}\,dx$
(b) $\displaystyle\int \dfrac{16}{(x+1)^2(x-3)}\,dx$
(c) $\displaystyle\int \dfrac{1}{(x+3)(x-5)}\,dx$

(d) $\displaystyle\int \dfrac{5x^2+2x+3}{(x+1)(x^2+1)}\,dx$
(e) $\displaystyle\int \dfrac{x-x^2}{(x^2+1)(x+1)}\,dx$
(f) $\displaystyle\int \dfrac{x^2+15}{(x+1)(x-3)^2}\,dx$

(g) $\displaystyle\int \dfrac{3}{(2x-1)(x+1)}$
(h) $\displaystyle\int \dfrac{x^2+3x-2}{(x^2+1)(x-1)}\,dx$
(i) $\displaystyle\int \dfrac{3-x}{(x^2-3)(x-1)}\,dx$

(j) $\displaystyle\int \dfrac{x^2+2x+6}{(x^2+4)(x+1)}\,dx$
# (k) $\displaystyle\int \dfrac{6}{(x^2-1)(x+2)}\,dx$

2. Evaluate the following definite integrals

(a) $\displaystyle\int_1^2 \dfrac{1}{(x+1)(2x-1)}\,dx$
(b) $\displaystyle\int_2^4 \dfrac{5+x}{(1+2x)(1-x)^2}\,dx$
(c) $\displaystyle\int_4^9 \dfrac{5x^2-8x+1}{2x(x-1)^2}\,dx$

(d) $\displaystyle\int_2^3 \dfrac{2(x+1)}{(x^2+1)(1-x)}\,dx$
(e) $\displaystyle\int_2^4 \dfrac{x^2-2x+3}{(x-1)(x^2+1)}\,dx$
#(f) $\displaystyle\int_0^2 \dfrac{x^2+6x+3}{(3-x)(x^2+1)}\,dx$

3. Show that $\displaystyle\int_2^4 \frac{5+x}{(1-x)(5+x^2)}\,dx = \ln\frac{\sqrt{21}}{9}$ and explain why $\displaystyle\int_0^2 \frac{5+x}{(1-x)(5+x^2)}\,dx$ cannot be evaluated.

## INTEGRATION BY SUBSTITUTION

Just as, when we were learning to differentiate complex functions, we discovered first the chain rule and then the product and quotient rules, so for integrating some complex functions we can use the integration equivalents of the chain and product rules.

Corresponding to the chain rule for differentiation is the method of integration by substitution.

Suppose you had to integrate $I = 2x(x^2 + 4)^5$.

You might realise that the derivative of $(x^2 + 4)^6$ is $12x(x^2 + 4)^5$ and deduce that

$I = \int 2x(x^2 + 4)^5\,dx = \frac{1}{6}(x^2 + 4)^6 + c.$

In doing this you would be using the chain rule in reverse; but it is not always easy to spot situations like this. Instead, you could make a suitable substitution in the same way as you did when using the chain rule.

$u = x^2 + 4$ looks like a suitable substitution; but if we substitute this in $I$ we get an expression with a mixture of $u$s and $x$s:

$\qquad I = \int 2xu^5\,dx.$

Since we are trying to make the new integrand a function of $u$ we need to integrate it with respect to $u$ rather than $x$.

However, since $u = x^2 + 4$, $\dfrac{du}{dx} = 2x$.

If we 'separate the variables', we get $du = 2x\,dx$.

$I$ then becomes $I = \int u^5\,du$ which is easily integrated as $I = \frac{1}{6}u^6$,

and replacing $u$ in terms of $x$ gives $I = \frac{1}{6}(x^2 + 4)^6 + c.$

> **!** The statement $du = 2x\,dx$ is not an equation – it is an equivalence of operators – it only has meaning when combined with integral signs (as in the solving of differential equations).

## Example 17.10

Find $I = \int 2x(x + 3)^4\,dx.$    ①

> ➤ Choose a substitution:     let $u = x + 3$,    then $x = u - 3$

> ➤ Differentiate $u$ w.r.t. $x$:     $\dfrac{du}{dx} = 1 \;\Rightarrow\; du = dx$

> ➤ Make substitutions in ①:     $I = \int 2(u - 3)u^4\,du = \int (2u^5 - 6u^4)\,du$

> ➤ Integrate:     $I = \frac{2}{6}u^6 - \frac{6}{5}u^5 = \frac{1}{15}u^5(5u - 18)$

> ➤ Give answer in terms of $x$:     $I = \frac{1}{15}(5x - 3)(x + 3)^5 + c$

Choosing the substitution in example 17.10 was straightforward, but it is not always that easy. Although with practice you will learn what substitutions are likely to work in various situations, for the CAPE exam you will be told what substitution to use when it is not really obvious.

Don't forget to differentiate the substitution, and replace the d$x$ with   ...d$u$.

## Example 17.11

Use the substitution $u = \cos x$ to find $I = \int \sin x \cos^7 x\, dx$

> Differentiate $u$ w.r.t. $x$:    $\dfrac{du}{dx} = -\sin x, \quad \therefore \quad -\sin x\, dx = du$

> Make substitutions:    $I = \int -u^7\, du$

> Integrate:    $I = -\tfrac{1}{8}u^8 + c$

> Give answer in terms of $x$:    $I = c - \tfrac{1}{8}\cos^8 x$

*(handwritten margin notes:)*
$dx = \dfrac{du}{-\sin x}$

$I = \int \sin x\, u^7 \cdot \dfrac{du}{-\sin x}$

$= -\int u^7\, du$

$= -\dfrac{u^8}{8} + c$

## Example 17.12

Use the substitution $u = \sin x$ to find $I = \int \cos^3 x\, dx$

> Comparing the integrand with the substitution you expect to have to use $\cos^2 x = 1 - \sin^2 x$

> Differentiate $u$ w.r.t. $x$:    $\dfrac{du}{dx} = \cos x, \quad \therefore \quad \cos x\, dx = du$

> Make substitutions:    $I = \int \cos x \left(1 - \sin^2 x\right) dx = \int (1 - u^2)\, du$

> Integrate:    $I = u - \tfrac{1}{3}u^3 + c$

> Give answer in terms of $x$:    $I = \sin x - \tfrac{1}{3}\sin^3 x + c$

## Example 17.13

Use the substitution $x = \tan u$ to find $\displaystyle\int \dfrac{1}{1 + x^2}\, dx$

> Differentiate the substitution :    $\dfrac{dx}{du} = \sec^2 u, \quad \therefore \quad dx = \sec^2 u\, du$

> Make substitutions:    $I = \displaystyle\int \dfrac{1}{1 + \tan^2 u}\sec^2 u\, du = \int \dfrac{1}{\sec^2 u}\sec^2 u\, du = \int 1\, du$

> Integrate:    $I = u + c$

> Give answer in terms of $x$:    $I = \tan^{-1} x + c$

## Definite Integrals using Substitution

The limits on a definite integral give the values of $x$ between which the integral is evaluated. If $u$ is substituted for some function of $x$ then allowance must be made for this in the limits. **Either** the integration is performed w.r.t. $u$ and then, as in indefinite integrals, $u$ is replaced by the function of $x$ before the limits are used to evaluate the integral, **or**, which is usually simpler, the substitution is used to replace the limits by values of $u$ and the integral, still in terms of $u$, is evaluated using the new limits

## Example 17.14

Use the substitution $u = x^2 + 1$, to evaluate $\displaystyle\int_0^1 \dfrac{3x}{\sqrt{(x^2 + 1)}}\, dx$

> Find limits in terms of $u$:    when $x = 0$, $u = 1$, when $x = 1$, $u = 2$

➤ Differentiate $u$ w.r.t. $x$:   $\dfrac{du}{dx} = 2x \Rightarrow du = 2x\,dx$

➤ Make substitutions:   $\displaystyle\int_0^1 \frac{3x}{\sqrt{(x^2+1)}}\,dx = \int_1^2 \frac{\frac{3}{2}}{u^{\frac{1}{2}}}\,du$

➤ Integrate and evaluate:   $= \left[3u^{\frac{1}{2}}\right]_1^2 = 3\sqrt{2} - 3$

Working with limits of $x$ rather than $u$, gives the same result as follows:

➤ Differentiate $u$ w.r.t. $x$:   $\dfrac{du}{dx} = 2x \Rightarrow du = 2x\,dx$

➤ Make substitutions:   $\displaystyle\int_0^1 \frac{3x}{\sqrt{(x^2+1)}}\,dx = \int_{x=0}^{x=1} \frac{\frac{3}{2}\,du}{u^{\frac{1}{2}}}$

➤ Integrate:   $= \left[3u^{\frac{1}{2}}\right]_{x=0}^{x=1}$

➤ Replace $u$ with function of $x$ and evaluate:   $= \left[3\sqrt{x^2+1}\right]_0^1 = 3\sqrt{2} - 3$

# EXERCISE 17 E

1.  Use the substitution given to find the following integrals:

(a) $\int x^2 \sqrt{(x^3+5)}\,dx$,   $u = x^3 + 5$

(b) $\int x\,(x^2-4)^5\,dx$,   $u = x^2 - 4$

(c) $\int x(x-2)^8\,dx$,   $u = x - 2$

(d) $\int 4x \cos(3-x^2)\,dx$,   $u = 3 - x^2$

(e) $\int 6x^2 \sin(x^3 - \tfrac{1}{3}\pi)\,dx$,   $u = x^3 - \tfrac{1}{3}\pi$

(f) $\int e^x\sqrt{(1+e^x)}\,dx$,   $u = 1 + e^x$

(g) $\int (x+2)(2x-3)^6\,dx$,   $u = 2x - 3$

(h) $\int \cos x \sin^5 x\,dx$,   $u = \sin x$

(i) $\int \sec^2 x \tan^4 x\,dx$,   $u = \tan x$

(j) $\int \sin x\,e^{\cos x}\,dx$,   $u = \cos x$

(k) $\int 3x^2 e^{x^3}\,dx$,   $u = x^3$

(l) $\displaystyle\int \frac{x}{\sqrt{x+2}}\,dx$,   $u = x + 2$

2.  By choosing a suitable substitution, integrate the following functions w.r.t. $x$

(a) $8x(2x^2+4)^5$

(b) $x(x-5)^4$

(c) $(2x+3)\sqrt{(x^2+3x)}$

(d) $\dfrac{2x}{\sqrt{1-x^2}}$

(e) $4x^3 e^{x^4}$

(f) $\dfrac{x}{(x+2)^5}$

3.  Using the substitution given, evaluate the following integrals:

(a) $\displaystyle\int_0^1 x^2\sqrt{x^3+3}\,dx$,   $u = x^3 + 3$

(b) $\displaystyle\int_0^{\frac{\pi}{2}} \cos^4 x \sin x\,dx$,   $u = \cos x$

(c) $\displaystyle\int_1^2 x^2(x-1)^5\,dx$,   $u = x - 1$

(d) $\displaystyle\int_0^{\frac{\pi}{2}} \cos x\sqrt{\sin x}\,dx$,   $u = \sin x$

(e) $\displaystyle\int_1^2 xe^{x^2}\,dx$,   $u = x^2$

(f) $\displaystyle\int_0^1 2x(2x-1)^4\,dx$,   $u = 2x - 1$

4.  Use the substitution $t^2 = x + 1$, where $x > -1$, $t > 0$, to find $\displaystyle\int \frac{x}{\sqrt{x+1}}\,dx$.

Hence evaluate $\displaystyle\int_0^3 \frac{x}{\sqrt{x+1}}\,dx.$

#5. (a) Find $\displaystyle\int \frac{1}{x(x+1)}\,dx, \quad x > 0.$

(b) Using the substitution $u = e^x$ and the answer to (a), find $\displaystyle\int \frac{1}{1+e^x}\,dx.$

### *Some Useful Rules that can be Deduced from the Preceding Exercises*

In many of the questions in Exercise 17 E you may have noticed that the integrand was of the form

$$f'(x)\,g\{f(x)\}$$

and that the suggested substitution was $u = f(x)$. From these examples we can deduce the following results, which can be checked by differentiation:

$$\int f'(x)[f(x)]^n \, dx = \frac{[f(x)]^{n+1}}{n+1} + c$$

$$\int \cos x \sin^n x \, dx = \frac{\sin^{n+1} x}{n+1} + c$$

$$\int \sin x \cos^n x \, dx = -\frac{\cos^{n+1} x}{n+1} + c$$

$$\int f'(x)e^{f(x)}\,dx = e^{f(x)} + c$$

These rules enable you to do some integrations by inspection, or help you to know what substitution to make, if the substitution is not given in the question. See if you can find the integrals in exercise 17F.

## EXERCISE 17 F

1. $\displaystyle\int 2\cos x \sin^3 x \, dx$

2. $\displaystyle\int x(2x^2 + 4)^5 \, dx$

3. $\displaystyle\int e^x (1 + 2e^x)^4 \, dx$

4. $\displaystyle\int xe^{x^2+3}\,dx$

5. $\displaystyle\int (3x^2 - 1)e^{x^3-x} \, dx$

6. $\displaystyle\int 3\sin x\,(1 + \cos x)^5 \, dx$

7. $\displaystyle\int \frac{4x-2}{x^2-x}\,dx$

8. $\displaystyle\int \frac{(1+\ln x)^3}{x}\,dx$

9. $\displaystyle\int \frac{\sin x}{1-\cos x}\,dx$

## INTEGRATION BY PARTS

Most of the problems in the previous section have involved the integration of products or quotients, but not all products that can be integrated can be integrated by substitution. Some (including one or two that are not normally regarded as products) can be integrated by the method known as integration by parts. This is derived from the product rule, as follows:

The product rule states that if $u$ and $v$ are functions of $x$, then $\displaystyle\frac{d}{dx}(uv) = v\frac{du}{dx} + u\frac{dv}{dx}$

Integrating both sides w.r.t. $x$ gives: $\quad uv = \int\left(v\dfrac{du}{dx}dx + u\dfrac{dv}{dx}dx\right)$

Rearranging the terms gives the integration by parts rule $\quad\boxed{\int u\dfrac{dv}{dx}dx = uv - \int v\dfrac{du}{dx}dx}$

The best way to see how this formula works is to look at some examples:

## Example 17.15

Find $\quad\int x\cos x\, dx$

Let $\quad u = x$ and $\dfrac{dv}{dx} = \cos x$

then $\left(\text{differentiating } u \text{ and integrating } \dfrac{dv}{dx}\right)\quad \dfrac{du}{dx} = 1$ and $v = \sin x$

so $\quad\int u\dfrac{dv}{dx}dx = x\sin x - \int\sin x\, dx$

$\qquad\qquad = x\sin x + \cos x + c$

$\boxed{!}$ Notice that
1) although the integration by parts formula looks as though it merely substitutes one hard integral for another, by wise choice of $u$ and $v$ the second integral is (as in this case) straightforward;

2) this happens because the function chosen for $u$ 'disappears' when it is differentiated;

3) the other function in the integrand becomes $\dfrac{dv}{dx}$, and so has to be integrated.

4) the arbitrary constant is added after the last integration has been performed.

In the next example, the first factor of the integrand does not disappear immediately, but it does if the formula is applied twice.

## Example 17.16

Find $\quad\int x^2\, e^x\, dx$

➢ choose substitutions: $\qquad\qquad$ let $u = x^2$ and $\dfrac{dv}{dx} = e^x$

➢ differentiate $u$ and integrate $\dfrac{dv}{dx}$: $\qquad \dfrac{du}{dx} = 2x,\quad v = e^x$

➢ substitute in the formula: $\qquad\qquad \int x^2 e^x\, dx = x^2 e^x - \int 2x e^x\, dx$

➢ choose new substitutions: $\qquad\qquad$ let $u = 2x$ and $\dfrac{dv}{dx} = e^x$

➢ differentiate $u$ and integrate $\dfrac{dv}{dx}$: $\qquad \dfrac{du}{dx} = 2,\quad v = e^x$

➢ substitute in the formula: $\qquad\qquad \int x^2 e^x dx = x^2 e^x - \left[2x e^x - \int 2e^x dx\right]$

➢ complete the integration: $\qquad\qquad \int x^2 e^x\, dx = x^2 e^x - 2x e^x + 2e^x + c$

## Example 17.17

Find $\int (x + 2)(2x - 3)^6 \, dx$  (this question was 1 (g) in exercise 17 E, where it was solved by substitution, the same result is arrived at here using integration by parts.)

➤ choose substitutions:      let $u = x + 2$ and $\dfrac{dv}{dx} = (2x - 3)^6$

➤ differentiate $u$ and integrate $\dfrac{dv}{dx}$:    $\dfrac{du}{dx} = 1, \quad v = \tfrac{1}{14}(2x - 3)^7$

➤ substitute in the formula:    $\int (x + 2)(2x - 3)^6 \, dx = \tfrac{1}{14}(x + 2)(2x - 3)^7 - \int \tfrac{1}{14}(2x-3)^7 dx$

➤ complete the integration:    $\int (x + 2)(2x - 3)^6 \, dx = \tfrac{1}{14}(x + 2)(2x - 3)^7 - \tfrac{1}{14.16}(2x - 3)^8 + c$

➤ simplify the result:    $\int (x + 2)(2x - 3)^6 \, dx = \tfrac{1}{14.16}(2x - 3)^7 \{16(x + 2) - (2x - 3)\} + c$

$$= \tfrac{1}{32}(2x + 5)(2x - 3)^7 + c$$

Care was taken in all these examples to choose $u$ as a function that 'dropped out' when the formula was applied, i.e. a power of $x$.

# EXERCISE 17 G

Use integration by parts to find the following integrals

1. $\int x \, e^x \, dx$
2. $\int 2x(x + 4)^5 dx$
3. $\int x \cos 3x \, dx$
4. $\int x^2 \sin x \, dx$

5. $\int x \, e^{-x} \, dx$
6. $\int x \sqrt{(x - 3)} \, dx$
7. $\int x^3 \cos x \, dx$
8. $\int e^x (2x + 1) \, dx$

## *Integration by Parts Applied to Integrals containing Logarithms*

We do not know how to integrate ln $x$, but we do know how to differentiate it, so even if the other factor in an integrand is a power of $x$ we choose the logarithmic function as $u$.

## Example 17.18

Find $\int x \ln x \, dx$

➤ Choose substitutions:      let $u = \ln x$ and $\dfrac{dv}{dx} = x$

➤ Differentiate $u$ and integrate $\dfrac{dv}{dx}$:    $\dfrac{du}{dx} = \dfrac{1}{x}, \quad v = \dfrac{x^2}{2}$

➤ Substitute in the formula:    $\int x \ln x \, dx = \dfrac{x^2}{2} \ln x - \int \dfrac{x^2}{2} \dfrac{1}{x} dx$

➤ Complete the integration:    $\int x \ln x \, dx = \dfrac{x^2}{2} \ln x - \dfrac{x^2}{4} + c$

This method also enables us to integrate ln $x$. Treat ln $x$ as a product: ln $x = 1 \times$ ln $x$

## Example 17.19

Find $\int \ln x \, dx$

- ➤ Choose substitutions:    let $u = \ln x$ and $\dfrac{dv}{dx} = 1$

- ➤ Differentiate $u$ and integrate $\dfrac{dv}{dx}$:    $\dfrac{du}{dx} = \dfrac{1}{x}, \quad v = x$

- ➤ Substitute in the formula:    $\int \ln x \, dx = x \ln x - \int x \dfrac{1}{x} \, dx$

- ➤ Complete the integration:    $\int \ln x \, dx = x \ln x - x + c$

## Integration by Parts Applied to Integrals like $\int e^x \cos x \, dx$

If neither factor of the product to be integrated will disappear, as with $e^x \cos x$, integration by parts may lead, after two applications, to the original integral returning.

## Example 17.20

Find $\int e^x \cos x \, dx$

- ➤ Call the integral $I$:    let $I = \int e^x \cos x \, dx$

- ➤ Choose substitutions:    let $u = \cos x$ and $\dfrac{dv}{dx} = e^x$

- ➤ Differentiate $u$ and integrate $\dfrac{dv}{dx}$:    $\dfrac{du}{dx} = -\sin x, \quad v = e^x$

- ➤ Substitute in the formula:    $I = \int e^x \cos x \, dx = e^x \cos x + \int e^x \sin x \, dx$

- ➤ Make new substitutions:    let $u = \sin x$ and $\dfrac{dv}{dx} = e^x$

- ➤ Differentiate $u$ and integrate $\dfrac{dv}{dx}$:    $\dfrac{du}{dx} = \cos x, \quad v = e^x$

- ➤ Substitute in the formula:    $I = e^x \cos x + e^x \sin x - \int e^x \cos x \, dx$

- ➤ This new integral $= I$, so    $2I = e^x \cos x + e^x \sin x$

     giving   $I = \tfrac{1}{2} e^x (\cos x + \sin x) + c$

!  It does not matter whether $e^x$ or the trig function is chosen as $u$, *but you must use the same type of function for u for both substitutions*, otherwise you end up proving $I = I$ !

# EXERCISE 17 H

1. Use the method of integration by parts to find the following integrals:

    (a) $\int x^2 \ln x \, dx$       (b) $\int e^x \sin 2x \, dx$       (c) $\int \dfrac{1}{x} \ln x \, dx$

    (d) $\int \ln 2x \, dx$       (e) $\int e^{2x} \cos x \, dx$       (f) $\int \cos^2 x \, dx$

2. Use the method of integration by parts to evaluate the following integrals:

(a) $\int_0^{\frac{\pi}{4}} x\cos 2x\,dx$    (b) $\int_1^4 \frac{\ln x}{x^2}\,dx$    (c) $\int_0^1 xe^x\,dx$

(d) $\int_0^{\pi} x^2\sin x\,dx$    (e) $\int_0^2 x(x-2)^4\,dx$

## INTEGRATION OF TRIG FUNCTIONS

The techniques we have learnt so far for integration enable us to integrate a variety of functions, but there are a lot more functions that we cannot integrate. Many functions cannot be integrated at all, and others can be integrated if you know what special techniques to apply. In a later section I will summarise all the techniques we have learnt, and how to decide which one to use, but first there are some special techniques that can often be used to help integrate trig functions.

How would you find $\int \cos^2 x\,dx$? There is no immediate straightforward answer. In exercise 17 H question 1 (f) you solved this problem using integration by parts, but the solution was long and tedious and therefore liable to error. So is there a simpler way?

Yes, there is. $\cos^2 x$ has no obvious integral, but $\cos 2x$ does; and we know an identity which will turn $\cos^2 x$ into a function of $\cos 2x$.

*Handwritten:* $\cos 2x = \cos(x+x)$
$= \cos x\cos x - \sin x\sin x$
$= \cos^2 x - \sin^2 x$
$= \cos^2 x - (1-\cos^2 x)$
$= \cos^2 x - 1 + \cos^2 x$
$\cos 2x = 2\cos^2 x - 1$
$\cos^2 x = \frac12 + \frac12\cos 2x$
$= \frac12(1+\cos 2x)$

### *Using Trig Identities to Solve some Trig Integrals*

**Example 17.21**

Find the integral $\int \cos^2 x\,dx$

➤ Use trig identity $\cos 2x = 2\cos^2 x - 1$:   $\int \cos^2 x\,dx = \int \frac12(1+\cos 2x)\,dx$

➤ Integrate:   $\cos^2 x = \frac12(1+\cos 2x) = \frac12 x + \frac14\sin 2x + c$

**Example 17.22**

By writing $\sin^3 x$ as $\sin x\sin^2 x$ and using the identity $\cos^2 x + \sin^2 x \equiv 1$, find $\int \sin^3 x\,dx$

Do this yourself! (You should get $\frac13\cos^3 x - \cos x + c$)

*Handwritten:* $\int \sin x\sin^2 x$, $\int(1-\cos^2 x)\sin x$, $-\int 1 - \cos^3 x$

The next examples use the product identities and half-angle identities which we derived in unit 1

$2\sin A\cos B \equiv \sin(A+B) + \sin(A-B)$    $2\cos A\sin B \equiv \sin(A+B) - \sin(A-B)$

$2\cos A\cos B \equiv \cos(A+B) + \cos(A-B)$    $-2\sin A\sin B \equiv \cos(A+B) - \cos(A-B)$

If $\tan\frac{\alpha}{2} = t$, then $\sin\alpha = \frac{2t}{1+t^2}$ and $\cos\alpha = \frac{1-t^2}{1+t^2}$

**Example 17.23**

Find the integral: $\int \sin 4x\cos 2x\,dx$

➤ Use trig identity $2\sin A\cos B \equiv \sin(A+B) + \sin(A-B)$: $\sin 4x\cos 2x = \frac12(\sin 6x + \sin 2x)$

$\int \sin 4x\cos 2x\,dx = \int \frac12(\sin 6x + \sin 2x)\,dx$

➤ Integrate:    $= -\frac{1}{12}\cos 6x - \frac14\sin 2x + c$

## Example 17.24

Find the integral: $\int \tan x \, dx$

- Use identity $\tan x = \dfrac{\sin x}{\cos x}$:  $\quad\quad \int \tan x \, dx = \int \dfrac{\sin x}{\cos x} \, dx$

- make numerator the derivative of denominator:  $\quad = -\int \dfrac{-\sin x}{\cos x} \, dx$

- Integrate:  $\quad = -\ln|\cos x|$

- Simplify:  $\quad = \ln|\cos x|^{-1} = \ln|\sec x| + c$

- State the answer:  $\int \tan x \, dx = \ln|\sec x| + c$

## Example 17.25

Find $I = \int \dfrac{dx}{1 + \cos x + \sin x}$

- Use half-angle identities, where $t = \tan \tfrac{1}{2}x$

  N.B. you need to substitute for $dx$ as well as $\sin x$ & $\cos x$

$$\frac{dt}{dx} = \tfrac{1}{2}\sec^2 \tfrac{1}{2}x = \tfrac{1}{2}(1 + \tan^2 \tfrac{1}{2}x) = \tfrac{1}{2}(1 + t^2) \quad \therefore \quad dx = \frac{2dt}{1 + t^2}$$

- Substitute for $x$:  $I = \int \dfrac{dx}{1 + \cos x + \sin x} = \int \dfrac{1}{\left(1 + \frac{1-t^2}{1+t^2} + \frac{2t}{1+t^1}\right)}\left(\dfrac{2dt}{1+t^2}\right)$

- Simplify:  $= \int \dfrac{2dt}{(1+t^2) + (1-t^2) + 2t} = \int \dfrac{dt}{1+t}$

- Integrate:  $I = \ln|1 + t| + c$

- Give answer in terms of $x$:  $I = \ln|1 + \tan \tfrac{1}{2}x| + c$  or  $I = \ln A|1 + \tan \tfrac{1}{2}x|$

## Integrating $\cos^n x$ and $\sin^n x$

When $n$ is small, the methods used in examples 17.21 and 17.22 extended can be used to integrate even and odd powers, respectively, of $\cos x$ and $\sin x$.

## Example 17.26 (*even powers*)

Find the integral: $\int \sin^4 x \, dx$

- Use trig identity for $\cos 2x$:  $\sin^4 x = (\sin^2 x)^2$

  $= \{\tfrac{1}{2}(1 - \cos 2x)\}^2$

  $= \tfrac{1}{4}(1 - 2\cos 2x + \cos^2 2x)$

- and again for $\cos 4x$:  $= \tfrac{1}{4} - \tfrac{1}{2}\cos 2x + \tfrac{1}{4}\cdot\tfrac{1}{2}(1 + \cos 4x)$

- Integrate:  $\int \sin^4 x \, dx = \tfrac{1}{4}x - \tfrac{1}{4}\sin 2x + \tfrac{1}{8}x + \tfrac{1}{8}\cdot\tfrac{1}{4}\sin 4x$

- State the answer:  $\int \sin^4 x \, dx = \tfrac{3}{8}x - \tfrac{1}{4}\sin 2x + \tfrac{1}{32}\sin 4x + c$

## Example 17.27 (odd powers)

Find the integral: $\int \cos^5 x \, dx$

➤ $\cos^5 x = \cos x . \cos^4 x$

$= \cos x \, (\cos^2 x)^2$

$= \cos x \, (1 - \sin^2 x)^2$

➤ $= \cos x \, (1 - 2\sin^2 x + \sin^4 x)$

➤ Integrate: $\int \cos^5 x \, dx = \int (\cos x - 2\sin^2 x \cos x + \sin^4 x \cos x) \, dx$

$= \sin x - \tfrac{2}{3}\sin^3 x + \tfrac{1}{5}\sin^5 x + c.$

*(handwritten annotations:)*
$\cos x . \cos^4 x$
$\cos x . (\cos^2 x)^2$
$\cos x . \{\tfrac{1}{2}(1 + \cos 2x)\}^2$
$\int [\cos x . \tfrac{1}{4}(1 + 2\cos 2x + \cos^2 2x)]$
$\int \cos x . \tfrac{1}{4} + \dfrac{\cos 2x}{2} + \dfrac{\cos^2 2x}{4}$

*(handwritten:)* $2\sin^2 x \cos x$
$2(1 - \cos^2 x)\cos x$
$2\cos x - \cos^3 x$

# EXERCISE 17 I

1. Integrate the following functions w.r.t. x, using trig identities

   (a) $\sin^2 x$     (b) $\cos^3 x$ *(handwritten: $2\sin x$)*     (c) $\cos 7x \cos 5x$     (d) $\sin 4x \sin 3x$

   (e) $\cos^4 x$     (f) $\tan^2 x$     (g) $\sin^5 x$     (h) $\dfrac{1}{1 + \sin x}$

2. Integrate the following functions w.r.t. x (mixed trig questions)

   (a) $3\cos^2 x \sin x$     (b) $\sec^2 5x$     (c) $\cos 5x \sin x$     (d) $\sec^2 x \tan^3 x$

   (e) $\sin^3 2x$     (f) $\sin 3x \sin x$     (g) $\tan 3x$     (h) $\sin^5 x \cos x$

   (i) $x \, (\cos x + \sin x)$     (j) $\sin 7x$     (k) $e^x \sin x$     (l) $\sin 2x \cos x$

## Integrating $\cos^n x$ and $\sin^n x$ when n is large (REDUCTION FORMULAE)

The methods used for integrating even and odd powers of $\cos x$ and $\sin x$ demonstrated in examples 17.21, 22, 26 and 27 work well for small values of $n$ (up to about 6), but for higher powers they become unwieldy. Is there some other way? Let us see what happens if we use integration by parts to find $\int \cos^n x \, dx$.

## Example 17.28

➤ Let $I_n = \int \cos^n x \, dx$

➤ Write $I_n$ as a product:     $I_n = \int \cos x . \cos^{n-1} x \, dx$   *(handwritten: → differentiate properly)*

➤ Choose substitutions:     let $u = \cos^{n-1} x$ and $\dfrac{dv}{dx} = \cos x$

➤ Differentiate $u$ and integrate $\dfrac{dv}{dx}$:     $\dfrac{du}{dx} = -(n-1)\cos^{n-2} x \sin x$ and $v = \sin x$

➤ Substitute in the formula:     $I_n = \cos^{n-1} x \sin x + \int (n-1)\sin^2 x . \cos^{n-2} x \, dx$

➤ Use trig identity:

$I_n = \cos^{n-1} x . \sin x + (n-1)\int (1 - \cos^2 x)\cos^{n-2} x \, dx$

$= \cos^{n-1} x . \sin x + (n-1)\left[ \int \cos^{n-2} x \, dx - \int \cos^n x \, dx \right]$

*(handwritten: $t = \dfrac{\sin \frac{1}{2}x}{\cos \frac{1}{2}x}$)*

This may look worse than the original integral, but if $\int \cos^p x \, dx$ is replaced by $I_p$ it leads to a useful formula

> Use notation for integrals $\qquad I_n = \cos^{n-1} x \sin x + (n-1)(I_{n-2} - I_n)$

> Rearrange: $\qquad I_n + (n-1)I_n = \cos^{n-1} x \sin x + (n-1)I_{n-2}$

or $\qquad I_n = \frac{1}{n}\cos^{n-1} x \sin x + \frac{n-1}{n}I_{n-2}$

This is known as a **reduction formula**, because it enables you to find the integral of a higher power in terms of a lower power. It also enables you to find the integral of any power of cos $x$, provided you know the value of $I_0$ if you want an even power, or $I_1$ if you want an odd power.

$I_0 = \int \cos^0 x \, dx = \int 1 \, dx = x$ and $I_1 = \int \cos x \, dx = \sin x$

so $I_2 = \frac{1}{2}\cos x \sin x + \frac{1}{2}I_0$

$\qquad = \frac{1}{4}\sin 2x + \frac{1}{2}x + c$

i.e. $\int \cos^2 x \, dx = \frac{1}{2}x + \frac{1}{4}\sin 2x + c$

also $I_3 = \frac{1}{3}\cos^2 x \sin x + \frac{2}{3}I_1$

$\qquad = \frac{1}{3}\cos^2 x \sin x + \frac{2}{3}\sin x + c$

i.e. $\int \cos^3 x \, dx = \frac{1}{3}\sin x (\cos^2 x + 2) + c$

rewriting this in terms of $\sin x$ gives $\int \cos^3 x \, dx = \sin x - \frac{1}{3}\sin^3 x + c$.

These results agree with the results found earlier (example 17.21 and exercise 17 I, 1 (b)) and the process can be continued indefinitely.

## Reduction Formulae for Definite Integrals

In example 17.28 we derived a reduction formula for an indefinite integral. It can clearly be used to find the integral of any power of cos $x$; but although there is no further integration to be done, the trig involved for higher powers is increasingly complex.

Reduction formulae are particularly useful for definite integrals, especially if the limits of integration mean that most of the terms drop out.

## Example 17.29

Derive the reduction formula $I_n = \left(\dfrac{n-1}{n}\right)I_{n-2}$ for the integral $I_n = \displaystyle\int_0^{\frac{\pi}{2}} \sin^n x \, dx$

and hence evaluate $\displaystyle\int_0^{\frac{\pi}{2}} \sin^5 x \, dx$

> Write $I_n$ as a product: $\qquad I_n = \displaystyle\int_0^{\frac{\pi}{2}} \sin x \sin^{n-1} x \, dx$

> Choose substitutions: $\qquad$ let $u = \sin^{n-1} x$ and $\dfrac{dv}{dx} = \sin x$

> Differentiate $u$ and integrate $\dfrac{dv}{dx}$ : $\dfrac{du}{dx} = (n-1)\sin^{n-2} x \cos x, \quad v = -\cos x$

➤ Substitute in the formula: $\quad I_n = \left[ -\cos x \sin^{n-1} x \right]_0^{\frac{\pi}{2}} + \int_0^{\frac{\pi}{2}} (n-1)\cos^2 x \sin^{n-2} x \, dx$

➤ $\cos\frac{\pi}{2} = 0$ and $\sin 0 = 0$, so : $\qquad \left[ -\cos x \sin^{n-1} x \right]_0^{\frac{\pi}{2}} = 0$

➤ Write $\cos^2 x$ in terms of $\sin x$: $\qquad I_n = 0 + \int_0^{\frac{\pi}{2}} (n-1)(1 - \sin^2 x)\sin^{n-2} x \, dx$

➤ Simplify: $\qquad I_n = \int_0^{\frac{\pi}{2}} (n-1)\{\sin^{n-2} x - \sin^n x\} \, dx$

➤ Use notation for integrals $\qquad I_n = (n-1)(I_{n-2} - I_n)$

➤ Rearrange: $\qquad I_n + (n-1)\,I_n = (n-1)I_{n-2}$

so $\qquad I_n = \left( \dfrac{n-1}{n} \right) I_{n-2}$

➤ We need an odd power, so find $I_1$: $\quad I_1 = \int_0^{\frac{\pi}{2}} \sin x \, dx = \left[ -\cos x \right]_0^{\frac{\pi}{2}} = 1$

➤ Use the reduction formula: $\qquad I_3 = \frac{2}{3} I_1 = \frac{2}{3}$

$I_5 = \frac{4}{5} I_3$

$= \frac{4}{5} \times \frac{2}{3} = \frac{8}{15}$

## Example 17.30

Find a reduction formula for $\quad I_n = \int_1^e \ln^n x \, dx$ and hence evaluate $\int_1^e \ln^3 x \, dx$

➤ Write $I_n$ as a product: $\qquad I_n = \int_1^e 1.\ln^n x \, dx$

➤ Choose substitutions: $\qquad$ let $u = \ln^n x$ and $\dfrac{dv}{dx} = 1$

➤ Differentiate $u$ and integrate $\dfrac{dv}{dx}$ : $\dfrac{du}{dx} = n\left( \dfrac{1}{x} \right)\ln^{n-1} x, \quad v = x$

➤ Substitute in the formula: $\qquad I_n = \left[ x \ln^n x \right]_1^e - \int_1^e n \ln^{n-1} x \, dx$

➤ $\ln 1 = 0$ and $\ln e = 1$, so : $\qquad \left[ x \ln^n x \right]_1^e = e$

➤ Use notation for integrals: $\qquad I_n = e - n\,I_{n-1}$

➤ Find $I_0$: $\qquad I_0 = \int_1^e \ln^0 x \, dx = \int_1^e 1 \, dx = \left[ x \right]_1^e = e - 1$

➤ Find $I_3$: $\qquad I_1 = e - I_0 = 1, \quad I_2 = e - 2I_1 = e - 2, \quad I_3 = e - 3I_2 = 6 - 2e$

## EXERCISE 17 J

1. If $I_n = \int_0^{\frac{\pi}{2}} \cos^n x \, dx$ find a reduction formula for $I_n$ in terms of $I_{n-2}$, and hence evaluate $\int_0^{\frac{\pi}{2}} \cos^4 x \, dx$.

2. Find a reduction formula for the integral $I_n = \int_0^{\frac{\pi}{4}} \tan^n x \, dx, n \geq 2$, and hence evaluate $\int_0^{\frac{\pi}{4}} \tan^4 x \, dx$
   (Hint: use the identity $\sec^2 x = 1 + \tan^2 x$)

3. If $I_n = \int x^n e^x \, dx$, show that $I_n = x^n e^x - n I_{n-1}$

## GENERAL METHODS FOR INTEGRATION

You have now learnt a number of different techniques for integration, each of which is appropriate under certain circumstances. Faced with an integration problem, how do you know what method to use? Recognising the best method comes with experience, but to gain that experience the best thing to do is to ask yourself a number of questions which will lead to either using or eliminating each method in turn. On the next page you will find a flow chart (figure 17.1) which takes you through the questions to ask. Then exercise 17 K provides a set of mixed questions to give you practice at selecting the right method.

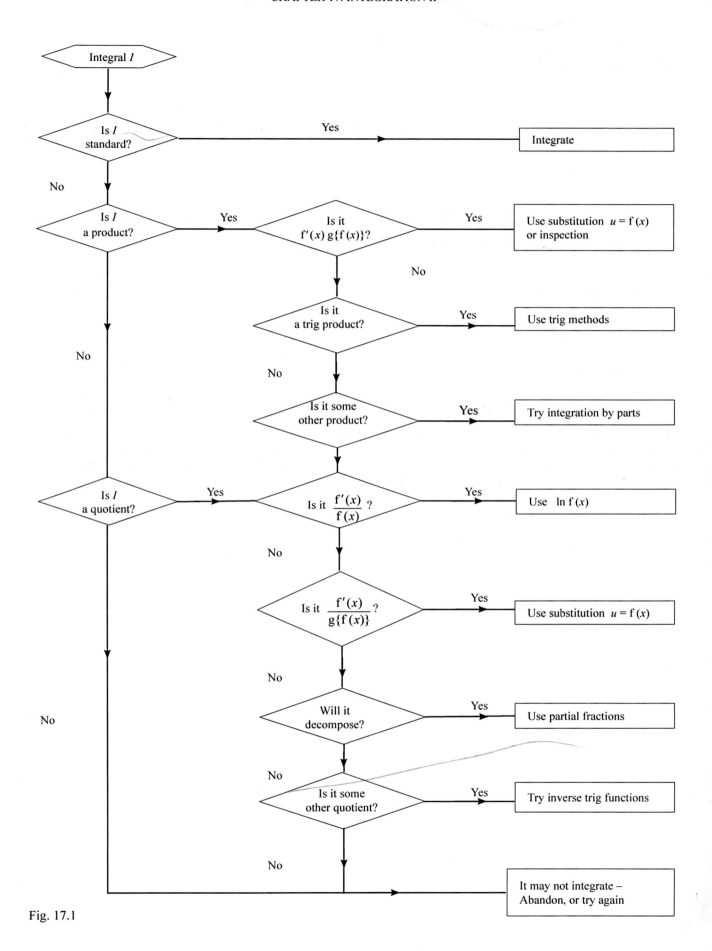

Fig. 17.1

# EXERCISE 17 K

1.  Find the following integrals

    (a) $\int e^{2x}\, dx$

    (b) $\int x \sin x\, dx$

    (c) $\int \cos x\, e^{\sin x}\, dx$

    (d) $\int 10 \cos 3x \sin x\, dx$

    (e) $\int (x-2)(x-3)dx$

    (f) $\int \dfrac{x+5}{x^2+10x-3}\, dx$

    (g) $\int \dfrac{7}{x}\, dx$

    (h) $\int x e^{x^2}\, dx$

    (i) $\int \dfrac{2}{x(x+1)}\, dx$

    (j) $\int \dfrac{2}{\sqrt{4x+5}}\, dx$

    (k) $\int 2\tan^2 x\, dx$

    (l) $\int \cos \tfrac{1}{2}x\, dx$

    (m) $\int \dfrac{4}{5x-2}\, dx$

    (n) $\int x e^{-x}\, dx$

    (o) $\int x^2 \sqrt{2x^3-7}\, dx$

    (p) $\int 3\cos 7x \cos 3x\, dx$

    (q) $\int \dfrac{2}{(1-x)^2}\, dx$

    (r) $\int \dfrac{2}{x^2-1}\, dx$

    (s) $\int \dfrac{2x}{x^2-1}\, dx$

    (t) $\int \dfrac{1}{\sqrt{1-x^2}}\, dx$

    (u) $\int x^3 \ln x\, dx$

    (v) $\int \cos^3 \theta\, d\theta$

    (w) $\int e^{5\ln x}\, dx$

    (x) $\int e^{2x}\,(1+e^{2x})^4\, dx$

    (y) $\int e^x \sin 2x\, dx$

    (z) $\int (x+3)e^x\, dx$

    (aa) $\int \dfrac{8}{4+x^2}\, dx$

    (bb) $\int \dfrac{\ln x}{\sqrt{x}}\, dx$

2.  Evaluate the following integrals

    (a) $\int_0^1 \dfrac{3}{2x+1}\, dx$

    (b) $\int_0^2 \dfrac{2x}{(x^2+1)^3}\, dx$

    (c) $\int_0^3 \dfrac{6}{x^2+9}\, dx$

    (d) $\int_{\frac{\pi}{3}}^{\frac{\pi}{2}} \dfrac{\sin 2x}{\cos x}\, dx$

    (e) $\int_1^2 \ln x\, dx$

    (f) $\int_0^2 \dfrac{8}{\sqrt{16-x^2}}\, dx$

    (g) $\int_0^{\frac{\pi}{3}} \sec^2 x\, dx$

    (h) $\int_1^3 \dfrac{6}{x^2}\, dx$

    (i) $\int_0^{\frac{\pi}{6}} \cos^2 x\, dx$

    (j) $\int_0^2 (x-2)^3\, dx$

# DIFFERENTIAL EQUATIONS

Differential equations form a very important subject in mathematics. They crop up in all sorts of practical situations in physics, chemistry, biology, engineering…. The first job when such a situation occurs is to translate the real life problem into mathematical language; this translation is what is known as mathematical modelling.

A reminder of what we learnt in Unit one about modelling with differential equations:

❖ Rate of change implies using a derivative, and if no independent variable is given it is w.r.t. $t$.

❖ BUT derivatives only exist for **continuous** functions

❖ Rate of increase means the variable is increasing and its derivative is therefore positive

❖ Rate of decrease means the variable is decreasing and its derivative is therefore negative

❖ Where the rate of change is proportional to something a constant of proportion must be introduced

❖ When the differential equation is solved an arbitrary constant is introduced

❖ Each of these constants can be found by using the initial and other conditions given in the question

❖ Any unit of time may be used, so long as the same unit is used throughout the problem

❖ Velocity is rate of change of displacement

❖ Acceleration is rate of change of velocity

## Example 17.31

The rate of change of a population is proportional to the population. Letting P be the population, form a differential equation in P.

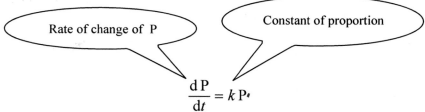

$$\frac{dP}{dt} = kP$$

# EXERCISE 17L

Form differential equations to express the following situations (use $t$ to represent time):

1.   The rate of decay of a radioactive material is proportional to the mass, $m$, of the material.

2.   The acceleration (rate of change of velocity) of a moving body is proportional to the velocity, $v$, of the body at a any given instant.

3.   The rate at which the height, $h$, of a tree increases is inversely proportional to the difference between its present height and its final height, $H$.

4.   The gradient of a graph at any point is proportional to the sum of the $x$- and $y$-coordinates at that point.

5.   In a chemical reaction, the rate of decrease in the concentration of the reagent is proportional to the square of the concentration, $c$.

## _First Order Differential Equations_

In unit 1 we learnt how to solve two types of first order differential equation:

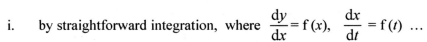

i.   by straightforward integration, where $\dfrac{dy}{dx} = f(x)$,  $\dfrac{dx}{dt} = f(t)$ ...

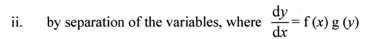

ii.   by separation of the variables, where $\dfrac{dy}{dx} = f(x)\,g(y)$

Now that we have learnt further integration techniques we can apply these methods in more complex situations. We will look at modelling and solving examples using these differential equations briefly before going on to a third type of first order differential equation.

## Example 17.32

The radioactive element carbon-14 has a half-life of 5750 years. If 200 g of this element are present initially, how much will be left after 1000 years.

❖      Assume that radioactive decay is exponential – proportional to mass (or number of atoms)

❖      The half-life is the time it takes for the mass to halve

> ➤ Form the differential equation: $\dfrac{dx}{dt} = -kx$

> ➤ Integrate: $\displaystyle\int \dfrac{dx}{x} = \int -k\,dt \;\to\; \ln x = -kt + c \;\to\; x = Ae^{-kt}$

> ➤ Use initial conditions: $200 = Ae^{0} \;\therefore\; A = 200, \;\; x = 200\,e^{-kt}$

> ➤ Use half-life: $100 = 200\,e^{-5750k} \;\;\therefore\; k = \frac{1}{5750}\ln 2$

> ➤ Give the formula: $x = 200\,e^{-kt}$ where $k = \frac{1}{5750}\ln 2$        ①

> ➤ Solve the problem: when $t = 1000$, $x = 200e^{-\frac{1000}{5750}\ln 2}$

$$= 200(2)^{-\frac{1000}{5750}}$$

$$= 200 \times 0.8864 = 177$$

---

① leads to the standard result for radioactive decay:   $x = x_{0}\,e^{-kt}$ where $k = \frac{1}{h}\ln 2$

                     ($x_{0}$ = initial mass, $h$ = half-life)

---

## Example 17.33

In a boarding school of 100 girls on a certain Caribbean island it is discovered that 3 girls have contracted an infectious disease. 5 days later 20 girls have the disease. It is assumed that the rate of infection is proportional to the product of the number of girls who already have the disease and the number of girls who have not yet succumbed to it. By creating a model based on this assumption, find out how many girls have been infected after ten days, and how long it will be before 99 of the girls are infected.

By considering how long it will take for the whole school to be infected, or otherwise, discuss the shortcomings of the model.

> ➤ Let N represent the number of girls infected at time $t$ days, and $k$ be the constant of proportion

> ➤ Write down the mathematical model: $\dfrac{dN}{dt} = kN(100 - N)$

> ➤ Separate the variables: $\displaystyle\int \dfrac{dN}{N(100 - N)} = \int k\,dt$

> ➤ Use partial fractions: $\displaystyle\int \left\{ \dfrac{\frac{1}{100}}{N} + \dfrac{\frac{1}{100}}{(100 - N)} \right\} dN = \int k\,dt$

> ➤ Replace $100k$ by $p$ and integrate: $\ln N - \ln(100 - N) = pt + c$

$$\text{i.e. } \ln\left( \dfrac{N}{100 - N} \right) = pt + c \quad \text{or} \quad \dfrac{N}{100 - N} = Ae^{pt}$$

> ➤ When $t = 0$, $N = 3$, $\therefore A = \frac{3}{97}$

> ➤ When $t = 5$, $N = 20$, $\therefore p = 0.41796$ (store the exact value in your calculator)

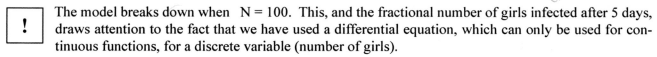

> model is: $\dfrac{N}{100-N} = \dfrac{3}{97}e^{pt}$

> Answers: when $t = 10$, $N = 67$ (to nearest whole number)
> when $N = 99$, $t = 19.3$

> [!] The model breaks down when $N = 100$. This, and the fractional number of girls infected after 5 days, draws attention to the fact that we have used a differential equation, which can only be used for continuous functions, for a discrete variable (number of girls).

In practice, as long as the numbers are large enough to be treated as if they were continuous, populations are often modelled using differential equations. However, in these cases great care needs to be taken when N (or some function of N like the $100 - N$ of this example) is close to 0.

## Example 17.34

According to Newton's law of cooling, the rate at which the temperature of a body falls is proportional to the amount by which its temperature exceeds that of its surroundings.

This situation can be modelled by the differential equation $\dfrac{dT}{dt} = -k(T - T_S)$ where $T$ is the temperature of

the body at time $t$, $T_S$ is the surrounding temperature and $k$ is the constant of proportion. Find the general solution to this equation.

> Separate the variables: $\displaystyle\int \dfrac{dT}{(T - T_S)} = \int -k\,dt$

> Integrate: $\ln(T - T_S) = -kt + c$ *

> Antilog: $T - T_S = Ae^{-kt}$

> Give the answer: $T = T_S + Ae^{-kt}$

\* $(T - T_S)$ cannot be negative, so no modulus sign is necessary

## Example 17.35

Find the general solution of the differential equation $(1 + x)\dfrac{dy}{dx} = 1 - \sin^2 y$

> Separate the variables: $\displaystyle\int \dfrac{dy}{1 - \sin^2 y} = \int \dfrac{dx}{1 + x}$ ①

> Apply trig identities: $\text{LHS} = \displaystyle\int \dfrac{dy}{\cos^2 y} = \int \sec^2 y\,dy$

so ① becomes $\displaystyle\int \sec^2 y\,dy = \int \dfrac{dx}{1 + x}$

> Integrate: $\tan y = \ln|1 + x| + c$

The solutions to the differential equations in examples 17.34 and 17.35 are **general solutions**. That is to say that they contain at least one arbitrary constant and represent a family of solutions. In particular cases, particular solutions can be found by using further information given in the question. For instance, in example 17.34 we might know the initial temperature of the body, and also it's temperature at a particular time.

# EXERCISE 17 M

In questions 1 – 8 find the general solution for each of the differential equations:

1. $\dfrac{dy}{dx} = 2xe^{x^2}$

2. $\dfrac{dy}{dx} = \dfrac{3}{\cos y}$

3. $x\dfrac{dy}{dx} = y(y+1)$

4. $(x+1)\dfrac{dy}{dx} = y$

5. $\dfrac{dy}{dx} = e^{x+y}$

6. $\dfrac{dy}{dx} = \cos^2 y \sin x$

7. $\dfrac{dy}{dx} = e^y \cos x$

8. $\dfrac{dy}{dx} = 4x\sqrt{1-y^2}$

9. Find the general solution to the differential equation $\dfrac{dy}{dx} = e^{2x+y}$.
   You may leave your solution in implicit form.

10. (a) Obtain the general solution to the differential equation $\dfrac{dy}{dx} = xy^2$, $y > 0$.

    (b) Given also that $y = 1$ at $x = 1$ show that $y = \dfrac{2}{3-x^2}$, $-\sqrt{3} < x < \sqrt{3}$, is a particular solution of the differential equation.

11. $\dfrac{dy}{dx} = \dfrac{2xy}{1+x^2}$, $y > 0$.

    (a) Find the general solution of this differential equation.

    (b) Given that $y = 2$ at $x = 0$, find the particular solution in the form $y = f(x)$.

12. (a) Find $\int xe^{-x}\,dx$.

    (b) Given that $y = \tfrac{1}{4}\pi$ at $x = 0$, solve the differential equation
    $$e^x \frac{dy}{dx} = \frac{x}{\sin 2y}$$

13. Find

    (a) $\int x\cos x\,dx$,

    (b) $\int \cos^2 y\,dy$.

    Hence find the general solution of the differential equation $\dfrac{dy}{dx} = x\cos x \sec^2 y$, $0 < y < \tfrac{1}{2}\pi$.

## Further First Order Differential Equations

Here we learn how to solve one further type of first order differential equation: those in the form

$$\frac{dy}{dx} + ky = f(x). \qquad \text{①}$$

 This is a first order linear differential equation and it is solved by means of an *integrating factor.* If the whole equation is multiplied by the factor $e^{kx}$ it can be integrated.

From the work we have done on implicit differentiation and the product rule we know that

$$\frac{d}{dx}(ye^{kx}) = e^{kx}\frac{dy}{dx} + y\frac{d}{dx}(e^{kx})$$

$$= e^{kx}\frac{dy}{dx} + kye^{kx}$$

$$= e^{kx}\left(\frac{dy}{dx} + ky\right) \qquad ②$$

so if ① is multiplied by $e^{kx}$ we get $\qquad e^{kx}\left(\frac{dy}{dx} + ky\right) = e^{kx}f(x)$

using ② and integrating gives $\qquad\qquad ye^{kx} = \int e^{kx}f(x)\,dx$

> The general solution of the differential equation $y' + ky = f(x)$ is
> $ye^{kx} = \int e^{kx}f(x)\,dx$ or $y = e^{-kx}\left[\int e^{kx}f(x)\,dx + C\right]$

If $k$ is negative, we can make the substitution $-k = p$, and the rule becomes

> The general solution of the differential equation $y' - py = f(x)$ is
> $ye^{-px} = \int e^{-px}f(x)\,dx$ or $y = e^{px}\left[\int e^{-px}f(x)\,dx + C\right]$

This is a simplified version of the rule for integrating differential equations in the form $y' + g(x)y = f(x)$ where the integrating factor is $e^{\int g(x)dx}$ and the general solution is $e^{\int g(x)dx}y = \int e^{\int g(x)dx}f(x)\,dx$

## Example 17.36

Solve the differential equation $\frac{dy}{dx} + 2y = e^{-x}$
- Identify $k$ and $f(x)$: $\qquad k = 2,\ f(x) = e^{-x}$
- Apply the formula: $\qquad ye^{2x} = \int e^{2x}e^{-x}\,dx = \int e^x\,dx = e^x + c$
- Give the answer: $\qquad y = e^{-2x}(e^x + c)$ or $y = e^{-x} + ce^{-2x}$

## Example 17.37

Solve the differential equation $\frac{dy}{dx} + 5y = x^2 + 2$
- Identify $k$ and $f(x)$: $\qquad k = 5,\ f(x) = x^2 + 2$
- Apply the formula: $\qquad ye^{5x} = \int e^{5x}(x^2 + 2)$

➤ Using integration by parts, $\int x^2 e^{5x}\, dx = \dfrac{1}{125} e^{5x}\{25x^2 - 10x + 2\}$

so $\quad ye^{5x} = \dfrac{1}{125} e^{5x}\{25x^2 - 10x + 2\} + \dfrac{2}{5} e^{5x} + C$

➤ Give the answer: $\quad y = \dfrac{1}{125}\{25x^2 - 10x + 52\} + Ce^{-5x}$

Here are two examples of how such an equation might arise in a real life situation

# Example 17.38

In a drive to diversify tourism, the government of Sun Island have decided to develop the local reservoir by providing angling facilities. This will require stocking the reservoir with the right number of extra fish so that the fish population remains stable while anglers remove fish. They require a model to help them determine what this stable population ($P$) should be, given the following data:

a) 30 fish enter the reservoir daily from the river that feeds it

b) 60 fish are eaten daily by predatory birds (pelicans and herons)

c) in the first few weeks they expect the number of anglers to increase steadily so that the number of fish caught each day will be $10t$, where $t$ is the time in days from the opening of the attraction (thereafter the number of anglers will be limited)

d) in the absence of other factors the fish population doubles every 50 days

Form a differential equation to model the situation during the initial period.

Assuming that population growth is proportional to the population, $\dfrac{dP}{dt} = kP$
$k$ can be calculated using d)

separating the variables $\int \dfrac{dP}{P} = \int k\, dt$

$\ln P = kt + c \quad$ or $\quad P = Ae^{kt}$

if $n_0$ is the population at time 0 then the population at time 50 is $2n_0$

when $t = 0 \qquad n_0 = A \times e^0 = A$

when $t = 50 \qquad 2n_0 = Ae^{50k}$

eliminate $n_0$: $\quad 2A = Ae^{50k} \quad \therefore\ e^{50k} = 2 \quad$ i.e. $k = \dfrac{1}{50} \ln 2$

Rate of change in $P$ = rate of increase in $P$ − rate of decrease in $P$

$$\dfrac{dP}{dt} \quad = \quad \underbrace{(kP + 30)}_{\text{d} \quad \text{a}} \quad - \quad \underbrace{(60 + 10t)}_{\text{b} \quad \text{c}}$$

so $\qquad \dfrac{dP}{dt} - kP = -30 - 10t$

this is of the form $P' - kP = f(t)$ where $k = \dfrac{1}{50} \ln 2$ and $f(t) = -30 - 10t$

it can be solved (using integration by parts) to give $P = \dfrac{10t + 30}{k} + \dfrac{10}{k^2} + Ce^{kt}$

When the government has decided how many anglers they will allow to fish each day, they can calculate a value for the constant $C$ ($k$ is known) and hence can work out how many fish need to be in the reservoir at the start of the process.

> ! Notice how many simplifications and assumptions have been made in creating this model. See if you can find at least 7.

## Example 17.38

The diagram, figure 17.2, shows an electric circuit containing an electromotive force (a battery or generator) producing a voltage of $E(t)$ volts and a current of $I_t$ amperes at time $t$, a resistor with resistance $R$ ohms, an inductor with an inductance of $L$ henries and a switch. The laws of physics lead to the differential equation

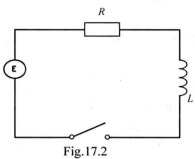

Fig.17.2

$$L\frac{dI}{dt} + RI = E(t)$$

In a particular circuit the resistance is 12 Ω, the inductance is 4 H, and the electromotive force is supplied by a generator producing a variable voltage of 60 sin 30$t$ volts. If the switch is closed when the generator is started at time $t = 0$, i.e. $I_0 = 0$, find an expression for the current $I$ at time $t$.

> ➤ Apply the model to this problem: $4\dfrac{dI}{dt} + 12I = 60\sin 30t$

> ➤ Divide through by 4: $\dfrac{dI}{dt} + 3I = 15\sin 30t$

> ➤ This is a differential equation of the form we have been working with where $k = 3$ and $f(t) = 15\sin 30t$
> At this point you can either multiply through by the integrating factor $e^{3t}$ or you can use the formula on page 77 either way it leads to $\quad Ie^{3t} = \int 15e^{3t}\sin 30t\,dt$

> ➤ Integrating by parts gives $Ie^{3t} = 15\dfrac{e^{3t}}{909}(3\sin 30t - 30\cos 30t) + C$
> i.e. $I = \frac{5}{101}(\sin 30t - 10\cos 30t) + Ce^{-3t}$

> ➤ Use the initial conditions to find $C$: $0 = \frac{5}{101}(-10) + C$ i.e. $C = \frac{50}{101}$

> ➤ Give the solution: $I_t = \frac{5}{101}(\sin 30t - 10\cos 30t + 10e^{-3t})$

## EXERCISE 17 N (There are more practical differential equation problems in exercise 14B)

1. Solve the following differential equations:

   (a) $\dfrac{dy}{dx} + y + 3 = x$     (b) $\dfrac{dy}{dx} + 2y = e^{-2x}\cos x$     (c) $\dfrac{dy}{dx} + y - \dfrac{1}{1+e^x} = 0$

   (d) $y' + 2y = e^{2x}$     (e) $y' + y = \cos(e^x)$     (f) $2y' - 4y = 1$

2. (a) In the circuit shown in figure 17.2, a battery supplies a constant voltage of 60 volts, the inductance is 4 H, the resistance is 12 Ω and $I_0 = 0$. Find (i) $I_t$ and (ii) the current after 1 second.

   (b) In the circuit shown in figure 17.2, a battery supplies a constant voltage of 40 volts, the inductance is 2 H, the resistance is 10 Ω and $I_0 = 0$. Find (i) $I_t$ and (ii) the current after 0.1 second.

   #(c) In the circuit shown in figure 17.2, a generator supplies a voltage of 40 sin 60$t$ volts, the inductance is 1 H, the resistance is 20 Ω and $I_0 = 1$ amp. Find $I_t$.

3. A population of rabbits is growing in such a way that at time $t$ weeks, the rate at which the population is increasing is proportional to the size, P, of the population at that time.

   Express this situation as a differential equation in P and $t$.

   If the size of the population is initially 8, show that $P = 8e^{kt}$, where $k$ is a positive constant.

   After 12 weeks the population has grown to 72. Show that $k = \frac{1}{12} \ln 9$.

   Assuming that no rabbits die, what will be the size of the population after 1 year?

4. Find a curve in the $xy$-plane that passes through $(0, 3)$ and whose tangent at a point $(x, y)$ has gradient $\dfrac{2x}{y^2}$.

5. In a chemical reaction a substance A is continuously transformed into a substance B. Throughout the reaction the sum of the masses of A and B remains constant and equal to $m$. The mass of B present at time $t$ after the start of the reaction is denoted by $x$. At any instant the rate of increase of the mass of B is $k$ times the mass of A, where $k$ is a positive constant. Write down the differential equation relating $x$ and $t$.

   Solve this differential equation, given that $x = 0$ when $t = 0$. Given also that $x = \frac{1}{2}m$ when $t = \ln 2$, determine the value of $k$, and show that, at time $t$, $x = m(1 - e^{-t})$.

   Hence find (i) the value of $x$ (in terms of $m$) when $t = 3 \ln 2$, (ii) the value of $t$ when $x = \frac{3}{4}m$.

6. A disease affects a proportion $p$ of a population. If the rate of infection is proportional to the product of the proportion already infected with the proportion not yet infected, form a differential equation connecting $p$ and $t$. If 5% of the population are infected initially, and 2 weeks later 15% are found to be infected, find how long it will be before 50% are infected.

7. The rate at which a body loses speed at any given instant as it travels through a resistive medium is given by $kv$ m s$^{-2}$ where $v$ is the speed of the body at that instant and $k$ is a positive constant. If its initial speed is $u$ m s$^{-1}$ show that the time taken for the body to decrease its speed to $\frac{1}{2}u$ m s$^{-1}$ is $\frac{1}{k} \ln 2$ seconds.

8. A chemical substance X decays at a rate equal to twice the quantity of X present. If $x$ is the quantity of X present at time $t$, write down a differential equation in $x$ and $t$.

   Given that initially $x = a$, find an expression for $x$ in terms of $a$ and $t$.

9. During a chemical reaction two substances A and B decompose. The number of grams, $x$ of substance A present at time $t$ is given by $x = \dfrac{10}{(1+t)^3}$.

   There are $y$ grams of B present at time $t$ and $\dfrac{dy}{dt}$ is directly proportional to the product of $x$ and $y$.

   (a) Given that $y = 20$ and $\dfrac{dy}{dt} = -40$ when $t = 0$, show that $\dfrac{dy}{dt} = \dfrac{-2y}{(1+t)^3}$.

   (b) Hence determine $y$ as a function of $t$.

   (c) Determine the amount of substance B remaining when the reaction is essentially complete.

10. (a) As he falls, a skydiver is subject to drag force proportional to the square of his velocity, $v$ ms$^{-1}$. This leads to the equation of motion $\dfrac{dv}{dt} = g - kv^2$ where $g \approx 10$ ms$^{-2}$ (the acceleration due to gravity).

    (i) Can this equation be solved to find an expression for $v$ in terms of $t$? (don't solve it, but state what method you could use)

    When the acceleration $\left(\dfrac{dv}{dt}\right)$ reaches 0 the diver continues to fall at a constant velocity, which is known as the terminal velocity.

    (ii) In the case where $k = .004$ find the diver's terminal velocity, V.

(b) If the diver now opens his parachute, he experiences an air resistance force which is proportional to his velocity so that the equation of motion becomes $\dfrac{dv}{dt} = g - cv$.

    (i)    If $c = 1.5$, find his new terminal velocity.

    (ii)   Assuming that he opens his parachute when his velocity is V, find an expression in terms of $t$ for his velocity $t$ seconds after he opens the parachute.

## *Second Order Differential Equations*

We are concerned here with equations of the form $ay'' + by' + cy = f(x)$           ①
where $a = 0$ and $f(x)$ is either a polynomial of degree $\leq 2$ or a trig function.

In unit 1 when learning about second derivatives you sometimes had to differentiate a function twice and hence show that it satisfied a particular (second order differential) equation

e.g. if $y = 3x + \cos 2x$ find $\dfrac{d^2y}{dx^2}$ and hence show that $\dfrac{d^2y}{dx^2} + 4y = 12x$        ②

or if $y = \dfrac{A}{x} + Bx$ where A and B are constants, show that $x^2\dfrac{d^2y}{dx^2} + x\dfrac{dy}{dx} = y$      ③

In each of these cases we see that a second order differential equation can be formed when a function is differentiated twice.

If $\dfrac{d^2y}{dx^2} = f(x)$ (i.e. $b = c = 0$ in ① above) the equation can be solved by integrating twice, but otherwise solving second order differential equations is not so straightforward. However examples ② and ③ above give us some useful hints:

1) since the equations come from differentiating twice we would expect the solution to be equivalent to integrating twice which introduces 2 arbitrary constants, as you can see from ③.

2) ② reminds us that if a cos or sin function is differentiated twice the second derivative contains the same trig function e.g. if $y = \sin 3x$, $y'' = 9\sin 3x$, and similarly if $y = e^{2x}$, $y' = 2e^{2x}$ and $y'' = 4e^{2x}$.

This suggests that the general solution to ① will contain 2 arbitrary constants and is likely to contain either an exponential or a trig function.

What follows is not a formal proof of the results you need to know for solving equations of the form ① but finds those results using deduction from observation.

## *Homogenous linear differential equations*

① is known as a ***second order linear*** differential equation. Linear because the derivatives and $y$ are to the first power (not squared etc) When $f(x) = 0$ it is also ***homogeneous***.

We will start with the simpler case where $f(x) = 0$, giving the homogenous differential equation

$$ay'' + by' + cy = 0 \quad \text{or} \quad a\frac{d^2y}{dx^2} + b\frac{dy}{dx} + cy = 0 \qquad ④$$

Let us assume that the solution to ④ is (or is similar to) $y = e^{ux}$.

then $y' = ue^{ux}$ and $y'' = u^2e^{ux}$

substituting in ④ we get $au^2e^{ux} + bue^{ux} + ce^{ux} = 0$

i.e. $e^{ux}(au^2 + bu + c) = 0$

    now $e^{ux}$ cannot be zero, so $au^2 + bu + c = 0$   ⑤

⑤ is known as the *auxiliary equation* (**or *characteristic equation***) for equation ④.

! The auxiliary equation is found from equation ④ by replacing $y''$ with $u^2$, $y'$ with $u$ and $y$ with 1. Notice also that this equation is a quadratic in $u$, so if you know $a$, $b$ and $c$ you can find $u$.

Now a quadratic equation normally has 2 roots which can be found either by factorisation or from the quadratic formula. If these roots are $\alpha$ and $\beta$ then we would expect the solution to ④ to be $y = Ae^{\alpha x} + Be^{\beta x}$. ⑥

This contains 2 arbitrary constants (as long as $\alpha$ and $\beta$ are distinct) and is therefore the solution we are looking for.

Taking a closer look at the auxiliary equation we see that $\alpha$ and $\beta$ are given by

$\alpha = \dfrac{-b + \sqrt{b^2 - 4ac}}{2a}$ and $\beta = \dfrac{-b - \sqrt{b^2 - 4ac}}{2a}$ and we need to consider what happens when the discriminant

$(b^2 - 4ac)$ is 0 or negative.

If $b^2 - 4ac = 0$, $\alpha = \beta$ and the solution is $y = e^{\alpha x}(A + Bx)$ ⑦

If $b^2 - 4ac < 0$, the auxiliary equation does not have real roots. The roots are in fact complex numbers (see chapter 25) The roots are in the form $\alpha = p + iq$ and $\beta = p - iq$ where $i = \sqrt{-1}$. You will discover in chapter 25 that there is a close relationship between exponential functions with imaginary indices and sine and cosine functions, but for now it is enough to know that when $b^2 - 4ac < 0$ the solution to ④ is $y = e^{px}(A\cos qx + B\sin qx)$

Before applying these results here is a summary:

> To solve the differential equation $ay'' + by' + cy = 0$
>
> find the roots $\alpha$ and $\beta$ of the auxiliary equation $au^2 + bu + c = 0$
>
> if $\alpha$ and $\beta$ are real and distinct (when $b^2 > 4ac$), the general solution is $y = Ae^{\alpha x} + Be^{\beta x}$
>
> if $\alpha$ and $\beta$ are real but equal (when $b^2 = 4ac$), the general solution is $y = e^{\alpha x}(A + Bx)$
>
> if $\alpha$ and $\beta$ are complex, namely $p \pm iq$, (when $b^2 < 4ac$), the general solution is $y = e^{px}(A\cos qx + B\sin qx)$

Example 17.39 will illustrate this method and checking the answers will show that the method works

## Example 17.39

Find the general solution to to following differential equations

    (a) $y'' - 4y' + 4y = 0$

    (b) $y'' + 3y' - 10y = 0$

    (c) $y'' - 8y' + 17y = 0$

(a)

➤ the auxiliary equation is $u^2 - 4u + 4 = 0$

➤ factorising:      $(u - 2)^2 = 0$ i.e. $\alpha = \beta = 2$

➤ this equation has real equal roots so the solution is $y = e^{2x}(A + Bx)$

➤ [check the answer: if $y = e^{2x}(A + Bx)$ then $y' = e^{2x}(2A + 2Bx + B)$ and $y'' = e^{2x}(4A + 4Bx + 2B + 2B)$

➤ we could eliminate A and B from these 3 equations, but it is easier to substitute into the LHS of (a)

➤ $y'' - 4y' + 4y = e^{2x}\{4A + 4Bx + 4B - 4(2A + 2Bx + B) + 4(A + Bx)\}$

➤ $= e^{2x}\{A(4 - 8 + 4) + B(4-4) + Bx(4 - 8 + 4)\} = 0]$

(b)

➢ the auxiliary equation is $u^2 + 3u - 10 = 0$

➢ factorising: $(u + 5)(u - 2) = 0$ i.e. $\alpha = -5$, $\beta = 2$

➢ this equation has distinct real roots so the solution is $\underline{y = Ae^{-5x} + Be^{2x}}$

➢ [check: if $y = Ae^{-5x} + Be^{2x}$ then $y' = -5Ae^{-5x} + 2Be^{2x}$ and $y'' = 25Ae^{-5x} + 4Be^{2x}$

➢ substitute in LHS of (b) $y'' + 3y' - 10y = Ae^{-5x}(25 - 15 - 10) + Be^{2x}(4 + 6 - 10) = 0]$

(c)

➢ the auxiliary equation is $u^2 - 8u + 17 = 0$

➢ this does not factorise, so using the quadratic formula $u = \dfrac{8 \pm \sqrt{64 - 68}}{2} = 4 \pm i$
   (i.e. $p = 4$, $q = 1$)

➢ the solution is $\underline{y = e^{4x}(A \cos x + B \sin x)}$

➢ [check: if $y = e^{4x}(A \cos x + B \sin x)$ then

➢ $y' = e^{4x}(4A\cos x + 4B\sin x - A\sin x + B\cos x) = e^{4x}\{(4A + B)\cos x + (4B - A)\sin x\}$

➢ and $y'' = e^{4x}\{(16A + 4B)\cos x + (16B - 4A)\sin x - (4A + B)\sin x + (4B - A)\cos x\}$
   $= e^{4x}\{(15A + 8B)\cos x + (15B - 8A)\sin x\}$

➢ substitute in LHS of (c)

➢ $y'' - 8y' + 17y = e^{4x}\{(15A + 8B)\cos x + (15B - 8A)\sin x - 8[(4A + B)\cos x + (4B - A)\sin x]$
   $+ 17(A \cos x + B \sin x)\}$
   $= e^{4x}\{(15 - 32 + 7)A\cos x + (8 - 8)B\cos x + (-8 + 8)A\sin x + (15 - 32 + 17)B\sin x\} = 0]$

 I have included the check for these examples to show that the rules really do work; in an exam you don't <u>need</u> to do the check, but if you have time at the end it is worth doing to make sure you haven't made a silly mistake.

## Nonhomogenous second order linear differential equations

When $f(x) \neq 0$ equation ① $ay'' + by' + cy = f(x)$ is a nonhomogenous second order linear differential equation. Its solution contains the solution of the simpler (homogeneous) equation $ay'' + by' + cy = 0$ ④ but with an extra term or terms.

 In this case the solution to ④ is known as the *complementary function* and the extra term/s is called the *particular integral*. The general solution then is $y = $ C.F. + P.I.

The particular integral takes a similar form to $f(x)$. If $f(x)$ is a constant the P.I. will be a constant, if $f(x)$ is a quadratic polynomial the P.I. will be a quadratic function, whose constants can be determined, and if $f(x)$ is a sin/cos function the P.I. will take the form $C\sin x + D\cos x$.

## Example 17.40

Find the general solution to the differential equation $y'' - 2y' + 5y = 10x + 1$. ①

➢ Write down and solve the auxiliary equation: $u^2 - 2u + 5 = 0$

   $u = \dfrac{2 \pm \sqrt{4 - 20}}{2}$ i.e. $u = 1 \pm 2i$

➢ the roots are complex, so the complementary function is $y = e^x(A\cos 2x + B\sin 2x)$

➢ Find the particular integral: since $f(x)$ is linear it should be $\underline{y = Cx + D}$ ②

➢ Find C and D: differentiate ② twice: $y' = C$, $y'' = 0$

➢ substitute in $y$, $y'$ & $y''$ in ①: $0 - 2C + 5(Cx + D) = 10x + 1$ i.e. $5Cx + 5D - 2C = 0$

$C(5x - 2) + 5D = 0$

➤ this is an identity, so equate coefficients: $5C = 10$, $5D - 2C = 1$

∴ $C = 2$, $D = 1$

$5x - 2 = x$

➤ particular integral is $y = 2x + 1$

➤ general solution is $y = $ C.F. + P.I. $= e^x(A\cos 2x + B\sin 2x) + 2x + 1$

> [!] Notice that there are two arbitrary constants, which both come from the complementary function. For a particular solution you will be given 2 pieces of information – either the values of both $y$ and $y'$ when $x = 0$, or the values of $y$ for 2 different values of $x$.

## Example 17.41

Find the general solution of the differential equation $y'' - 2y' = 5\sin x$ ①

Given that when $x = 0$, $y = 0$ and $y' = 3$, find the corresponding particular solution.

➤ Write down and solve the auxiliary equation: $u^2 - 2u = 0$ ②

$u(u - 2) = 0$ i.e. $\alpha = 0$, $\beta = 2$

the complementary function is therefore $y = A + Be^{2x}$ ③

➤ Find the particular integral: it should be $\underline{y = C\cos x + D\sin x}$ ④

➤ Find C and D: differentiate ④ twice: $y' = -C\sin x + D\cos x$, $y'' = -C\cos x - D\sin x$

➤ substitute for $y$, $y'$ & $y''$ in ①: $-C\cos x - D\sin x - 2(-C\sin x + D\cos x) = 5\sin x$

this is an identity, so equate coefficients of $\cos x$ & $\sin x$: $-C - 2D = 0$, $-D + 2C = 5$

∴ $C = 2$, $D = -1$

$-C\cos x - D\sin x + 2C\sin x - 2D\cos x = 5\sin x$

$\cos x(-C - 2D) + \sin x(-D + 2C) = 5\sin x + 0\cos x$

$-D + 2C = 5$

$-C - 2D = 0$

➤ particular integral is $y = 2\cos x - \sin x$

➤ general solution is $y = $ C.F. + P.I. $\underline{= A + Be^{2x} + 2\cos x - \sin x}$ ⑤

➤ to find the particular solution for the given initial conditions we need to find the constants A and B. We therefore need to write down expressions for $y$ and $y'$. (differentiate ⑤)

$y = A + Be^{2x} + 2\cos x - \sin x$

$y' = 2Be^{2x} - 2\sin x - \cos x$

➤ using initial conditions: $A + B + 2 = 0$, $2B - 1 = 3$ ∴ $B = 2$, $A = -4$

➤ particular solution is $\underline{y = 2e^{2x} + 2\cos x - \sin x - 4}$

## Example 17.42

Find the particular solution of the differential equation $y'' + 16y = 9$ ①, given that when $x = 0$, $y' = 0$, and $y'' = 32$ ②

➤ Write down and solve the auxiliary equation: $u^2 + 16 = 0$

$u = \pm 4i$ i.e. $p = 0$, $q = 4$

the complementary function is $y = A\sin 4x + B\cos 4x$

➤ Find the particular integral: it should be $\underline{y = C}$ ③

➤ Find C: differentiate ③ twice: $y' = 0$, $y'' = 0$

substitute in ①: $0 + 16C = 9$ ∴ $C = \frac{9}{16}$

➤ particular integral is $y = \frac{9}{16}$

- general solution is $y = \text{C.F.} + \text{P.I.} = \text{A}\sin 4x + \text{B}\cos 4x + \frac{9}{16}$ ④
- to find the constants A and B write down expressions for $y'$ and $y''$. (differentiate ④ twice)

  $y' = 4\text{A}\cos 4x - 4\text{B}\sin 4x, \quad y'' = -16\text{A}\sin 4x - 16\text{B}\cos 4x$
- using initial conditions ②: $0 = 4\text{A}$, $\therefore$ A = 0; $32 = -16\text{B}$, $\therefore$ B = −2
- particular solution is $y = \frac{9}{16} - 2\cos 4x$

Second order differential equations occur widely, especially in physics. ***Simple harmonic motion*** describes the motion of a body oscillating about a fixed point without additional forces other than that due to gravity. In SHM the acceleration of the body is towards the centre of motion and proportional to the displacement. The equation of motion is therefore $\dfrac{\mathrm{d}^2x}{\mathrm{d}t^2} = -n^2x$ (a homogeneous second order differential equation).

Common examples of SHM are a mass oscillating on an elastic string or spring and a simple pendulum. The rise and fall of the tides follows the same pattern.

Where a resisting force proportional to the velocity (such as friction) is applied to SHM, the motion becomes damped harmonic motion and the equation of motion is in the form $\dfrac{\mathrm{d}^2x}{\mathrm{d}t^2} + k\dfrac{\mathrm{d}x}{\mathrm{d}t} + n^2x = 0$, which is again a homogeneous second order differential equation.

The further addition of a force which is a constant or function of time leads to a nonhomogeneous differential equation.

Similar differential equations also arise in electric circuits.

## Example 17.43

The equation of motion of a mass attached to a spring immersed in a fluid is $\dfrac{\mathrm{d}^2x}{\mathrm{d}t^2} + 20\dfrac{\mathrm{d}x}{\mathrm{d}t} + 64x = 0$. Initially the mass is set in motion from the equilibrium position by a push giving it a velocity of 0.6 ms$^{-1}$. (i.e. when $t = 0$, $x = 0$ and $\dfrac{\mathrm{d}x}{\mathrm{d}t} = 0.6$ ). Find the displacement of the mass at time $t$ seconds.

PARTICULAR INTEGRALS

$y = C\cos x + D\sin x$

$y = Cx + D$

$y = C$

- The auxiliary equation is $u^2 + 20u + 64 = 0$ with roots $u = -4$ and $-16$

  so $x = \text{A}e^{-4t} + \text{B}e^{-16t}$
- Differentiate: $\dfrac{\mathrm{d}x}{\mathrm{d}t} = -4\text{A}\,e^{-4t} - 16\text{B}\,e^{-16t}$
- Apply initial conditions : when $t = 0$ $x = \text{A} + \text{B} = 0$, $\dfrac{\mathrm{d}x}{\mathrm{d}t} = -4\text{A} - 16\text{B} = 0.6$

  so A = 0.05 and B = −0.05
- Answer: $x = 0.05(e^{-4t} - e^{-16t})$ m

# EXERCISE 17O

1. Differential equations that can be solved by integrating twice

   (a) Find $y$ as a function of $x$ given that $\dfrac{\mathrm{d}^2y}{\mathrm{d}x^2} = 4 - 6x$ and that when $x = 2$, $\dfrac{\mathrm{d}y}{\mathrm{d}x} = -4$ and $y = 7$.

   (b) Find the equation of the curve for which $\dfrac{\mathrm{d}^2y}{\mathrm{d}x^2} = 6x - 4$ and which passes through the points (1, 4) and (−1, 2).

   (c) Find the equation of the curve for which $\dfrac{\mathrm{d}^2y}{\mathrm{d}x^2} = 30x$ and whose gradient is 5 at the point (1, −1)

2. Homogeneous differential equations

    (a)   Find the general solution of the differential equation $y'' + y' - 6y = 0$

    (b)   Find the general solution of the differential equation $y'' - 2y' + y = 0$

    (c)   Find the general solution of the differential equation $y'' - 6y' + 13y = 0$

    (d)   Find the general solution of the differential equation $y'' + 8y' + 41y = 0$

    (e)   Find the general solution of the differential equation $y'' - 6y' + 8y = 0$

    (f)   Find the general solution of the differential equation $y'' + 2y' + y = 0$

    (g)   Solve the differential equation $y'' + 6y' + 13y = 0$, given that when $x = 0$, $y = 4$ and $y' = 0$

    (h)   Solve the differential equation $y'' - 8y' + 16y = 0$, given that when $x = 0$, $y = 2$ and $y' = 11$

    (i)   Solve the differential equation $y'' + y' - 6y = 0$, given that when $x = 0$, $y = 5$ and $y' = 0$

3. Nonhomogeneous differential equations

    (a)   Find the general solution of the differential equation $y'' + 2y' + 2y = 8$

    (b)   Find the general solution of the differential equation $y'' + y' - 2y = 4\cos x$

    (c)   Find the general solution of the differential equation $y'' - 4y' + 4y = 12$

    (d)   Find the general solution of the differential equation $y'' - 2y' + 5y = 10x + 1$

    (e)   Find the general solution of the differential equation $y'' + 3y' + 2y = 4x^2$

          (Hint: although RHS only contains $x^2$, PI could contain $x$ and constant as well)

    (f)   Find the general solution of the differential equation $y'' + 6y' + 9y = 9x + 24$

    (g)   Find the general solution of the differential equation $y'' - 2y' = 40\sin 4x$

    (h)   Find the general solution of the differential equation $y'' + 4y' = 2x^2 + x - 5$

4. Mixed questions

    (a)   Find the general solution of the differential equation $y'' + 9y = 18$

    (b)   Solve the differential equation $y'' + y = 0$, given that when $x = 0$, $y = 2$ and $y' = 3$

    (c)   Find the general solution of the differential equation $y'' + 8y' + 16y = 32x^2 - 5$

    (d)   Solve the differential equation $y'' - 4y' + 3y = 0$, given that when $x = 0$, $y = 0$ and $y' = 2$

    (e)   Find the general solution of the differential equation $y'' + 3y' - 4y = 5 - 4x$

    (f)   Find the general solution of the differential equation $y'' + 2y' + 4y = 0$

    (g)   Solve the differential equation $y'' + 4y = 2x^2 - 3$, given that when $x = 0$, $y = 0$ and $y' = 4$

    (h)   Find the general solution of the differential equation $y'' - 2y' = 0$

    (i)   Find the general solution of the differential equation $y'' + y = \cos 2x$

    (j)   Solve the differential equation $y'' - 10y + 25 = 0$, given that when $x = 0$, $y = 2$ and $y' = 9$

    (k)   Find the general solution of the differential equation $y'' - 3y' - 10y = 0$

    (l)   Solve the differential equation $y'' - 9y = 18$, given that when $x = 0$, $y = -2$ and $y' = 6$

    (m)   Find the general solution of the differential equation $y'' - 4y = \sin 2x$

# EXERCISE 17P

1. The equation of motion of a body oscillating on the end of a spring is $\dfrac{d^2x}{dt^2} + 100x = 0$, where $x$ is the displacement in metres for the body from its equilibrium position after time $t$ seconds. Determine $x$ in terms of $t$, given that at time $t = 0$, $x = 2$ m and $\dfrac{dx}{dt} = 0$.

2.  The oscillations of a heavily damped pendulum satisfy the differential equation $\dfrac{d^2x}{dt^2} + 6\dfrac{dx}{dt} + 8x = 0$, where $x$ cm is the displacement of the bob at time $t$ seconds. The initial displacement is equal to $+4$ cm and the initial velocity is 8 cms$^{-1}$. solve the equation for $x$.

3.  In a galvanometer the deflection $\theta$ satisfies the differential equation $\dfrac{d^2\theta}{dt^2} + 4\dfrac{d\theta}{dt} + 4\theta = 8$. Solve the equation for $\theta$ given that when $t = 0$, $\theta = \dfrac{d\theta}{dt} = 2$.

4.  The equation of motion for a damped oscillator is $\dfrac{d^2x}{dt^2} + 6\dfrac{dx}{dt} + 9x = 0$. Given that at time $t = 0$, $x = s$ and $\dfrac{dx}{dt} = u$, find an expression for $x$ as a function of $t$.

5.  The equation of motion of a forced oscillation is $\dfrac{d^2x}{dt^2} + 2\dfrac{dx}{dt} + 5x = 4\sin 5t$. Solve the equation given that when $t = 0$, $x = 0$ and $\dfrac{dx}{dt} = 0$.

# APPROXIMATING INTEGRALS – THE TRAPEZIUM RULE

Not all functions can be integrated, but since we know that definite integrals represent areas, we can often find an approximation to an integral by calculating an approximation to the corresponding area. There are a number of ways of doing this, of which the simplest is the trapezium rule.

This uses the idea that we met earlier (in unit 1), that an area can be divided into strips whose areas are summed. In the case of the trapezium rule the strips are approximated to trapezia.

Here is the proof when an area is divided into 5 strips, and then the formula is generalised for $n$ strips.

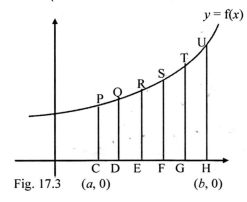

Fig. 17.3    $(a, 0)$        $(b, 0)$

Suppose we need to evaluate $\displaystyle\int_a^b f(x)\,dx$.

This integral can be represented by the area PUHC in fig.17.3, where C is the point $(a, 0)$ and H the point $(b, 0)$

Divide this area into 5 strips of equal width, so each strip will be of

width $h = \dfrac{b-a}{5}$.

Notice that 5 strips means 6 ordinates, PC, QD, RE, SF, TG and UH

If the coordinates of P, Q, …U are $(x_0, y_0)$, $(x_1, y_1)$…. $(x_5, y_5)$ then $x_0 = a$, $x_1 = a + h$, $x_2 = a + 2h$… and $y_0 = f(x_0)$  $y_1 = f(x_1)$  $y_2 = f(x_2)$ ……

Taking each strip to be approximately a trapezium we get

$A = \tfrac{1}{2}h(y_0 + y_1) + \tfrac{1}{2}h(y_1 + y_2) + \tfrac{1}{2}h(y_2 + y_3) + \tfrac{1}{2}h(y_3 + y_4) + \tfrac{1}{2}h(y_4 + y_5)$

$= \tfrac{1}{2}h\{y_0 + 2(y_1 + y_2 + y_3 + y_4) + y_5\}$

Generalizing the result we get

> **The Trapezium Rule**
> If $f(x)$ is a continuous function on $[a, b]$
>
> $$\int_a^b f(x)\,dx \approx \tfrac{1}{2}h\{y_0 + 2(y_1 + y_2 + \ldots + y_{n-1}) + y_n\}$$
>
> where $n$ is the number of strips into which the area is divided,
>
> $$h = \frac{b-a}{n}, \quad x_r = a + rh \quad \text{and} \quad y_r = f(x_r)$$
>
> The greater the value of $n$, the more accurate the approximation.

This may look a bit daunting like this, but is really very straightforward, if tedious, to apply.

## Example 17.44

Use the trapezium rule with 5 ordinates to estimate the value of the integral

$$\int_0^1 \sqrt{1 - x^2}\,dx$$

➤ find $h$           $n = 4, \;\; \therefore\; h = (1 - 0)/4 = 0.25$

➤ find $x_r, y_r$ for $r = $   0,1,2,3,4:

| $x_r$ | 0 | 0.25 | 0.5 | 0.75 | 1 |
|-------|---|------|-----|------|---|
| $y_r$ | 1 | 0.968246 | 0.866025 | 0.661438 | 0 |

➤ substitute values in the formula

$$\int_0^1 \sqrt{1 - x^2}\,dx = \tfrac{1}{2} \times 0.25\{1 + 2(0.968246 + 0.866025 + 0.661438) + 0\}$$

$$= 0.749$$

(the exact answer, which can be found by integration using the substitution $u = \sin x$, is actually $\tfrac{1}{4}\pi \approx 0.785$)

## Example 14.45

(a) Find the value of $\displaystyle\int_0^{\frac{\pi}{2}} \cos x\,dx$ by integration.

(b) Use the trapezium rule with 3 trapezia to find an approximation to the integral $\displaystyle\int_0^{\frac{\pi}{2}} \cos x\,dx$

(a) $\displaystyle\int_0^{\frac{\pi}{2}} \cos x\,dx = \left[\sin x\right]_0^{\frac{\pi}{2}} = 1 - 0 = 1$

(b) ➤ find $h$ :        $n = 3, \;\; \therefore\; h = \dfrac{b-a}{n} = \dfrac{\frac{\pi}{2} - 0}{3} = \dfrac{\pi}{6}$

➤ find $x_r, y_r$ for $r = 0,1,2,3$:

| $x_r$ | 0 | $\frac{1}{6}\pi$ | $\frac{1}{3}\pi$ | $\frac{1}{2}\pi$ |
|-------|---|------|------|------|
| $y_r$ | 1 | 0.866025 | 0.5 | 0 |

➤ substitute values in formula:    $\displaystyle\int_0^{\frac{\pi}{2}} \cos x\,dx \approx \tfrac{1}{2} \times \tfrac{1}{6}\pi \{1 + 2(0.866025 + 0.5) + 0\}$

$$= 0.977$$

> **!** You will give your answer correct to 3 s.f. This means that you must either use the memory keys on your calculator (preferable) or work with 6 s.f. for the calculation.

## EXERCISE 17Q

1. Find an approximate value for $\int_1^2 \frac{1}{x}\,dx$ using the trapezium rule with 5 strips.

2. (a) Use the trapezium rule with 4 ordinates to estimate the value of the integral $\int_0^3 \sqrt{x}\,dx$.
   (b) Use integration to show that this estimate is correct to the nearest whole number.
   (c) How could you use the trapezium rule to get a better approximation?

3. Use the trapezium rule with 3 trapezia to find an approximate value for $\int_0^\pi x\sin x\,dx$.

4. Use the trapezium rule with 5 ordinates to estimate the value of the integral $\int_0^1 \sqrt{x^3+1}\,dx$.

5. Find an approximate value for $\int_0^{\frac{\pi}{4}} \tan x\,dx$ using the trapezium rule with 3 strips.

6. Use the trapezium rule with 5 ordinates to estimate the value of the integral $\int_0^2 \frac{1}{\sqrt{x+5}}\,dx$

7. Donna and Tanya want to find an approximation to $\int_0^2 x^4\,dx$.
   (a) Donna uses the trapezium rule with 2 strips of equal width. What result does she get?

   Tanya thinks that a more accurate approximation can be found using 2 trapezia of different widths with ordinates at $x = 0$, $1\frac{1}{4}$ and 2.
   (b) Draw a sketch to illustrate Tanya's method, and calculate her approximation.

   (c) Use integration to calculate $\int_0^2 x^4\,dx$ exactly. Which approximation is the more accurate?

## EXERCISE 17 X

1. Express $\dfrac{x^4}{(x+1)(x^2+1)}$ as the sum of a polynomial and partial fractions. [5]

2. By means of the substitution $u = 1 + x^3$, show that
$$\int_0^2 \frac{x^2}{\sqrt{(1+x^3)}}\,dx = \int_1^9 \frac{u^{-\frac{1}{2}}}{3}\,du$$ [3]

$= \int (x-1) + \int \frac{1}{2(x+1)} - \int \frac{x-1}{2(x^2+1)}$

$\frac{-\frac{1}{2}x + \frac{1}{2}}{x^2+1} = \frac{-(x-1)}{2(x^2+1)}$

3. Express $\dfrac{x^2}{(x-1)(x^2+2)}$ in partial fractions. [6]

4. By means of the substitution $u^2 = x - 3$, show that

$$\int \frac{2x+1}{(x-3)^{\frac{3}{2}}} dx = \int \left(4 + \frac{14}{u^2}\right) du.$$ [6]

5. Express $\dfrac{1}{x(x+2)}$ in partial fractions and hence show that $\displaystyle\int \frac{dx}{x(x+2)} = \tfrac{1}{2}\ln\left(\frac{x}{x+2}\right) + k$,

where $k$ is constant and $x > 0$. [5]

✳ 6. Express $\dfrac{3x^2+2}{(2x+1)(x-2)}$ in partial fractions. $u=$ [6]

7. Evaluate the following integrals

(i) $\displaystyle\int x^3 e^x dx$ [6]

(ii) $\displaystyle\int x^3 \ln 4x \, dx, \ x > 0$ [6]

8. Find

(i) $\displaystyle\int 2xe^x dx$ [4]

(ii) $\displaystyle\int \frac{x}{(x+4)(6-x)} dx$ [5]

9. Given that $\dfrac{d}{dx}(e^{-x}) = -e^{-x}$, evaluate $\displaystyle\int x^2 e^{-x} dx$. [4]

10. Express in partial fractions $\dfrac{1-x^2}{x(x^2+1)}$ ← employ simultaneous equation [8]

11. Use the substitution $u = \cos x$ to find $\displaystyle\int \sin^3 x \, dx$. [6]

12. Use the substitution $u = \sin x$ to find $\displaystyle\int \cos^3 x \, dx$. [6]

13. Express $\dfrac{1}{x^2(x+1)}$ in partial fractions. Hence, find the indefinite integral $\displaystyle\int \dfrac{dx}{x^2(x+1)}$ [6]

14. Use the substitution $u = x - 2$ to evaluate

   (a) $\displaystyle\int_2^3 (x-2)^{\frac{3}{2}} dx$ [3]

   (b) $\displaystyle\int_2^3 x\sqrt{(x-2)}\, dx.$ [3]

15. Find the real constants A, B and C such that

   $$\dfrac{3x^2 + 4x + 1}{(x+2)(x^2+1)} = \dfrac{A}{x+2} + \dfrac{Bx+C}{x^2+1}$$

   Hence evaluate $\displaystyle\int \dfrac{3x^2 + 4x + 1}{(x+2)(x^2+1)}\, dx.$ [13]

16. (a) Express in partial fractions $\dfrac{x+1}{x(x+2)}$ [5]

   (b) Hence, find $\displaystyle\int \dfrac{x+1}{x(x+2)}\, dx,\ x > 0.$ [3]

17. Find $\displaystyle\int x^2 \ln x\, dx,\ x > 0$ [5]

18. Use the substitution $u = \sin x$ to find $\displaystyle\int e^{\sin x} \cos x\, dx.$ [5]

19. Express $\dfrac{x-1}{(x+1)(x^2+5)}$ in partial fractions, [6]

20. Find the following integrals

   (a) $\displaystyle\int \dfrac{1 + \tan^2\theta}{\tan\theta}\, d\theta$ [3]

   (b) $\displaystyle\int x^2 e^{3x}\, dx$ [6]

21. (a) (i) Using the fact that $e^{-x} = \dfrac{1}{e^x}$ or otherwise, show that $\dfrac{d}{dx}(e^{-x}) = -e^{-x}$ [3]

   (ii) hence evaluate $\displaystyle\int x^2 e^{-x}\, dx.$ [4]

   (b) (i) Given that $f(x) = \dfrac{x}{(x^2+4)}$, show that $f'(x) = \dfrac{4 - x^2}{(x^2+4)^2}.$ [4]

   (ii) Hence, evaluate $\displaystyle\int_0^2 \dfrac{12 - 3x^2}{(x^2+4)^2}\, dx.$ [4]

91

22. Use the substitution indicated in EACH case to find the following integrals

    (a) $\int \sin^8 x \cos x \, dx$ ;  $u = \sin x$ [4]

    (b) $\int x\sqrt{2x+1} \, dx$ ;  $u = 2x+1$ [6]

23. Given that $\dfrac{d}{d\theta}(\tan\theta) = \sec^2\theta$, find $\int \tan^3\theta \sec^2\theta \, d\theta$, [3]

24. (a) Find $\displaystyle\int \frac{6x}{x^2+1} \, dx$ [3]

    (b) By using the substitution $u = x^2$, or otherwise, find $\int xe^{-x^2} \, dx$. [4]

25. (a) Express $\dfrac{2x+1}{x^2(x+1)}$ in the form $\dfrac{A}{x} + \dfrac{B}{x^2} + \dfrac{C}{x+1}$, where A, B and C are constants. [7]

    (b) Hence, evaluate, $\displaystyle\int_1^2 \frac{2x+1}{x^2(x+1)} \, dx$. [7]

26. The rate of increase of a population P is proportional to P with the constant of proportionality being 0.03. Write down a differential equation model for P.

    If the initial population is $P_0$, find P at time $t$. [3]

*27. The rate of change of volume of a sphere is proportional to the volume of the sphere. Obtain a differential equation for the radius. [4]

28. The rate of decline of an insect population due to the application of a certain type of insecticide can be modelled by means of the differential equation
    $$\frac{dx}{dt} = -\frac{600}{1+6t}$$
    where $x$ is the number of insects alive $t$ hours after the application of the insecticide.

    If there were 1000 insects initially, calculate

    (a) the number of insects alive after 24 hours [4]
    (b) how long the population of insects will survive. [2]

*29. In a certain chemical reaction, the rate of production of a substance is directly proportional to the amount, $x$ grams, already produced.

    (a) (i) Write down a differential equation relating $x$ and the time $t$, measured in seconds. [1]
        (ii) By solving the differential equation in (i), show that $x = Ae^{kt}$, where A and $k$ are constants. [4]
        (iii) Given that $k = 0.005$, calculate the time taken for the amount of the substance to triple. [3]

    The substance is now removed at a constant rate of $h$ grams per second.

(b) (i) Show that the differential equation for this present condition is given by

$$\frac{dx}{dt} = kx - h, \text{ where } h \text{ is a constant.}$$  [1]

(ii) Given that at the beginning of the reaction there were $p$ grams of the substance, determine the amount of the substance at time $t$ seconds.  [6]

(iii) In one sentence, write what happens to the amount of substance if $kp \leq h$.  [1]

(c) Sketch graphs on the same rectangular coordinates system showing the variation with respect to time $t$, of the amount of substance $x$, for the cases,

(i) $kp \leq h$.  [2]

(ii) $kp > h$.  [2]

30. The rate of increase of a population of insects is directly proportional to the size of the population at time $t$, given in days. Initially the population is $p_0$ and it doubles its size in 3 days.

(i) Show that $p = p_0\, e^{kt}$, where $p$ is the size of the population after $t$ days and $k = \frac{1}{3} \ln 2$.  [7]

(ii) Find the proportional increase in population at the end of

(a) the first day

(b) the second day.  [3]

31. The price $x$ of a particular commodity is decreasing with time at a rate proportional to the square root of the price at that time. Write down a differential equation model for $x$ involving $x$ and $t$, where $t$ is the time since the decrease in price started.  [3]

32.

The diagram, **not drawn to scale**, shows the variation in the volume, $V$ cm³, of an air freshener block with time, $t$ weeks.

The variation can be written as $\dfrac{dV}{dt} = -kV$

(a) Describe clearly in words the variation shown above.  [2]

(b) Show that $V = A\, e^{-kt}$, where $k$ is a constant.  [5]

(c) Initially an air freshener block has a volume of 64 cm³. It loses half its volume after 6 weeks. Show that $V = 64e^{\,(\ln\frac{1}{2})\frac{1}{6}t}$  [8]

(d) The air freshener block becomes ineffective when its volume reaches 6 cm³. Calculate the time, to the nearest week, at which the block should be replaced.  [5]

33. Write down a differential equation to represent each of the following. Carefully define all variables used.

(i) The population of a Caribbean island increases by 3.5 % every year.  [2]

(ii) The number of rabbits on a farm would be twice as many every 40 days except for the fact that 5 rabbits are caught and sold every day.  [3]

34. The rate at which atoms in a mass of radioactive material are disintegrating is proportional to $n$, the number of atoms present at any time, $t$, measured in days. Initially the number of atoms is $m$.

    (i) Form and solve the differential equation which represents this data. [8]

    (ii) Given that half of the original mass disintegrates in 76 days, evaluate the constant of proportion in the differential equation. [4]

35. Once a person ingests medication, the human body eliminates the medication from the body in such a way that a fixed percent of the amount is removed daily. A person had a daily dose of Q units of a certain medication and his body eliminated 60% of the medication daily.

    (a) Write down, in terms of Q, the amount of the medication **not** eliminated

        (i) at the end of the first day [1]

        (ii) at the end of the second day [2]

    (b) What is the value of Q, if the amount of medication in the body at the end of the second day is 56 units. [1]

36. A fish pond is constructed for breeding groupers only. The number of groupers in the fish pond at time $t$ (measured in days), can be modelled by

    $$F(t) = \frac{500}{(1 + 49e^{-0.2t})}.$$

    (i) How many groupers were in the pond initially? [2]

    (ii) How quickly is the population of groupers increasing after 5 days? [3]

    (iii) What is the greatest number of groupers the pond can support? [3]

37. A disease spreads through an urban population.

    At time $t$, the proportion of the population who have the disease is $p$, where $0 < p < 1$.

    The rate of change of $p$ with respect to time is proportional to the product of $p$ and $(1 - p)$.

    (i) Form a differential equation in terms of $p$ and $t$ which models the situation described above. [2]

    (ii) Solve the differential equation. [8]

    (iii) Given that when $t = 0$, $p = \frac{1}{10}$ and when $t = 2$, $p = \frac{1}{5}$, show that

    $$\frac{9p}{1-p} = \left(\frac{3}{2}\right)^t$$ [10]

    (iv) Find $p$ when $t = 4$. [3]

38. The rate of change of the volume, V, of a sphere of radius $r$ with respect to time, $t$, is directly proportional to the volume of the sphere.

    (a) Obtain a differential equation involving the radius, $r$, of the sphere. [5]

    (b) Hence, show that $r = Ce^{\frac{kt}{3}}$, $C, r \in \mathbb{R}$ [4]

    $$\left[ V = \frac{4}{3}\pi r^3 \right]$$

39. The enrolment pattern of membership of a country club follows an exponential logistic function $N$,

$$N = \frac{800}{1 + ke^{-rt}}, \quad k \in \mathbb{R}, \quad r \in \mathbb{R},$$

where $N$ is the number of members enrolled $t$ years after the formation of the club, The initial membership was 50 persons, and after one year there were 200 persons enrolled in the club.

    (i)    What is the LARGEST number reached by the membership of the club? [2]

    (ii)    Calculate the EXACT value of $k$ and of $r$. [6]

    (iii)    How many members will there be in the club 3 years after its formation? [2]

40. (a) Express $\dfrac{1+x}{(x-1)(x^2+1)}$ in partial fractions. [6]

    (b) Hence, find $\displaystyle\int \frac{1+x}{(x-1)(x^2+1)}\,dx$ . [3]

41. Given that $I_n = \displaystyle\int_0^1 x^n e^x\,dx$ , where $n \in \mathbb{N}$.

    (a) Evaluate $I_1$ . [4]

    (b) Show that $I_n = e - nI_{n-1}$ [4]

    (c) Hence, or otherwise, evaluate $I_3$ , writing your answer in terms of e. [3]

# What you have learnt in Chapter 17

*Integration of Some Standard Functions*

| $f(x)$ | $\int f(x)dx$ |
|---|---|
| $e^x$ | $e^x + c$ |
| $e^{(ax+b)}$ | $\frac{1}{a}e^{(ax+b)} + c$ |
| $\sin x$ | $-\cos x + c$ |
| $\sin(ax+b)$ | $-\frac{1}{a}\cos(ax+b) + c$ |
| $\cos x$ | $\sin x + c$ |
| $\cos(ax+b)$ | $\frac{1}{a}\sin(ax+b) + c$ |
| $\sec^2 x$ | $\tan x + c$ |
| $\sec^2(ax+b)$ | $\frac{1}{a}\tan(ax+b) + c$ |
| $\dfrac{1}{\sqrt{1-x^2}},\ |x|<1$ | $\sin^{-1} x + c$ |
| $\dfrac{1}{\sqrt{a^2-x^2}},\ |x|<a$ | $\sin^{-1}\left(\dfrac{x}{a}\right) + c$ |
| $\dfrac{1}{1+x^2}$ | $\tan^{-1} x + c$ |
| $\dfrac{a}{a^2+x^2}$ | $\tan^{-1}\left(\dfrac{x}{a}\right) + c$ |
| $(ax+b)^n\ \ n \neq -1$ | $\dfrac{(ax+b)^{n+1}}{a(n+1)} + c$ |
| $\frac{1}{x} = x^{-1}\ (x \neq 0)$ | $\ln|x| + c$ or $\ln A|x|$ |
| $\dfrac{f'(x)}{f(x)}$ | $\ln|f(x)| + c$ or $\ln A|f(x)|$ |

## *Partial Fractions*

Fractions of the form $F(x) = \dfrac{f(x)}{g(x)}$, where $g(x)$ is the product of linear or quadratic factors, can be decomposed into partial fractions if $F(x)$ is a proper fraction (i.e. if $f(x)$ is of a lower order than $g(x)$).

Where $f(x)$ is not a proper fraction, $F(x)$ must first be reduced to a polynomial plus proper fraction, so that $F(x) = G(x) + \dfrac{h(x)}{g(x)}$

$$\frac{f(x)}{(ax+b)(cx+bd)} \equiv \frac{A}{ax+b} + \frac{B}{cx+d}$$

$$\frac{f(x)}{(ax^2+b)(cx+d)} \equiv \frac{Ax+B}{ax^2+b} + \frac{C}{cx+d}$$

$$\frac{f(x)}{(ax+b)^2(cx+d)} \equiv \frac{A}{(ax+b)^2} + \frac{B}{ax+b} + \frac{C}{cx+d}$$

## Integration by Substitution

Many functions can be integrated by making the substitution $u = f(x)$ especially if the integrand contains both $f'(x)$ and a function of $f(x)$.

Remember that <u>all</u> occurrences of $x$ must be substituted, including $dx$ and limits of integration.

The following results can be found using substitution, or by inspection

$$\int f'(x)[f(x)]^n \, dx = \frac{[f(x)]^{n+1}}{n+1} + c$$

$$\int f'(x)e^{f(x)} dx = e^{f(x)} + c$$

$$\int \cos x \sin^n x \, dx = \frac{\sin^{n+1} x}{n+1} + c$$

$$\int \sin x \cos^n x \, dx = -\frac{\cos^{n+1} x}{n+1} + c$$

## Integration by Parts

$$\int u \frac{dv}{dx} dx = uv - \int v \frac{du}{dx} dx$$

## Some Special Techniques

$\ln x$ should be substituted by $u$ not by $\dfrac{dv}{dx}$ in integration by parts

many trig integrals can be solved by using such identities as

$\cos^2 x + \sin^2 x = 1$, $\sec^2 x = 1 + \tan^2 x$

$\cos 2x = 2\cos^2 x - 1 = 1 - 2\sin^2 x$

$\sin A \cos B \equiv \frac{1}{2}\{\sin(A+B) + \sin(A-B)\}$    $\cos A \sin B \equiv \frac{1}{2}\{\sin(A+B) - \sin(A-B)\}$

$\cos A \cos B \equiv \frac{1}{2}\{\cos(A+B) + \cos(A-B)\}$    $\sin A \sin B \equiv \frac{1}{2}\{\cos(A-B) - \cos(A+B)\}$

If $t = \tan\dfrac{x}{2}$, then $\sin x = \dfrac{2t}{1+t^2}$, $\cos x = \dfrac{1-t^2}{1+t^2}$ and $dx = \dfrac{dt}{1+t^2}$

## First Order Differential Equations

if $\dfrac{dy}{dx} = f(x)$, then $y = \int f(x)\,dx$        (direct integration)

if $\dfrac{dy}{dx} = f(x)g(y)$ then $\int \dfrac{dy}{g(y)} = \int f(x)\,dx$      (separation of the variables)

if $y' + ky = f(x)$, then $ye^{kx} = \int e^{kx} f(x)\,dx$      (using the integrating factor $e^{kx}$)

if $y' - py = f(x)$, then $ye^{-px} = \int e^{-px} f(x)\,dx$      (using the integrating factor $e^{-px}$)

## Second Order Differential Equations

If $\dfrac{d^2 y}{dx^2} = f(x)$, then $y$ can be found by integrating twice

To solve the homogeneous differential equation $ay'' + by' + cy = 0$

find the roots $\alpha$ and $\beta$ of the auxiliary equation $au^2 + bu + c = 0$

if $\alpha$ and $\beta$ are real and distinct ($b^2 > 4ac$), the general solution is $y = Ae^{\alpha x} + Be^{\beta x}$

if $\alpha$ and $\beta$ are real but equal ($b^2 = 4ac$), the general solution is $y = e^{\alpha x}(A + Bx)$

if $\alpha$ and $\beta$ are complex, namely $p \pm iq$, ($b^2 < 4ac$),

     the general solution is $y = e^{px}(A\cos qx + B\sin qx)$

To solve the nonhomogeneous differential equation $ay'' + by' + cy = f(x)$

find a particular integral (PI) − if $f(x)$ is quadratic the PI will be $Cx^2 + Dx + E$

                if $f(x)$ is linear the PI will be $Cx + D$

                if $f(x)$ is a constant the PI will be $C$

                if $f(x)$ is sin or cos $ax$ the PI will be $C\sin ax + D\cos ax$

find the complementary function (CF) (the solution to the corresponding homogeneous equation)

the solution is $y = CF + PI$

## The Trapezium Rule

If $f(x)$ is a continuous function on $[a, b]$

$$\int_a^b f(x)\,dx \approx \tfrac{1}{2}h\{y_0 + 2(y_1 + y_2 + \dots + y_{n-1}) + y_n\}$$

where $n$ is the number of strips into which the area is divided,

$$h = \frac{b-a}{n}, \quad x_r = a + rh \text{ and } y_r = f(x_r)$$

The greater the value of $n$, the more accurate the approximation.

# CHAPTER 18
# 2.2a Sequences

## DEFINITIONS AND NOTATION

📖 A **sequence** is a set of numbers, in a particular order, which follows a pattern or rule, e.g.

| | |
|---|---|
| 1, 4, 7, 10, 13 | ① |
| 1, 4, 9, 16, 25, ... 441 | ② |
| 3, 6, 12, 24, 48, ... | ③ |
| 1, 1, 2, 3, 5, 8, 13, 21 ... | ④ |

📖 Each number of the sequence is called a **term**.

📖 A sequence can be **finite** or **infinite**.

In the examples above ① and ② are **finite** sequences because they finish after a certain number of terms (① after 5 terms, ② after 21 terms). ③ and ④ are **infinite** sequences because they continue for an infinite number of terms.

For a finite sequence the number of terms is usually denoted by *n*, and for all sequences the general, or *r*th term by $a_r$ or $u_r$. (*n* or *k* may also be used instead of *r* for the general term.)

Depending on the pattern the sequence follows, the general term can usually be found

either from its position in the sequence, e.g. in ② $u_r = r^2$,

or from the term or terms that precede it, e.g. in ④, the Fibonacci sequence, where each term is found by adding the two preceding terms i.e. $u_{r+2} = u_r + u_{r+1}$, where $u_1 = u_2 = 1$ and $r \geq 1$

This is known as a **recurrence relation** (or **recursive formula**).

In ① and ③ both methods are possible.

In ① $u_{r+1} = u_r + 3, u_1 = 1, r \geq 1$ or $u_r = 3r - 2$;

in ③ $u_{r+1} = 2u_r, u_1 = 1, r \geq 1$ or $u_r = 3 \times 2^{r-1}$.

> Since the value of each term depends on its position in the sequence, the sequence can be seen as a function mapping the positive integers onto the real numbers.
> if $S = \{a_n\}$ then $a_n = f(n)$, for some function f, where $f : \mathbb{N} \rightarrow \mathbb{R}$

## *Convergent, divergent and periodic sequences*

Consider the sequences:

| | | | |
|---|---|---|---|
| $S = \{a_n\}$ | where $a_n = 1 + 3(n-1)$ | $S = \{1, 4, 7, 10, 13, ... \}$ | ① |
| $S = \{a_n\}$ | where $a_n = n^2$ | $S = \{1, 4, 9, 16, 25, ... \}$ | ② |
| $S = \{a_n\}$ | where $a_n = 3 \times 2^{n-1}$ | $S = \{3, 6, 12, 24, 48, ... \}$ | ③ |

$S = \{a_n\}$  where $a_n = 5 - 2(n-1)$          $S = \{5, 3, 1, -1, -3 \ldots \}$          ④

$S = \{a_n\}$  where $a_n = 3 \times (-2)^{n-1}$          $S = \{3, -6, 12, -24, 48, \ldots \}$          ⑤

$S = \{a_n\}$  where $a_n = 3 \times (\tfrac{1}{2})^{n-1}$          $S = \{3, \tfrac{3}{2}, \tfrac{3}{4}, \tfrac{3}{8}, \ldots \}$          ⑥

$S = \{a_n\}$  where $a_n = \dfrac{n+1}{n}$          $S = \left\{ \dfrac{2}{1}, \dfrac{3}{2}, \dfrac{4}{3}, \dfrac{5}{4}, \dfrac{6}{5}, \ldots \right\}$          ⑦

$S = \{a_n\}$  where $a_n = \sin(\tfrac{1}{2}n\pi)$          $S = \{1, 0, -1, 0, 1, 0, -1, 0, \ldots \}$          ⑧

Sequences ① to ⑤ are **divergent**. In ① to ③ the terms get larger and larger as $n$ gets larger, in ④ they get large and negative, while in ⑤ they oscillate between positive and negative, but getting larger all the time. In other words $|a_n| \to \infty$ as $n \to \infty$.

Sequences ⑥ and ⑦, on the other hand, are **convergent**. If we consider $a_n$ as $n \to \infty$ we see that for ⑥

$$\lim_{n\to\infty} a_n = 3 \lim_{n\to\infty} (\tfrac{1}{2})^n = 0 \quad \text{and for ⑦} \quad \lim_{n\to\infty} a_n = \lim_{n\to\infty} \frac{n+1}{n} = \lim_{n\to\infty}\left(1 + \frac{1}{n}\right) = 1.$$

So sequence ⑥ is convergent, and it converges to the limit 0; sequence ⑦ is convergent and it converges to the limit 1.

Sequence ⑧ is an example of a **periodic** sequence: the same group of numbers recur regularly. $a_{n+k} = a_n$ for some integer $k$ (in this case $k = 4$) Periodic sequences do not converge.

An **arithmetic sequence** is a sequence in which the difference between consecutive terms is constant.

e.g. sequence ① is arithmetic since $a_2 - a_1 = 4 - 1 = 3$, $a_3 - a_2 = 7 - 4 = 3$, $a_4 - a_3 = 11 - 7 = 3$, ...
$a_{n+1} - a_n = 1 + 3n - \{1 + 3(n-1)\} = 3$. 3 is called the **common difference** ($d$) for the sequence.

A **geometric sequence** is a sequence in which the quotient of consecutive terms is constant.

e.g. sequence ③ is geometric since $a_2 \div a_1 = 6 \div 3 = 2$, $a_3 \div a_2 = 12 \div 6 = 2$, $a_4 \div a_3 = 24 \div 12 = 2$, ...
$a_{n+1} \div a_n = (3 \times 2^n) \div (3 \times 2^{n-1}) = 2$. 2 is called the **common ratio** ($r$) for the sequence.

# Example 18.1

Write down the first four terms in the sequences defined by:

(a) $a_n = 4n - 3$          (b) $a_n = 3 - n^2$          (c) $a_{n+1} = \tfrac{1}{2} a_n$, $a_1 = 4$          (d) $a_n = 5(-1)^n$

Of these sequences, which are (i) arithmetic sequences, (ii) geometric sequences, (iii) periodic, (iv) divergent?

(a)  ➢  substitute $n = 1, 2, 3, 4$ in the formula: $a_1 = 4 - 3 = 1$, $a_2 = 8 - 3 = 5$,
$a_3 = 12 - 3 = 9$, $a_4 = 16 - 3 = 13$

➢  sequence is 1, 5, 9, 13, ...

➢  this sequence is (i) arithmetic, with common difference 4; it is also (iv) divergent.

(b)  ➢  substitute $n = 1, 2, 3, 4$ in the formula: $a_1 = 3 - 1 = 2$, $a_2 = 3 - 4 = -1$,
$a_3 = 3 - 9 = -6$, $a_4 = 3 - 16 = -13$

➢  sequence is 2, -1, -6, -13, ...

➢  this sequence does not have a common difference (not arithmetic), does not have a common ratio (not geometric) and is (iv) divergent.

(c)  ➢  substitute $n = 1$ in the formula: $a_2 = \tfrac{1}{2} \times a_1 = 2$,

➢  substitute $n = 2$ in the formula: $a_3 = \tfrac{1}{2} \times a_2 = 1$,

➢  substitute $n = 3$ in the formula: $a_4 = \tfrac{1}{2} \times a_3 = \tfrac{1}{2}$

➤ sequence is $4, 2, 1, \frac{1}{2}, \ldots$

➤ this sequence is (ii) geometric with common ratio $\frac{1}{2}$

(d)  ➤ substitute $n = 1, 2, 3, 4$ in the formula:  $a_1 = 5(-1) = -5, \ a_2 = 5(-1)^2 = 5,$

$a_3 = 5(-1)^3 = 9, \ a_4 = 5(-1)^4 = 5$

➤ sequence is $-5, 5, -5, 5, \ldots$

➤ this sequence is (iii) periodic, with period 2, it is also (ii) geometric with common ratio $-1$.

## Example 18.2

Show that the sequence defined by $a_n = 3\cos\frac{2}{3}n\pi$ is periodic.

Find the period, and hence find the values of $a_9$, $a_{17}$ and $a_{28}$.

➤ substitute $n = 1, 2, 3, 4 \ldots$ in the formula:  $a_1 = 3\cos\frac{2}{3}\pi = -\frac{3}{2}, \quad a_2 = 3\cos\frac{4}{3}\pi = -\frac{3}{2},$

$a_3 = 3\cos 2\pi = 3, \qquad a_4 = 3\cos\frac{8}{3}\pi = -\frac{3}{2}, \qquad a_5 = 3\cos\frac{10}{3}\pi = -\frac{3}{2}, \qquad a_6 = 3\cos 4\pi = 3$

➤ sequence is $-\frac{3}{2}, -\frac{3}{2}, 3, -\frac{3}{2}, -\frac{3}{2}, 3 \ldots$  *#3 repeats after every third time*

➤ this sequence is periodic with period 3.

> [!] notice that when proving periodicity you must find enough terms to show the repetition clearly.

➤ when period is 3,  $a_1 = a_{3+1} = a_{6+1} = a_{9+1} = \ldots$

$a_2 = a_{3+2} = a_{6+2} = a_{9+2} = \ldots$

$a_3 = a_{3+3} = a_{6+3} = a_{9+3} = \ldots$

so  $a_9 = a_{6+3} = a_3 = 3$

$a_{17} = a_{15+2} = a_2 = -\frac{3}{2}$

$a_{28} = a_{27+1} = a_1 = -\frac{3}{2}$

## Example 18.3

A sequence is generated by the recurrence relation $u_{r+1} = \dfrac{5}{u_r - 4}, \ r \geq 1.$
Given that $u_2 = u_1$, find the possible values of $u_1$.

➤ Use formula to write $u_2$ in terms of $u_1$:  $u_2 = \dfrac{5}{u_1 - 4}$

➤ Put $u_2 = u_1$:  $u_1 = \dfrac{5}{u_1 - 4}$

➤ Rearrange and simplify:  $u_1(u_1 - 4) = 5 \ \rightarrow \ u_1^2 - 4u_1 - 5 = 0$

➤ Solve for $u_1$:  $(u_1 - 5)(u_1 + 1) = 0 \quad \therefore \ u_1 = -1 \text{ or } u_1 = 5$

# EXERCISE 18 A

1. Write down the first four terms in the sequence defined by

   (a) $a_n = n^2 - 1$      (b) $a_n = (-1)^n$      (c) $a_n = \dfrac{n+3}{n}$      (d) $a_n = 3n - 7$

   (e) $a_n = (\frac{1}{3})^n$      (f) $a_n = (-3)^{n-1}$      (g) $a_n = n^3$      (h) $a_n = 8 - 3n$

   (i) $a_n = \dfrac{n+2}{3n-5}$      (j) $a_n = \cos(\frac{1}{3}n\pi)$      (k) $a_n = \dfrac{1}{n^3}$      (l) $a_n = n \tan \frac{1}{3}\pi$

2. Write the $n$th term of each of the given sequences in the form $a_n = f(n)$

   (a) 3, 7, 11, 15, ...      (b) 2, 6, 18, 54. ...      (c) $\dfrac{1}{1.2}, \dfrac{1}{2.3}, \dfrac{1}{3.4}, \dfrac{1}{4.5}\cdots$

   (d) 15, 12, 9, 6, ...      (e) 192, 96, 48, 24 ...      (f) 2, 5, 10, 17, 26, ...

   (g) 2, –2, 2, –2, ...      (h) $\dfrac{1}{3}, \dfrac{2}{4}, \dfrac{3}{5}, \dfrac{4}{6},\cdots$      (i) $-\frac{1}{8}, \frac{1}{4}, -\frac{1}{2}, 1, \ldots$

3. Write down the first four terms of the sequences generated by the given recurrence relations, using the given first term

   (a) $u_{r+1} = 2u_r - 1, \quad u_1 = 3$      (b) $u_{r+1} = \dfrac{1}{u_r + 1}, \quad u_1 = 0$      (c) $u_{r+1} = -3u_r, \quad u_1 = 2$

   (d) $u_{r+1} = (1 + u_r)^2, \quad u_1 = 2$      (e) $u_{r+1}.u_r = r, \quad u_1 = 1$      (f) $u_{r+1} = r^2 - u_r, \quad u_1 = 0$

4. Write down recurrence relations which generate the given sequences

   (a) 6, 8, 10, 12, ...      (b) 2, 6, 18, 54, ...      (c) 2, 5, 7, 12, 19, 31, ...

   (d) 1, 3, 7, 15, 31, ...      (e) 25, 21, 17, 13, ...      (f) 1, 2, 5, 14, 41, ...

5. Show that the sequence defined by $a_n = \sqrt{2} \sin \frac{2n-1}{4}$ is periodic.

   Find the period, and hence find the values of $a_9$, $a_{30}$ and $a_{52}$.

6. The sequence $\{a_n\}$ is generated by the recurrence relation $a_{n+2} = a_{n+1} - a_n; \ a_1 = 1, a_2 = 3, n \geq 1$

   (a) Show that the sequence is periodic.
   (b) State the period, and hence find the values of $a_{13}$, $a_{27}$ and $a_{48}$.

7. A sequence is generated by the recurrence relation $u_{r+1} = 2u_r^2 - 1$.

   Given that $u_2 = u_1$, find the two possible values of $u_1$.

   For each of these values of $u_1$, find the corresponding value of $u_3$.

8. A sequence is generated by the recurrence relation $u_{r+1} = \dfrac{4}{u_r - 3}, \ r \geq 1$.

   Given that $u_2 = u_1$, find the possible values of $u_1$.

9. A sequence is generated by the recurrence relation $u_{r+1} = u_r^2 - 1$.

   Given that $u_3 = u_1$, show that $u_1^4 - 2u_1^2 - u_1 = 0$.

10. A sequence of terms $u_1, u_2, u_3, \ldots,$ where $u_1$ is a given positive real number, is defined by

$$u_{n+1} = 1 + \frac{1}{u_n}.$$

   (a) For the case $u_1 = 1$, write down the values of $u_2$ and $u_3$.

   (b) Find the value of $u_1$ for which $u_2 = u_1$, giving your answer in an exact form.

#11. The $n$th term of a sequence is given by the formula $u_n = \frac{1}{\sqrt{5}}\left[\left(\frac{1+\sqrt{5}}{2}\right)^n - \left(\frac{1-\sqrt{5}}{2}\right)^n\right]$

   Using a calculator, write down the first 5 or 6 terms of this sequence.
   Do you recognise the sequence?
   Can you describe the sequence by means of a recurrence relation?

## Convergent Sequences

A sequence $\{a_n\}$ **converges** if its terms converge as $n$ tends to infinity, i.e. if $\lim\limits_{n\to\infty} a_n = L$ for some $L \in \mathbb{R}$. In Unit 1 we learnt rules for working with limits of continuous functions. The same rules apply to limits of functions of $n$, $n \in \mathbb{N}$, as $n$ tends to infinity. They are

If $\lim\limits_{n\to\infty} f(n) = F$, $\lim\limits_{n\to\infty} g(n) = G$ and $k$ is a constant,

then $\lim\limits_{n\to\infty} k\,f(n) = kF$, $\lim\limits_{n\to\infty} f(n)\,g(n) = FG$, $\lim\limits_{n\to\infty}\{f(n) \pm g(n)\} = F + G$

and, provided $G \neq 0$, $\lim\limits_{n\to\infty}\dfrac{f(n)}{g(n)} = \dfrac{F}{G}$.

Also if $h$ is a continuous function, $\lim\limits_{n\to\infty} h\{f(n)\} = h\{\lim\limits_{n\to\infty} f(n)\} = h(F)$

Remember also that
$$\begin{cases} \lim\limits_{n\to\infty} n^b = \infty, & b > 0 \quad \text{e.g. } \lim\limits_{n\to\infty} n = \infty, \ \lim\limits_{n\to\infty} n^2 = \infty \ \ldots \\[2mm] \lim\limits_{n\to\infty} n^0 = 1 \\[2mm] \lim\limits_{n\to\infty} n^b = 0, & b < 0 \quad \text{e.g. } \lim\limits_{n\to\infty}\left(\frac{1}{n}\right) = 0, \ \lim\limits_{n\to\infty}\left(\frac{1}{n^2}\right) = 0 \ \ldots \\[2mm] \lim\limits_{n\to\infty} k^n = 0, & |k| < 1 \end{cases}$$

# Example 18.4

Determine whether the sequence defined by $a_n = \dfrac{n+1}{2n^2 - 3}$ converges, and, if so, state the limit to which it converges.

➤ Divide numerator and denominator by the highest power of $n$ in the denominator: $a_n = \dfrac{\frac{1}{n} + \frac{1}{n^2}}{2 - \frac{3}{n^2}}$

➤ Apply the limit theorems: $\lim\limits_{n\to\infty} a_n = \lim\limits_{n\to\infty} \dfrac{\frac{1}{n} + \frac{1}{n^2}}{2 - \frac{3}{n^2}} = \dfrac{0+0}{2-0} = 0$

➤ Give answer:  the sequence converges to the limit 0

---

## Example 18.5

The sequence $\{x_n\}$, is defined by the recurrence relation $x_{n+1} = \sqrt{2x_n + 5}$.

(a) Prove that if $l = \lim_{n \to \infty} x_n$ exists, then $l$ satisfies the equation $l^2 - 2l - 5 = 0$.

(b) Use the recurrence relation with $x_1 = 3$ to find one value of $l$ to 2 decimal places.

(a) ➤ If $l$ exists, then $\lim_{n \to \infty} x_{n+1} = \lim_{n \to \infty} x_n = l$

➤ $x_{n+1} = \sqrt{2x_n + 5} \rightarrow \lim_{n \to \infty} x_{n+1} = \lim_{n \to \infty} \sqrt{2x_n + 5}$

➤ $\therefore$ (using limit theorems) $l = \sqrt{2l + 5}$

➤ $l^2 = 2l + 5, \rightarrow l^2 - 2l - 5 = 0$

(b) ➤ $x_1 = 3 \rightarrow x_2 = \sqrt{11} = 3.317 \rightarrow x_3 = \sqrt{(2 \times 3.317 + 5)} = 3.411$

$\rightarrow x_4 = \sqrt{(2 \times 3.411 + 5)} = 3.438 \rightarrow x_5 = \sqrt{(2 \times 3.438 + 5)} = 3.446$

$\rightarrow x_6 = \sqrt{(2 \times 3.446 + 5)} = 3.448$

➤ $l = 3.45$ to 2 d.p.

## EXERCISE 18 B

1. Evaluate $\lim_{n \to \infty} a_n$ for each of the following

(a) $a_n = \dfrac{n+1}{n}$  (b) $a_n = \dfrac{2n^2 - 3}{4n^2 + n}$  (c) $a_n = \dfrac{4n-1}{n^2 + 5}$  (d) $a_n = \dfrac{1}{\sqrt{n}}$  (e) $a_n = \dfrac{6n^5 + 3n^2 - 5}{2n^5 - 2n^4 + 1}$

2. Determine whether each of the following sequences $\{a_n\}$ is convergent or divergent. Where the sequence is convergent, state the limit to which it converges.

(a) $a_n = \dfrac{2n-3}{3n}$  (b) $a_n = \dfrac{n+1}{n^2}$  (c) $a_n = \dfrac{3n^2 + 2}{n+4}$  (d) $a_n = \sin n\pi$  (e) $a_n = 5^n$

(f) $a_n = \dfrac{(-1)^n}{n}$  (g) $a_n = n + \dfrac{1}{n}$  (h) $a_n = \dfrac{2n^3 - 7n^2 + 3}{5n^3 + n^2 - 1}$  (i) $a_n = \dfrac{\sqrt{n}}{n+1}$  (j) $a_n = \dfrac{1}{3^n}$

3. Which of the sequences in exercise 18A question 1 are

(i) divergent?

(ii) convergent? find the limit to which they converge;

(iii) periodic? what is the period?

(iv) arithmetic sequences? state the common difference;

(v) geometric sequences? state the common ratio.

4. Repeat question 3 for the sequences in exercise 18A question 2.

5.  The $n$th terms of two sequences are defined as follows:

    (a) $t_n = 1 - \dfrac{1}{n}$,  (b) $u_n = 1 - \dfrac{1}{u_{n-1}}$, where $u_1 = 2$.

    Decide in each case whether the sequence is convergent, divergent or periodic, giving reasons for your answers.

6.  A sequence is defined by $u_{n+1} = \dfrac{1}{1 - u_n}$, with $u_1 = a$, where $a \neq 1$. Find $u_2$, $u_3$ and $u_4$ in terms of $a$, simplifying your answers.

    Hence determine the behaviour of the sequence as $n$ increases.

7.  A sequence is generated by the recurrence relation $u_{r+1} = \dfrac{2}{u_r}$, $r \geq 1$, $u_1 = 5$

    Write down the first four terms of the sequence. Is the sequence convergent, divergent, periodic or none of these?

8.  Three sequences are defined, for $n = 1, 2, 3, \ldots$, as follows

    $$u_n = \frac{n}{n+1}, \quad v_n = 2n(n+2), \quad w_n = 2 + 3\left(\tfrac{1}{2}\right)^n.$$

    Describe the behaviour of each of these sequences as $n \to \infty$.

9.  The sequence $\{x_n\}$, is defined by the recurrence relation $x_{n+1} = \sqrt{3 + x_n}$, $x_1 = \sqrt{3}$

    Prove that if $l = \lim\limits_{n \to \infty} x_n$ exists, then $l$ satisfies the equation $l^2 - l - 3 = 0$.

10. The Fibonacci sequence is defined by the recurrence relation $a_{n+2} = a_n + a_{n+1}$ for $n \geq 1$, $a_1 = 1$, $a_2 = 1$.

    Prove that if $l = \lim\limits_{n \to \infty} \dfrac{a_{n+1}}{a_n}$ exists, then $l$ satisfies the equation $l^2 - l - 1 = 0$.

    Solve this equation for $l$, and state, giving reasons, which solution gives the true value for $l$.

$$\lim_{n \to \infty} \frac{a_{n+1}}{a_n} = \lim_{n \to \infty} \frac{a_{n+2}}{a_{n+1}}$$

## *Sequences and Proof by Induction*

This section assumes that you are confident at using proof by induction. Mathematical induction was introduced in Unit 1 (Chapter 1) – but you may have forgotten it. Chapter 19 contains an extended section on it, and you may prefer to leave this section until you have completed chapter 19.

Here are two examples of proofs by induction applied to sequences:

## Example 18.6

The sequence $\{u_n\}$ is defined by the recurrence relation $u_{n+1} = 2u_n(1 - u_n)$, $n \geq 0$, $u_0 = \tfrac{1}{4}$.  ①

Prove by induction that $u_n = \dfrac{1}{2}\left(1 - \dfrac{1}{2^{2^n}}\right)$

Let $u_n = \dfrac{1}{2}\left(1 - \dfrac{1}{2^{2^n}}\right)$ be the statement $P_n$.  ②

From ② putting $n = 0$, $u_0 = \tfrac{1}{2}(1 - \tfrac{1}{2}) = \tfrac{1}{4}$ which agrees with ①, so $P_0$ is true  ③

Assume $P_n$ is true for some $k$, where $k \geq 0$, then

105

➤ substituting $k$ for $n$ in ①: $\qquad u_{k+1} = 2u_k(1 - u_k)$

➤ using ② with $n = k$: $\qquad u_{k+1} = \left(1 - \dfrac{1}{2^{2^k}}\right)\left(1 - \dfrac{1}{2}\left(1 - \dfrac{1}{2^{2^k}}\right)\right)$

$$= \left(1 - \dfrac{1}{2^{2^k}}\right)\left(\dfrac{1}{2} + \dfrac{1}{2}\cdot\dfrac{1}{2^{2^k}}\right)$$

$$= \dfrac{1}{2}\left(1 - \dfrac{1}{2^{2^k}}\right)\left(1 + \dfrac{1}{2^{2^k}}\right) \qquad \text{(these are the factors of a difference of 2 squares)}$$

$$= \dfrac{1}{2}\left(1 - \dfrac{1}{2^{2^{k+1}}}\right)$$

➤ but this is $P_{k+1}$, so $P_k$ implies $P_{k+1}$ $\qquad\qquad\qquad\qquad\qquad\qquad\qquad$ ④

➤ Combining ③ and ④ $P_n$ is true for $n \geq 0$

## Example 18.7

The Fibonacci sequence $\{f_n\}$ is defined by the recurrence relation $f_{n+2} = f_n + f_{n+1}$ for $n \geq 1, f_1 = 1, f_2 = 1$. ①
Prove, by mathematical induction that $f_{4n}$ is divisible by 3 for all $n \geq 1$

➤ Let $P_n$ be the proposition that $f_{4n}$ is divisible by 3 for all $n \geq 1$

➤ Assume $P_n$ is true for some $k \geq 1$, so that $f_{4k}$ is divisible by 3 $\qquad\qquad\qquad$ ②

➤ from ①: $f_3 = 1 + 1 = 2$, $f_4 = 1 + 2 = 3$, which is divisible by 3, so $P_1$ is true. $\qquad$ ③

➤ Using ① repeatedly: $f_{4(k+1)} = f_{4k+4} = f_{4k+2} + f_{4k+3}$
$$= f_{4k} + f_{4k+1} + f_{4k+1} + f_{4k+2}$$
$$= f_{4k} + 2f_{4k+1} + f_{4k} + f_{4k+1}$$
$$= 2f_{4k} + 3f_{4k+1} \qquad\qquad\qquad ④$$

➤ Now from ② $f_{4k}$ is divisible by 3, and $3f_{4k+1}$ is also divisible by 3, so ④ is divisible by 3.

➤ $P_{k+1}$ therefore follows from $P_k$ $\qquad\qquad\qquad\qquad\qquad\qquad\qquad\qquad$ ⑤

➤ Combining ⑤ and ③, $f_{4n}$ is divisible by 3 for all $n \geq 1$

See also example 19.15

## EXERCISE 18C

1. The sequence of triangle numbers $\{u_n\}$ is defined by the recurrence relation $u_{n+1} = u_n + n + 1$, $n \geq 1$, $u_1 = 1$. Prove by mathematical induction that $u_n = \frac{1}{2}n(n + 1)$.

2. The sequence $\{u_n\}$ is defined by the recurrence relation $u_{n+1} = 2u_n + 1$, $n \geq 0$, $u_0 = 0$. Prove by mathematical induction that $u_n = 2^n - 1$.

3. The sequence $\{u_n\}$ is defined by the recurrence relation $u_{n+1} = \dfrac{2u_n - 1}{3}$, $n \geq 1$, $u_1 = 1$. Write down the values of $u_2$, $u_3$ and $u_4$. Prove by mathematical induction that $u_n = 3(\frac{2}{3})^n - 1$

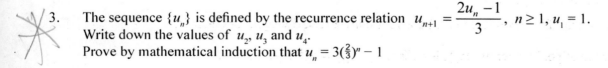

4.  Show, by mathematical induction, that for the Fibonacci sequence $\{f_n\}$ (see example 18.7) $f_{3n}$ is divisible by 2 for every $n \geq 1$

# EXERCISE 18 X

1.  The sequence $\{a_n\}$ is such that, for each positive integer $n$,

    $$a_n = \frac{3n^2 - 5n + 1}{2n^2 + n + 3},$$

    Evaluate $\lim_{n \to \infty} a_n$.                                                         [3]

2.  A sequence is generated by the recurrence relation $u_{r+1} = \dfrac{3}{u_r - 2}$, $r \geq 1$.

    Given that $u_2 = u_1$, find the possible values of $u_1$.                                  [5]

3.  The sequence $\{x_n\}$, is defined by the recurrence relation

    $$x_{n+1} = \frac{1}{2}\left(x_n + \frac{a}{x_n}\right), \quad a > 0, a \in \mathbb{R}, \text{ and } x_1 = 1.$$

    Prove that if $l = \lim_{n \to \infty} x_n$ exists, then $l = \sqrt{a}$.                     [8]

4.  Express $\dfrac{1}{n}\sqrt{2n^2 + 3}$ in the form $\sqrt{p + \dfrac{q}{n^2}}$ where $p, q \in \mathbb{R}$.

    Hence show that $\lim_{n \to \infty} \dfrac{2n + 3}{\sqrt{2n^2 + 3}} = \sqrt{2}$             [8]

5.  Three sequences are given below

    (i)   1, 4, 7, 10 ...

    (ii)  $1, -\frac{1}{4}, \frac{1}{7}, -\frac{1}{10}, \ldots$

    (iii) $(-1)^1, (-1)^4, (-1)^7, (-1)^{10}, \ldots$

    Determine which of the sequences is convergent, divergent or periodic and state which of the sequences is an arithmetic sequence.                                                             [12]

6.  A sequence of real numbers $\{u_n\}$ satisfies the recurrence relation:
    $$u_1 = 1, \quad u_n u_{n+1} = 2.$$

    (a)  Show that $u_{n+2} = u_n$.                                                             [2]

    (b)  Given that $a_n = u_{n+1} + u_n$ and $b_n = u_{n+1} - u_n$, write down the first FOUR terms of each of the sequences $\{u_n\}$, $\{a_n\}$ and $\{b_n\}$                                    [6]

    (c)  State which of the sequences in (b) above is convergent, divergent or periodic.        [3]

7.  The sequence of positive terms $\{x_n\}$, is defined by $x_{n+1} = x_n^2 + \frac{1}{4}$, $x_1 < \frac{1}{2}$.

Show by mathematical induction, or otherwise, that

(a)  $x_{n+1} - \frac{1}{2} < 0$                                          [4]

(b)  $x_{n+1} > x_n$, for all positive integers              [2]

     (You may use the results $x + y \geq 2\sqrt{(xy)}$, $x > 0$, $y > 0$ where appropriate.)

[Mathematical induction is explained more fully in chapter 19]

8.    A sequence $\{u_n\}$ is generated by the recurrence relation $u_{n+1} = \dfrac{8}{u_n - 3}$, $n \geq 1$.

    (a)   Given that $u_2 = 2u_1$, find the possible values of $u_1$.             [4]

    (b)   Find $u_3$ corresponding to EACH value of $u_1$.              [4]

9.    A sequence $\{u_n\}$ of real numbers satisfies $u_{n+1} u_n = 3(-1)^n$ ; $u_1 = 1$.

    (a)   Show that

        (i)   $u_{n+2} = -u_n$                                       [3]

        (ii)  $u_{n+4} = u_n$                                       [1]

    (b)   Write the FIRST FOUR terms of this sequence.            [3]

10.   The sequence of positive terms, $\{x_n\}$, is defined by $x_{n+1} = x_n^2 + \frac{1}{4}$, $x_1 < \frac{1}{2}$.

    (a)   Show, by mathematical induction, or otherwise, that $x_n < \frac{1}{2}$ for all positive integers $n$.    [7]

    (b)   By considering $x_{n+1} - x_n$ , or otherwise, show that $x_n < x_{n+1}$.         [3]

# What you have learnt in Chapter 18

## *Sequences*

A sequence is a function mapping the positive integers onto the real numbers.

i.e. if $S = \{a_n\}$ then $a_n = f(n)$, for some function f, where $f : \mathbb{N} \to \mathbb{R}$

Sometimes it is easier to express the terms of a sequence using a **recurrence relation**, where $a_n$ is given as a function of one or more preceding terms.

## *Convergent, Divergent and Periodic Sequences*

A sequence $\{a_n\}$ is **convergent** if its terms converge to a limit L as $n$ tends to infinity.

i.e. $\lim\limits_{n \to \infty} a_n = L$ for some $L \in \mathbb{R}$.

A sequence $\{a_n\}$ is **periodic** if the same group of numbers recur regularly throughout the sequence.

i.e. $a_{n+k} = a_n$ for some integer $k$.

A sequence $\{a_n\}$ is **divergent** if it is neither convergent nor periodic.

## *Arithmetic and Geometric Sequences*

An **arithmetic sequence** is a sequence in which the difference ($d$) between consecutive terms is constant.

A **geometric sequence** is a sequence in which the quotient ($r$) of consecutive terms is constant.

# CHAPTER 19

# 2.2b Series

## Example 19.1

Suppose you wish to make a journey of 1 km. You travel for $\frac{1}{2}$ km, and then pause; then you travel a further $\frac{1}{4}$ km and pause again. After a further $\frac{1}{8}$ km you pause again, and so on, each time halving the remaining distance between pauses.

The journey can be represented graphically:

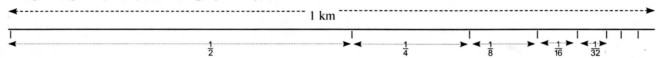

The distances travelled, in km, between pauses form the geometric sequence

$$G = \{\tfrac{1}{2}, \tfrac{1}{4}, \tfrac{1}{8}, \tfrac{1}{16}, \dots (\tfrac{1}{2})^r \dots\}$$

To find how far you had walked altogether at each pause you could add the distances travelled so far:

$$S_1 = \tfrac{1}{2} \qquad S_2 = \tfrac{1}{2} + \tfrac{1}{4} = \tfrac{3}{4} \qquad S_3 = \tfrac{1}{2} + \tfrac{1}{4} + \tfrac{1}{8} = \tfrac{7}{8} \qquad S_4 = \tfrac{1}{2} + \tfrac{1}{4} + \tfrac{1}{8} + \tfrac{1}{16} = \tfrac{15}{16}$$

$$S_n = \sum_{r=1}^{n} (\tfrac{1}{2})^r = (\tfrac{1}{2})^1 + (\tfrac{1}{2})^2 + (\tfrac{1}{2})^3 + (\tfrac{1}{2})^4 + \cdots + (\tfrac{1}{2})^n$$

The sums $S_1$, $S_2$, $S_3$, ... $S_n$ are known as the first, second, third, ... $n$th partial sums of the terms of G.

As $n \to \infty$, the distance covered each time, $(\tfrac{1}{2})^n$, tends to 0, while the total distance travelled approaches 1.

The partial sums themselves form a sequence $B = \{S_1, S_2, S_3, S_4, \dots \} = \left\{ \dfrac{1}{2}, \dfrac{3}{4}, \dfrac{7}{8}, \dfrac{15}{16}, \dfrac{31}{32}, \dots \dfrac{(\tfrac{1}{2})^n - 1}{(\tfrac{1}{2})^n} \dots \right\}$

This example helps set the scene for some

## DEFINITIONS

Given the sequence $A = \{a_r\} = \{ a_1, a_2, a_3, a_4, a_5, \dots \qquad \}$

When the first $n$ terms of this sequence are added together, the result is known as the **$n$th partial sum** of the sequence, denoted by $S_n$. So

$$S_1 = a_1 \qquad S_2 = a_1 + a_2 \qquad S_3 = a_1 + a_2 + a_3 \qquad S_4 = a_1 + a_2 + a_3 + a_4$$

$$S_n = \sum_{r=1}^{n} a_r = a_1 + a_2 + a_3 + \cdots + a_n$$

The sum, S, of the terms of an infinite sequence, $S = \displaystyle\sum_{r=1}^{\infty} a_r = a_1 + a_2 + a_3 + \cdots + a_n + \cdots$ is known as an **infinite series**. Some books, especially American ones, only use the term 'series' for such an infinite series, but more usually the term **series** can be used of a finite series $S = \sum_{r=1}^{n} a_r = a_1 + a_2 + a_3 + \cdots + a_n$, which is like a partial sum – the sum of the terms of a finite sequence.

The partial sums themselves form a sequence $B = \{S_n\} = \{S_1, S_2, S_3, S_4, \dots\}$

If, as in example 19.1, the terms of B converge to a limit, so that $\lim_{n\to\infty} S_n = L$, then $S = \sum_{r=1}^{\infty} a_r = L$. The series is said to **converge** to the limit L. If the sequence B does not converge to a limit the series is said to **diverge**, and the series does not have a sum.

A series will only converge if $\lim_{n\to\infty} a_n = 0$, but $\lim_{n\to\infty} a_n = 0$ does not guarantee that the series converges (see the note on the harmonic series at the end of this chapter).

At this point it is important to be clear about the difference between a sequence and a series.

---

**!**

A **sequence** is a set or *succession* of terms, written either as $\{a_r\}$ or $\{a_1, a_2, a_3, a_4, a_5, \dots\}$

A **series** is a *sum* of terms, written either as $\sum_{r=1}^{n} a_r$ or $a_1 + a_2 + a_3 + a_4 + a_5 + \dots + a_n$

The **sum of the series** is the result of adding the terms of the series, and is either a formula involving *n*, the number of terms in the series, or, where *n* is known, a single number.

---

## Example 19.2

A is the arithmetic sequence $A = \{2, 5, 8, 11, 14, 17, 20, 23\}$

(a)    Write down a formula for $a_r$, the general term of A, as a function of *r*.

(b)    Write down the first 4 partial sums of A.

(c)    Express the finite series, S, corresponding to A, in $\Sigma$ notation

(d)    Calculate the sum of the series S.

    (a)    $a_r = 3r - 1$

    (b)    $S_1 = 2$, $S_2 = 2 + 5 = 7$, $S_3 = 2 + 5 + 8 = 15$, $S_4 = 2 + 5 + 8 + 11 = 26$

    (c)    $S = 2 + 5 + 8 + 11 + 14 + 17 + 20 + 23 = \sum_{r=1}^{8} (3r - 1)$

    (d)    $S = 100$

In the previous chapter, we introduced the idea of arithmetic and geometric sequences. The sums of the terms of arithmetic and geometric sequences are **Arithmetic** and **Geometric Series or Progressions**.

## THE ARITHMETIC SERIES OR ARITHMETIC PROGRESSION (A.P.)

$3 + 7 + 11 + 15 + \dots$     ①     is an example of an **arithmetic progression**.

In an A.P. each term is found by adding a constant amount to the previous term.

📖    The amount added is known as the **common difference** and is denoted by *d*.
    $u_r = u_{r-1} + d$.    i.e.    $d = u_r - u_{r-1}$.

The common difference can be found by taking any term and subtracting the previous term from it.

e.g. ① can be shown to be an A.P. since $11 - 7 = 4,\ 7 - 3 = 4,\ 15 - 11 = 4$, etc. i.e. $d = 4$.

Clearly, an arithmetic series can be written

$$a + (a + d) + (a + 2d) + (a + 3d) + (a + 4d) + \ldots\ldots \quad \text{where the \textbf{first term} is denoted by } a$$

and we can see from this that the general term is $u_r = a + (r - 1)d$ and the last term is $u_n = a + (n - 1)d$

To find the sum of the first $n$ terms of an A.P. we write the series down twice, first forwards and then backwards:

$$S_n = \quad a \quad + \quad [a+d] \quad + \quad [a+2d] \quad + \ldots\ldots + [a+(n-3)d] + [a+(n-2)d] + [a+(n-1)d]$$

$$S_n = [a+(n-1)d] + [a+(n-2)d] + [a+(n-3)d] + \ldots\ldots + [a+2d] \quad + \quad [a+d] \quad + \quad a$$

adding gives

$$2S_n = [2a+(n-1)d] + [2a+(n-1)d] + [2a+(n-1)d] + \ldots\ldots + [2a+(n-1)d] + [2a+(n-1)d] + [2a+(n-1)d]$$

since there are $n$ terms, which are all the same we get

$$2S_n = n \times [2a+(n-1)d] \qquad \text{or} \qquad S_n = \tfrac{1}{2}n[2a+(n-1)d]$$

When the last term of the A.P. is known, this formula can be written in a simpler form. If the last term is $l$, then $l = a + (n - 1)d$ and the formula becomes $S_n = \tfrac{1}{2}n[a + l]$

For an arithmetic progression we therefore have

$$\boxed{\begin{array}{c} u_n = a + (n - 1)d \\[4pt] S_n = \tfrac{1}{2}n[2a + (n-1)d] \ \text{ or } \ S_n = \tfrac{1}{2}n[a + l] \end{array}}$$

## Example 19.3

For the following series find $a$ and $d$ and the term and sum stated

(i)   $11 + 25 + 39 + 53 + \ldots$        7th term, sum of first 20 terms

(ii)   $95 + 90 + 85 + 80 + \ldots$        21st term    , sum of first 12 terms

(iii)   $5 + 9 + 13 + 17 + \ldots + 65$        sum

(i)   $a = 11,\ d = 39 - 25 = 14$.

      $u_7 = a + (7 - 1)d = 11 + 6 \times 14$

         $= 11 + 84 = 95$

      $S_{20} = \tfrac{1}{2}n[2a + (n - 1)d]$

           $= 10[22 + 19 \times 14] = 2880$

(ii)   $a = 95,\ d = 85 - 90 = -5$.

      $u_{21} = a + (21 - 1)d = 95 + 20 \times (-5)$

         $= 95 - 100 = -5$

      $S_{12} = \tfrac{1}{2}n[2a + (n - 1)d] = \tfrac{1}{2} \times 12[190 + 11 \times (-5)]$

           $= 6[190 - 55] = 810$

(iii) $a = 5$, $d = 13 - 9 = 4$

we need to find $n$. We know the $n$th term, it's 65, so

$u_n = a + (n-1)d = 60$

$\therefore 5 + (n-1)4 = 65$

i.e. $4(n-1) = 60$

$\Rightarrow n - 1 = 15 \Rightarrow n = 16$

$S_n = \tfrac{1}{2}n\,[a + l] \Rightarrow S_{16} = 8\,[5 + 65]$

$\qquad\qquad = 8 \times 70 = 560$

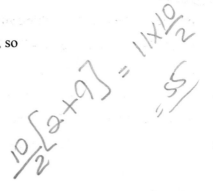

## Example 19.4

In an arithmetic series the fifth term is 34 and the eighth term is 52. Find $a$ and $d$ and the sum of the first 15 terms

$u_5 = 34 \Rightarrow a + 4d = 34$

$u_8 = 52 \Rightarrow a + 7d = 52$

Solving the simultaneous equations gives $3d = 52 - 34 = 18$

$\therefore d = 6, \ a = 10$

$S_{15} = \tfrac{1}{2} \times 15\,(20 + 14 \times 6)$

$\qquad = 7.5 \times 104 = 780$

## Example 19.5

The sum $S_n$, of the first $n$ terms of a series is given by $S_n = n(3 - 2n)$.

(i)    Find the $n$th term of the series

(ii)   Show that the terms of the series are in arithmetic progression with common difference $-4$.

(i)    the $n$th term can be found by subtracting the sum of $n - 1$ terms from the sum of $n$ terms

$\boxed{u_n = S_n - S_{n-1}}$

$\qquad = n(3 - 2n) - (n - 1)\{3 - 2(n - 1)\}$

$\qquad = 3n - 2n^2 - (n - 1)(5 - 2n)$

$\qquad = 5 - 4n$

(ii)  (using sums)

$a = u_1 = S_1 = 1,$

$u_2 = S_2 - S_1 = -2 - 1 = -3,$

$u_3 = S_3 - S_2 = -9 - (-2) = -7$

$u_4 = S_4 - S_3 = -20 - (-9) = -11$

or  (using the general term)

$u_1 = 5 - 4 \times 1 = 1$

$u_2 = 5 - 4 \times 2 = -3$

$u_3 = 5 - 4 \times 3 = -7$

$u_4 = 5 - 4 \times 4 = -11$

$\left.\begin{array}{l} u_2 - u_1 = -3 - 1 = -4 \\[4pt] u_3 - u_2 = -7 - (-3) = -4 \\[4pt] u_4 - u_3 = -11 - (-7) = -4 \end{array}\right\}$ $\therefore$ terms are in A.P. with common difference $-4$.

113

## Example 19.6

In an arithmetic series the first term is 12 and the common difference is 15. Find the least value of $n$ such that the sum of the first $n$ terms is greater than 1000.

**N.B.** $n$ is the number of terms and must therefore be an integer. However, in this type of question, to be able to do the calculation it is easier to solve an equation (A) which will usually give a non-integral solution than to work with inequalities The answer to the problem will be the nearest integer above or below the solution to A, depending on how the question is worded.

Let $S_n = 1000$.                 (A)

Then $\frac{1}{2} n\,[2a + (n-1)d\,] = 1000$    i.e. $\frac{1}{2} n\,[24 + 15(n-1)] = 1000$

$n(15n + 9) = 2000 \Rightarrow 15n^2 + 9n - 2000 = 0$

Using the quadratic formula:

$\qquad n \approx 11.25$ or $n \approx -11.85$.

$n$ cannot be negative, so taking the positive value, $n$ is either 11 or 12 (the nearest integers to 11.25). Common sense suggests that when $n$ is 11 the sum will be just less than 1000 and when $n$ is 12 it will be just greater than 1000 (this can be checked by calculation) The answer is therefore $n = 12$.

## EXERCISE 19 A

1. Write out the following series in full:

(a) $\displaystyle\sum_{r=1}^{5} r(r-1)$    (b) $\displaystyle\sum_{r=3}^{7} (2r-1)$ (c) $\displaystyle\sum_{r=1}^{4} \sqrt{r^2 + 1}$ (d) $\displaystyle\sum_{r=1}^{\infty} \frac{1}{r}$

2. Express the following series in $\Sigma$ notation:

(a) $3 + 5 + 7 + 9 + \ldots + 33$

(b) $1 + 8 + 27 + 64 + \ldots$

(c) $2\cdot3 + 4\cdot5 + 6\cdot7 + 8\cdot9 + 10\cdot11 + 12\cdot13$

3. The sequence $A = \{a_n\}$ is such that, for each positive integer $n$, $a_n = n^2 - n$, and the sequence $B = \{b_n\}$ is such that, for each positive integer $n$, $b_n = \dfrac{a_n}{a_{n+1}}$

    (a) Write down the first four **partial sums** of sequence A,

    (b) Write down the first four **terms** of sequence B,   $(b_{n+1} - b_n) < 0$

    (c) Show that the sequence B converges to a limit, and evaluate that limit.

4. Determine which of the following series are arithmetic. For those that are, (i) write down the value of $a$ and of $d$, (ii) find the 7th term, and (iii) find the sum of the first 12 terms.

(a) $5 - 7 + 9 - 11 + 13 - 15 + \ldots$         (b) $7 + 15 + 23 + 31 + \ldots$

(c) $25 + 17 + 9 + 1 - 7 \ldots$               (d) $12 + 24 + 48 + 96 + \ldots$

(e) $2 + 3 + 5 + 8 + 12 + \ldots$           (f) $-12 - 9 - 6 - 3 \ldots$

5. Find how many terms there are in, and hence the sum of, each of the following series

(a) $7 + 10 + 13 + \ldots + 61$           (b) $170 + 163 + 156 + \ldots + 93$

(c) $-3 - 8 - 13 - 18 - \ldots - 53$       (d) $49 + 45 + 41 + \ldots - 3 - 7$

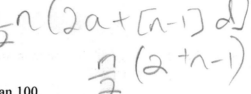

$$\tfrac{1}{2}n(2a+[n-1]d)$$
$$\tfrac{n}{2}(2+n-1)$$

6. Find (a) $\displaystyle\sum_{r=1}^{12} r$  (b) $\displaystyle\sum_{r=1}^{n} r$

7. Find the sum of all the odd numbers less than 100.

8. Find the sum of all the even numbers up to and including 100.

9. Find the sum of all the numbers between 100 and 400 (inclusive) that
   (a) are divisible by 3,
   (b) are **not** divisible by 3.

10. The 4th term of an A.P. is 18 and the 7th term is 30. Find the first term and common difference.

11. The 3rd term of an A.P. is 12 and the 10th term is 47. Find the sum of the first 10 terms.

12. The 5th term of an A.P. is 2 and the 11th term is –13. Find the first term and common difference.

13. The 12th term of an A.P. is –8 and the 20th term is –32. Find the first term and common difference.

14. The 3rd term of an A.P. is –3 and the 7th term is 5. Find the twentieth term .

15. The third term of an arithmetic series is 20 and the sum of the first 7 terms is 182. Find the first term, the common difference, the twenty-first term and the sum of the first 21 terms.

16. The sum of the first 6 terms of an arithmetic series is –3 and the sixth term is –8. Find the first term, the common difference, the twenty-fifth term and the sum of the first 25 terms.

#17. The sum of the first 5 terms of an arithmetic progression is 50, and the ninth term is 4 times the second. Find the first term, the common difference and the sum of the *second* five terms.

#18. The first and third terms of an A.P. are respectively 21 and 15. Find the two values of $n$ for which the sum of the first $n$ terms is 75.

19. The $n$th term of a series is given by $u_n = 30 - 4n$. Find the first 4 terms of the series and hence show that the series is arithmetic. State the values of $a$ and $d$, and find the sum of the first 15 terms.

20. The $n$th term of a series is given by $u_n = 3n + 5$. Find the $(n + 1)$th term of the series and hence show that $u_{n+1} - u_n$ is independent of $n$. Find the sum of the first 20 terms of the series.

21. Find the value of the first term of the A.P. $99 + 93 + 87 + \dots$ to be negative.

22. Find the least value of $n$ such that the sum of the first $n$ terms of the series $8 + 12 + 16 + \dots$ exceeds 200.

23. Find the greatest possible value of $n$ for which the sum of $n$ terms of the series $2 + 5 + 8 + 11 + \dots$ does not exceed 500.

24. By finding the common difference, show that the series
    $$\log_a 5 + \log_a(5)^2 + \log_a(5)^3 + \log_a(5)^4 + \log_a(5)^5 + \dots$$
    is an A.P., and hence find the $n$th term and the sum to $n$ terms.

#25. Find $\displaystyle\sum_{r=1}^{n} \log_a(bc^r)$

26. The $n$th partial sum of a series is given by $S_n = 3n^2 + 5n$.

(a) Find an expression for (i) $S_{n-1}$ (ii) the $n$th term of the series;

(b) show that the series is an A.P., and find the first term and common difference.

## _Divergence of Arithmetic Sequences and Series_

For arithmetic sequences and series, $a_n = a + (n-1)d = (a-d) + nd$.

When $d \neq 0$, $|a_n| \to \infty$ as $n \to \infty$, so both the sequence and the series diverge.

When $d = 0$, $a_n = a$. The sequence therefore converges to $a$, but the series is $\Sigma a = na$ which $\to \infty$ as $n \to \infty$.

> All arithmetic series are divergent
>
> All arithmetic sequences are divergent except when the common difference is 0.

# THE GEOMETRIC SERIES OR GEOMETRIC PROGRESSION (G.P.)

The series $3 + 6 + 12 + 24 + 48 \ldots$  ②  is known as a geometric series or **geometric progression**.

In a G.P. each term is found by multiplying the previous term by a fixed amount.

The amount multiplied by is known as the **common ratio** and is denoted by $r$.

$$u_n = u_{n-1} \times r. \quad \text{i.e. } r = \frac{u_n}{u_{n-1}}.$$

The common ratio can be found by taking any term and dividing it by the previous term.

e.g. ② can be shown to be a G.P. since $6 \div 3 = 2$, $12 \div 6 = 2$, $48 \div 24 = 2$, etc. i.e. $r = 2$.

Clearly, a geometric series can be written

$$a + ar + ar^2 + ar^3 + ar^4 + \ldots\ldots$$

and we can see from this that the general ($n$th) term $u_n = ar^{n-1}$

To find the sum of the first $n$ terms of a G.P., we write down the series $S_n$ and then subtract $r \times S_n$ from it

$$S_n = a + ar + ar^2 + ar^3 + ar^4 + \ldots\ldots\ldots + ar^{n-1}$$

$$r\,S_n = ar + ar^2 + ar^3 + ar^4 + ar^5 + \ldots\ldots\ldots + ar^{n-1} + ar^n$$

$$(1-r)\,S_n = a \ldots\ldots\ldots\ldots\ldots\ldots\ldots\ldots\ldots\ldots\ldots - ar^n = a(1 - r^n)$$

giving $\quad S_n = \dfrac{a(1 - r^n)}{1 - r}$

Multiplying numerator and denominator by $-1$ gives the alternative version $S_n = \dfrac{a(r^n - 1)}{r - 1}$

For a geometric progression we therefore have

$$u_n = ar^{n-1}$$

$$S_n = \frac{a(1-r^n)}{1-r}, \quad r < 1$$

$$\text{or} \quad S_n = \frac{a(r^n-1)}{r-1}, \quad r > 1$$

## Example 19.7

For the following series find $a$ and $r$ and the term and sum stated

(i)  $2 + 6 + 18 + 54 + \ldots$       10th term, sum of first 10 terms

(ii)  $8 - 12 + 18 - 27 + \ldots$       12th term, sum of first 12 terms

(iii)  $128 + 64 + 32 + 16 + \ldots$     10th term, $S_{15}, S_{24}$

     (i)  $r = 18 \div 6 = 3, \ a = 2,$

         $u_{10} = ar^9 = 2 \times 3^9 = 39366,$

         $S_{10} = a(r^{10} - 1) \div (r - 1)$

              $= 2(3^{10} - 1) \div 2 = 59048$

     (ii)  $r = -12 \div 8 = -1.5$ (notice the negative sign ), $a = 8,$

         $u_{12} = ar^{11} = 8 \times (-1.5)^{11} \approx -691.98$

         $S_{12} = a(1 - r^{12}) \div (1 - r)$

              $= 8(1 - (-1.5)^{12}) \div (1 + 1.5) \approx -51.5$

     (iii)  $r = 64 \div 128 = \frac{1}{2}$  (notice that you must divide the term on the right by the term on its left, which in this case gives a fraction, not the larger by the smaller) $a = 128$

         $u_{10} = ar^9 = 128 \times (\frac{1}{2})^9 = \frac{1}{4}$

         $S_{15} = a(1 - r^{15}) \div (1 - r) = 128(1 - (\frac{1}{2})^{15}) \div \frac{1}{2} = 255.9922$

         $S_{24} = a(1 - r^{24}) \div (1 - r) = 128(1 - (\frac{1}{2})^{24}) \div \frac{1}{2} = 255.999985$

## Example 19.8

In a geometric series the third term is 24 and the sixth term is 192. Find $a$ and $r$ and the sum of the first 12 terms.

     $u_3 = ar^2 = 24$    ①            $u_6 = ar^5 = 192$    ②

     ② ÷ ① gives   $\dfrac{ar^5}{ar^2} = \dfrac{192}{24}$   $\therefore r^3 = 8 \Rightarrow r = 2$

     substituting in ① gives   $a = 24 \div 2^2 = 6$

     $S_{12} = a(r^{12} - 1) \div (r - 1)$

         $= 6(2^{12} - 1) \div (2 - 1) = 24570$

## Example 19.9

In a geometric series the fourth term is 189 and the sixth term is 1701. Find the possible value(s) of $r$, the common ratio.

$$u_4 = ar^3 = 189 \qquad ① \qquad\qquad u_6 = ar^5 = 1701 \qquad ②$$

② ÷ ① gives $r^2 = 9$, ∴ $r = \pm 3$

> ! notice that there are 2 possible values for $r$.

## Example 19.10

Find the least number of terms needed so that the sum of the geometric series $1 + 2 + 4 + \ldots$ exceeds $10^6$.

$$S_n = \frac{a(r^n - 1)}{r - 1} = \frac{2^n - 1}{2 - 1} = 2^n - 1$$

If $2^n - 1 = 1\,000\,000$, then $2^n = 1\,000\,001$

taking logs of both sides: $n \log 2 = \log 1\,000\,001 \rightarrow n \approx 19.93$

If sum exceeds $10^6$, then $n > 19.93$, ∴ $n = 20$, since $n$ must be an integer.

## Example 19.11

Given the series $\dfrac{1}{2} + \dfrac{1}{2^3} + \dfrac{1}{2^5} + \dfrac{1}{2^7} + \ldots$

(a)   show that the series is geometric

(b)   find the sum of the series to $n$ terms

(c)   show that as $n$ approaches infinity, the sum of the series approaches $\frac{2}{3}$.

(a)  ➤ To show that series is geometric, show that it has a common ratio:

$$\frac{a_2}{a_1} = \frac{\frac{1}{2^3}}{\frac{1}{2}} = \frac{2}{2^3} = \frac{1}{4},$$

$$\frac{a_3}{a_2} = \frac{\frac{1}{2^5}}{\frac{1}{2^3}} = \frac{2^3}{2^5} = \frac{1}{4}$$

$$\frac{a_4}{a_3} = \frac{\frac{1}{2^7}}{\frac{1}{2^5}} = \frac{2^5}{2^7} = \frac{1}{4}$$

∴ the series has a common ratio of $\frac{1}{4}$, so is a G.P..

(b)  ➤ Use $S_n = \dfrac{a(1 - r^n)}{1 - r}$,

➤ putting $a = \frac{1}{2}$ and $r = \frac{1}{4}$, $S_n = \dfrac{\frac{1}{2}(1 - (\frac{1}{4})^n)}{1 - \frac{1}{4}} = \dfrac{2}{3}(1 - (\frac{1}{4})^n)$

(c)  ➤ as $n \rightarrow \infty$, $(\frac{1}{4})^n \rightarrow 0$, ∴ $S_n \rightarrow \frac{2}{3}$

### Convergent Geometric Series

Part (c) of example 19.11 shows that it is possible for a geometric series to converge to a limit (as does example 19.1, with which we started the chapter). This is because $r^n \rightarrow 0$ as $n \rightarrow \infty$ whenever $|r| < 1$. We therefore have the result for a geometric series

$$\text{if } |r| < 1$$
$$\lim_{n \to \infty} S_n = \frac{a}{1-r}, \quad \text{or} \quad S_\infty = \frac{a}{1-r}$$

! Notice again the importance of the distinction between a sequence and a series. A sequence is convergent if the *terms* converge to a limit, while a series is convergent if the ***sum of the terms*** converges to a limit.

For examples of sequences and series where the sequence converges but the corresponding series diverges, see the end of the chapter, page 138.

## Example 19.12

Find the range of values of $x$ for which the series $1 + 2x + 4x^2 + 8x^3 + \ldots$ converges.

➤ Show this is a G.P. :
$$2x \div 1 = 2x,$$
$$4x^2 \div 2x = 2x,$$
$$8x^3 \div 4x^2 = 2x$$
$$\therefore \text{ G.P. with } r = 2x$$

➤ State condition for convergence: $|r| < 1 \rightarrow |2x| < 1,$
$$\therefore \text{ series converges when } |x| < \tfrac{1}{2}$$

## EXERCISE 19 B

1.  Determine which of the following series are in geometric progression, and state the common ratio.

    (a)  $5 + 10 + 20 + 40 + \ldots$

    (b)  $2 + 6 + 24 + 120 + 720 + \ldots$

    (c)  $162 + 54 + 18 + 6 + \ldots$

    (d)  $7 - 14 + 28 - 56 + \ldots$

    (e)  $12 + 15 + 18 + 21 + \ldots$

    (f)  $12 - 6 + 3 - 1.5 + \ldots$

    (g)  $\displaystyle\sum_{k=1}^{n} 5(3)^k$

    (h)  $\displaystyle\sum_{k=1}^{n} 2(5k)^k$

2.  For each of the following geometric series find $r$, the common ratio, and hence find the term stated and the sum stated.

    (a)  $1 + \tfrac{1}{2} + \tfrac{1}{4} + \tfrac{1}{8} + \ldots$     12th term, sum of 15 terms

    (b)  $7 + 21 + 63 + 189 + \ldots$     9th term, sum of 11 terms

    (c)  $5 - 10 + 20 - 40 + \ldots$     11th term, sum of 20 terms

    (d)  $450 + 150 + 50 + \ldots$     10th term, sum of 12 terms

    (e)  $-1000 + 200 - 40 + \ldots$     7th term, sum of 15 terms

    (f)  $\displaystyle\sum_{k=1}^{n} 3(2)^{k-1}$     9th term, sum of 9 terms

    (g)  $\displaystyle\sum_{k=1}^{n} 5(\tfrac{1}{5})^k$     8th term, sum of 20 terms

    (h)  $\displaystyle\sum_{k=1}^{n} 64(-1.5)^{k-1}$     5th term, sum of 12 terms

3.    Which of the series in question 2 above are convergent?  Find their sum to infinity.

4.    Find the number of terms in the following G.P.s:
      (a)  $81 + 27 + 9 + \ldots + \frac{1}{27}$
      (b)  $4 + 8 + 16 + \ldots + 1024$
      (c)  $-1 + 2 - 4 + \ldots - 256$

5.    The third term of a G.P. is 4 and the sixth term is 108.  Find the common ratio, the first term and the sum of the first 12 terms.

6.    $a$, $b$ and $9a$ are three consecutive terms of a G.P..  Find the value(s) of the common ratio.

7.    Find the range of values of $x$ for which the following series converge:
      (a)  $1 + x + x^2 + x^3 + \ldots$
      (b)  $(3 + x) + (3 + x)^2 + (3 + x)^3 + \ldots$
      (c)  $x^9 + x^7 + x^5 + x^3 + \ldots$

#8.   The sum of the first 3 terms of a G.P. is 14 and the sum of the first 2 terms is –4.  Find the two possible values of $r$, the common ratio, and the corresponding values for the first term.

9.    Find the sum of the first $n$ terms of the G.P.  $3 + 15 + 75 + \ldots$

      What is the smallest number of terms of this series whose total is more than $10^9$?

10.   Show that the series $\displaystyle\sum_{r=0}^{\infty} \frac{(x-1)^r}{2^r}$ is a G.P.  Find the range of values of $x$ for which the series converges.

11.   Given that the sum of the first $n$ terms of a series, S, is  $2^{n+3} - 8$
      (a)  find the $n$th term of  S
      (b)  show that  S is a geometric progression
      (c)  does S converge?  If so, find the sum to infinity of  S.

12.   Find the sum of the convergent geometric series  $S = 3 + 1 + \frac{1}{3} + \frac{1}{9} + \ldots$

13.   A geometric progression with first term  $p$ and common ratio  $q$ has sum to infinity equal to $3p$.  Find the common ratio  $q$.

## THE MACLAURIN SERIES

In chapter 15 we found that it was possible to write the exponential function as a polynomial – a series of ascending powers of $x$.  This is known as a power series, and many other functions, like $\sin x$, $\cos x$, $\ln x$ etc. can be written as power series.  Before finding series for these specific functions we will derive a general formula for finding a power series, known as Maclaurin's Theorem.

Let $f(x)$ be a function of $x$ and suppose that $f(x)$ can be written as a series of ascending powers of $x$,

so that     $f(x) = a_0 + a_1 x + a_2 x^2 + a_3 x^3 + a_4 x^4 + a_5 x^5 + \ldots$     ①

putting $x = 0$ in ① gives     $f(0) = a_0$

differentiating ① gives $\qquad$ $f'(x) = a_1 + 2a_2x + 3a_3x^2 + 4a_4x^3 + 5a_5x^4 + \ldots$ $\qquad$ ②

putting $x = 0$ in ② gives $\qquad$ $f'(0) = a_1$

differentiating ② gives $\qquad$ $f''(x) = 2a_2 + 2\cdot3a_3x + 3\cdot4a_4x^2 + 4\cdot5a_5x^3 + \ldots$ $\qquad$ ③

putting $x = 0$ in ③ gives $\qquad$ $f''(0) = 2a_2$

differentiating ③ gives $\qquad$ $f'''(x) = 2\cdot3a_3 + 2\cdot3\cdot4a_4x + 3\cdot4\cdot5a_5x^2 + \ldots$ $\qquad$ ④

putting $x = 0$ in ④ gives $\qquad$ $f'''(0) = 2\cdot3\, a_3 = 3!\, a_3$

differentiating ④ gives $\qquad$ $f''''(x) = 2\cdot3\cdot4a_4 + 2\cdot3\cdot4\cdot5a_5x + \ldots$ $\qquad$ ⑤

putting $x = 0$ in ⑤ gives $\qquad$ $f''''(0) = 2\cdot3\cdot4a_4 = 4!\, a_4$

substituting for $a_0, a_1, a_2 \ldots$ in ① gives

---

**Maclaurin's Theorem**

$$f(x) = f(0) + f'(0)x + \frac{f''(0)}{2!}x^2 + \frac{f'''(x)}{3!}x^3 + \frac{f''''(0)}{4!}x^4 + \ldots$$

---

Maclaurin's Theorem can be used to find series for $f(x)$, a function of $x$, provided that

❖ it is possible to find all the derivatives of $f(x)$

❖ all the derivatives are defined when $x = 0$

❖ the series is convergent – this may mean placing certain restrictions on the range of $x$.

## Example 19.13

Use Maclaurin's theorem to show that $\sin x$ can be expressed as a series of ascending powers of $x$, and hence find $\sin \frac{1}{3}\pi$ correct to 3 d.p.

➤ check that conditions are satisfied:

if $f(x) = \sin x$, $f'(x) = \cos x$, $f''(x) = -\sin x$, $f'''(x) = -\cos x$, $f''''(x) = \sin x \ldots$ ✔

$f(0) = \sin 0 = 0$, $f'(0) = \cos 0 = 1$, $f''(0) = -\sin 0 = 0$, $f'''(0) = -\cos 0 = -1$, $f''''(0) = \sin 0 = 0 \ldots$ ✔

➤ apply Maclaurin's theorem: $\sin x =$

$$f(x) = f(0) + f'(0)x + \frac{f''(0)}{2!}x^2 + \frac{f'''(x)}{3!}x^3 + \frac{f''''(0)}{4!}x^4 + \ldots$$

$$= 0 + x + 0x^2 - \frac{x^3}{3!} + 0x^4 + \frac{x^5}{5!} + 0x^6 - \frac{x^7}{7!} + \ldots$$

$$= x - \frac{x^3}{3!} + \frac{x^5}{5!} - \frac{x^7}{7!} + \cdots + (-1)^{r+1}\frac{x^{2r-1}}{(2r-1)!} + \cdots$$

➤ substitute for $x$: $\qquad$ $\sin \frac{1}{3}\pi = 1.04719 - 0.19140 + 0.01049 - 0.00027 + \ldots$ ①

$$= 0.86601 \approx 0.866$$

➤ ① is clearly convergent, and higher powers of $x$ will not affect the 3rd d.p.

# EXERCISE 19 C

1. Use Maclaurin's theorem to find series expansions for each of the following, up to the term containing the given power of $x$.

   (a) $e^x$; $x^4$

   (b) $\cos x$; $x^6$

   (c) $\ln(1+x)$; $x^4$; for $-1 < x \le 1$

   (d) $\sin 2x$; $x^5$

   (e) $e^{-x}$; $x^4$

   (f) $(1-x)^{-1}$; $x^4$; for $|x| < 1$

2. Explain why it is not possible to expand $\ln x$ using Maclaurin's theorem.

3. Use the appropriate Maclaurin series, up to the term in $x^n$, to find approximate values (correct to 3sf) for the following (you may use the results from question 1 and example 19.13 where appropriate).

   (a) $e^{1.5}$      $n = 4$

   (b) $\ln 1.1$      $n = 3$

   (c) $\ln 0.7$      $n = 4$

   (d) $\cos \frac{1}{6}\pi$      $n = 4$

   (e) $0.98^{-1}$      $n = 2$

   (f) $\sin 1$ *      $n = 3$

   * what unit are you using for this angle? why?

# PROOF BY INDUCTION

Proof by induction was introduced for the summation of series in Chapter 1 (module 1.1) It also has much wider use in proving a number of mathematical statements, so we will now take another look at it, both as it relates to series and in other situations.

It is important to understand the three steps that make up the proof.

Suppose you are asked to show the truth of a proposition $P_n$. (e.g. $2^n > n$, $n \in \mathbb{N}$, $n \ge 1$)

The steps are:

❖ (1)    show that $P_n$ is true for a particular value of $n$, usually $n = 1$.

❖ (2)    show that $P_k$ is true implies that $P_{k+1}$ is true

❖    combining (1) and (2) shows that $P_n$ must be true for all $n$ since (1) proves it true for $n = 1$, and then applying (2) successively proves it true for $n = 2$, $n = 3$, … etc.

It might help to understand and remember the method if you think if a set of dominoes.

You are probably familiar with the fact that if you stand the dominoes on their ends at a suitable distance apart you can knock them all down by toppling the first one.

To be sure that all the dominoes will fall you need to be sure of two things:

1. The first domino will fall

2. Whenever a domino falls, its next neighbour will also fall,

Examples 19.14 – 19.16 illustrate the method in 3 different situations, then examples 19.17 and 19.18 show why both steps 1 and 2 are necessary.

Handwritten (top right margin):
$f(x) = e^x$
$f(0) = 1$
$f'(x) = e^x$
$f'(0) = 1$
$f''(x) = e^x$
$f''(0) = 1$
$f'''(0) = 1$

$f'''(x) e^x$
$f'''(0) = 1$
$f^{iv}(x) = e^x$
$f^{iv}(0) = 1$

## Example 19.14

Prove, by mathematical induction, that $\sum_{r=1}^{n} r^2 = \frac{1}{6}n(n+1)(2n+1)$       $P_n$

➢   when $n = 1$,   $\sum_{r=1}^{n} r^2 = 1^2 = 1$

and $\frac{1}{6}n(n+1)(2n+1) = \frac{1}{6} \times 1 \times 2 \times 3 = 1$
∴ $P_1$ is true       ①

➢   Assume $P_k$ is true. Then   $\sum_{r=1}^{k} r^2 = \frac{1}{6}k(k+1)(2k+1)$       ②

➢   (we want to show that $P_{k+1}$ is true i.e. that $\sum_{r=1}^{k+1} r^2 = \frac{1}{6}(k+1)(k+2)(2k+3)$       ③)

| ! | We cannot assume ③, we must use ② to prove ③ |

➢   now,  $\sum_{r=1}^{k+1} r^2 = \sum_{r=1}^{k} r^2 + u_{k+1}$

$= \frac{1}{6}k(k+1)(2k+1) + (k+1)^2$

$= \frac{1}{6}(k+1)\{k(2k+1) + 6(k+1)\}$

$= \frac{1}{6}(k+1)\{2k^2 + 7k + 6\}$

$= \frac{1}{6}(k+1)(k+2)(2k+3)$,  which is  ③, which is what we wanted to prove

➢   so, $P_k$ is true implies $P_{k+1}$ is true       ④

➢   combining statements ① and ④,  $P_n$ is true for $n = 1, 2, 3,\dots$ i.e. for all $n$.

Handwritten (right margin):
$f(x) = f(0) + f'(0)x + \frac{f''(0)x^2}{2!} + \frac{f'''(0)x^3}{3!} + \frac{f^{iv}(0)x^4}{4!}$

$f(x) = 1 + x + \frac{x^2}{2!} + \frac{x^3}{3!} + \frac{x^4}{4!}$

## Example 19.15

Prove, **by mathematical induction**, that, if $n$ is a positive integer, $7^n - 1$ is always divisible by 6.   $P_n$

➢   putting $n = 1$, $7^1 - 1 = 7 - 1 = 6$, which is divisible by 6.       ∴ $P_1$ is true   ①

➢   Assume $P_k$ is true,  then $7^k - 1$ is divisible by 6

i.e. $7^k - 1 = 6Z$ for some integer Z

➢   (we want to show that it follows that $P_{k+1}$ is true, i.e. that $7^{k+1} - 1$ is divisible by 6)

➢   $7^{k+1} - 1 = 7 \times 7^k - 1$

$= 7 \times (6Z + 1) - 1$

$= 42Z + 7 - 1$

$= 42Z + 6$

$= 6(7Z + 1)$  which is divisible by 6

➢   ∴ if $7^k - 1$ is divisible by 6, it follows that $7^{k+1} - 1$ is also divisible by 6       ②

➢   ① states that $7^n - 1$ is divisible by 6 when $n = 1$,

➢   applying statement ② repeatedly shows that $7^n - 1$ is divisible by 6 must also be true for $n = 2, 3 \dots$

 notice how we replaced $7^{k+1}$ by $7 \times 7^k$. This is often a useful technique in such questions.

# Example 19.16

Prove, by mathematical induction, that $2^n > n$, $n \in \mathbb{N}, n \geq 1$     $(P_n)$

- ➤   when $n = 1$, $2^n = 2$ and $n = 1$   $\therefore 2^n > n$. $\therefore$   $P_1$ is true       ①

- ➤   Assume $P_k$ is true. Then $2^k > k$            ②

(We want to show that $P_{k+1}$ is true, i.e. that $2^{k+1} > k + 1$)

- ➤     $2^{k+1} = 2.\,2^k$ so from ②, $2^{k+1} > 2k$

         i.e. $2^{k+1} > k + k$. But $k \geq 1$,   $\therefore 2^{k+1} > k + 1$

- ➤    $\therefore P_k$ is true implies $P_{k+1}$ is true.        ③

- ➤   combining statements ① and ③, $P_n$ is true for $n = 1, 2, 3, \ldots$ i.e. for all $n$.

# Example 19.17

$\boxed{!}$   Step (2) alone is not enough

Consider the proposition $P_n$: $\displaystyle\sum_{r=1}^{n} r = \tfrac{1}{8}(2n+1)^2$

Assume that $P_n$ is true when $n = k$, i.e. $P_k$ is true and $\displaystyle\sum_{r=1}^{k} r = \tfrac{1}{8}(2k+1)^2$

Now $\displaystyle\sum_{r=1}^{k+1} r = \left(\sum_{r=1}^{k} r\right) + (k+1) = \tfrac{1}{8}(2k+1)^2 + (k+1)$

$$= \tfrac{1}{8}\left\{(4k^2 + 4k + 1) + 8(k+1)\right\} = \tfrac{1}{8}\left\{4k^2 + 12k + 9\right\}$$

$$= \tfrac{1}{8}(2k+3)^2 = \tfrac{1}{8}\left(2(k+1)+1\right)^2$$

So $P_k$ is true implies $P_{k+1}$ is true.

However, if we substitute any value of $n$ in $P_n$, we find that the proposition is not true:

     when $n = 1$, LHS $= 1$   RHS $= 1\tfrac{1}{8}$

     when $n = 2$, LHS $= 3$   RHS $= 3\tfrac{1}{8}$

     when $n = 3$, LHS $= 6$   RHS $= 6\tfrac{1}{8}$

# Example 19.18

$\boxed{!}$   Step (1) alone is not enough

Consider the proposition $P_n$: $n^2 - n + 41$ is prime, $n \geq 1$

     when $n = 1$, $n^2 - n + 41 = 41$, which is prime

     when $n = 2$, $n^2 - n + 41 = 43$, which is prime

     when $n = 3$, $n^2 - n + 41 = 47$, which is prime

     when $n = 4$, $n^2 - n + 41 = 53$, which is prime

     when $n = 5$, $n^2 - n + 41 = 61$, which is prime

     when $n = 6$, $n^2 - n + 41 = 71$, which is prime

     when $n = 7$, $n^2 - n + 41 = 83$, which is prime

so $P_n$ is true not only for $n = 1$, but also for $n = 2, 3, 4, 5, 6, 7 \ldots$

It might seem reasonable to suppose that $P_n$ is true for all $n$, but an attempt to prove the proposition by exhaustion (which can't be done anyway because $n$ can take an infinity of values) fails after $n = 40$

when $n = 41$, $n^2 - n + 41 = 41^2$, which is **NOT** prime

## EXERCISE 19 D

1.  Prove, by mathematical induction, that the sum of the first $n$ integers is $\frac{1}{2}n(n+1)$

2.  Prove, by mathematical induction, that the sum of the first $n$ odd integers is $n^2$.

3.  Prove, by mathematical induction, that $\sum_{r=1}^{n} r^3 = \frac{1}{4}n^2(n+1)^2$ for all positive integers $n$.

4.  Prove, by mathematical induction, that $11^n - 1$ is always divisible by 10.

5.  Prove, by mathematical induction, that $\sum_{r=1}^{n} r(r+1) = \frac{1}{3}n(n+1)(n+2)$.

6.  Prove, by mathematical induction, that $\sum_{r=1}^{n} \frac{1}{r(r+1)} = \frac{n}{n+1}$.

7.  Prove, by mathematical induction, that $\sum_{r=1}^{n} 4^r = \frac{4}{3}(4^n - 1)$.

8.  Prove, by mathematical induction, that
    $(1 \times 2) + (2 \times 3) + (3 \times 4) + \ldots + n(n+1) = \frac{1}{3}n(n+1)(n+2)$

9.  Prove, by mathematical induction, that 2 is always a factor of $n^2 + n$.

10. Prove, by mathematical induction, that $n^3 - n$ is always divisible by 6.

11. Prove, by mathematical induction, that $\sum_{r=1}^{n} 2^{r-1} = 2^n - 1$.

12. Prove, by mathematical induction, that
    $(1 \times 4) + (2 \times 7) + (3 \times 10) + \ldots + n(3n+1) = n(n+1)^2$ for all positive integers $n$.

13. Prove, by mathematical induction, that 4 is always a factor of $5^n - 1$.

14. Prove, by mathematical induction, that 3 is always a factor of $n^3 + 2n$.

15. Prove, by mathematical induction, that $\sum_{r=1}^{n} \frac{1}{r(r-1)} = 1 - \frac{1}{n}$.

16. Prove, by mathematical induction, that $\sum_{r=1}^{n} \frac{r}{2^r} = 2 - \left(\frac{1}{2}\right)^n (2+n)$.

## SUMMATION OF SERIES BY THE METHOD OF DIFFERENCES

### Example 19.19

Consider the series $\dfrac{1}{2} + \dfrac{1}{6} + \dfrac{1}{12} + \dfrac{1}{20} + \cdots$  ①

The series can be written as $\dfrac{1}{1 \cdot 2} + \dfrac{1}{2 \cdot 3} + \dfrac{1}{3 \cdot 4} + \dfrac{1}{4 \cdot 5} + \ldots + \dfrac{1}{n(n+1)}$  or  $\displaystyle\sum_{r=1}^{n} \dfrac{1}{r(r+1)}$

We have already proved by induction (see Exercise 19 C, question 6) that the sum of this series, $S_n$, is

given by $\ S_n = \dfrac{n}{n+1}$, but is there another way of finding this sum?

We know that $\dfrac{1}{r(r+1)} \equiv \dfrac{1}{r} - \dfrac{1}{r+1}$,

so $\ S_n = \displaystyle\sum_{r=1}^{n} \dfrac{1}{r(r+1)} = \sum_{r=1}^{n} \left\{ \dfrac{1}{r} - \dfrac{1}{r+1} \right\}$

Using the rules for combining sums (see chapter 1, module 1.1) this can be written

$$S_n = \sum_{r=1}^{n} \dfrac{1}{r} - \sum_{r=1}^{n} \dfrac{1}{r+1}$$

Now, writing this expression in full in two rows we see that

$$S_n = \dfrac{1}{1} + \dfrac{\cancel{1}}{\cancel{2}} + \dfrac{\cancel{1}}{\cancel{3}} + \dfrac{\cancel{1}}{\cancel{4}} + \dfrac{\cancel{1}}{\cancel{5}} + \cdots\cdots + \dfrac{\cancel{1}}{\cancel{n}}$$

$$- \left\{ \dfrac{\cancel{1}}{\cancel{2}} + \dfrac{\cancel{1}}{\cancel{3}} + \dfrac{\cancel{1}}{\cancel{4}} + \dfrac{\cancel{1}}{\cancel{5}} + \cdots\cdots + \dfrac{\cancel{1}}{\cancel{n}} + \dfrac{1}{n+1} \right\}$$

$$= 1 - \dfrac{1}{n+1}$$

$$= \dfrac{n}{n+1}$$

When terms cancel out like this, the method is known as the **method of differences**, which can also be used when the terms of the sum to be found can be expressed as the difference between two functions.

### Example 19.20

(a) Show that if $f(r) \equiv r(r+1)(r+2)$ then

$\qquad f(r) - f(r-1) \equiv 3r(r+1).$

(b) Hence find the sum of the series $2 + 6 + 12 + \ldots + n(n+1)$  ②

$\qquad$ (a) $\ \succ\ \ f(r) - f(r-1) \equiv r(r+1)(r+2) - (r-1)r(r+1)$

$$\equiv r(r+1)\{(r+2) - (r-1)\}$$

$$\equiv r(r+1)3$$

$$\equiv 3r(r+1)$$

(b) ➤ $S_n = \sum\limits_{r=1}^{n} r(r+1)$

$\quad = \frac{1}{3} \sum\limits_{r=1}^{n} 3r(r+1)$

$\quad = \frac{1}{3} \sum\limits_{r=1}^{n} \{f(r) - f(r-1)\}$

$\qquad = \frac{1}{3}\{f(1) - f(0) + f(2) - f(1) + f(3) - f(2) + \dots + f(n) - f(n-1)\}$

$\qquad = \frac{1}{3}\{f(n) - f(0)\}$

$\qquad = \frac{1}{3}n(n+1)(n+2)$

> ! Notice that in example 19.19 the solution was set out as a difference of sums, while in 19.20 the sum of differences was used. Either method will work, so long as you can see which terms cancel out and therefore which are left.

## *Convergence of Series summed by the Method of Differences*

Once $S_n$ has been found as a function of $n$ it is easy to know whether or not the series converges.

In example 19.19 $\lim\limits_{n\to\infty} S_n = 1$ while in example 19.20 $\lim\limits_{n\to\infty} S_n = \infty$, so series ① converges and series ② does not.

$\underline{A(2r+1) + B(2r-1) = 2}$

$\quad A$

## EXERCISE 19 E

1. Express $\dfrac{2}{(2r-1)(2r+1)}$ in partial fractions and hence find an expression for $\sum\limits_{r=1}^{n} \dfrac{2}{(2r-1)(2r+1)}$

2. Express $\dfrac{2}{r(r+2)}$ in partial fractions and hence find an expression for $\sum\limits_{r=1}^{n} \dfrac{1}{r(r+2)}$

3. Find the $n$th partial sum of the series $S = \dfrac{1}{1\times 4} + \dfrac{1}{2\times 5} + \dfrac{1}{3\times 6} + \cdots$
   Does S converge? If so, find the limit to which it converges.

4. Given that $2r+1 \equiv (r+1)^2 - r^2$ obtain an expression for $S = \sum\limits_{r=1}^{n} (2r+1)$
   Does S converge as $n \to \infty$? If so, find the limit to which it converges.

## PROBLEMS INVOLVING SERIES

Many problems, especially financial ones, can be solved using arithmetic or geometric progressions. Although time is a continuous variable, if deposits are made or interest paid at regular intervals only, then time is being treated as a discrete variable.

## Example 19.21

Sherwin finds he is able to save $3000 in 2005, his first year in work. Each year he is able to save $250 more than the previous year. Find

(a)  the first year in which he saves $5000,

(b)  how many years it takes to save $60 000.

| Year | 2005, year 1 | 2006, year 2 | 2007, year 3 | 2008, year 4 | year $n$ |
|------|------|------|------|------|------|
| Saving | 3000 | 3250 | 3500 | 3750 | |

The numbers in the savings row form an AP, with $a = 3000$ and $d = 250$

(a)  Find $n$ when $a_n = 5000$: $3000 + 250(n-1) = 5000 \rightarrow n - 1 = 8,\ n = 9$

He will save $5000 in 2012

(b)  Find $n$ when $S_n = 60\ 000$:  $120\ 000 = n[6000 + 250(n-1)]$

$$n^2 + 23n - 480 = 0$$

$n = 13.24$  $\therefore$ it will take 14 years to save $60 000

## Example 19.22

Michelle decides that she can save $S every month. She invests the money in a deposit account which pays $x$% interest on the last day of every month, starting in January.

Write down expressions for the balance in her account on the 1st of February, March and April.

Deduce a formula for the balance at the start of the $n$th month.

If her monthly deposits are of $1000 and the interest rate is $\frac{1}{2}$% per month, find how long it will take for her to have saved $100 000.

$S + Sx$
$S(1+x)$

➤  Let $p = 1 + x \times 10^{-2}$   $[(100 + x)\%]$

| $n$ | month | start | end |
|-----|-------|-------|-----|
| 1 | Jan | S | $S \times p$ |
| 2 | Feb | $S + Sp = S(1 + p)$ | $S(1 + p)p$ |
| 3 | Mar | $S + S(1 + p)p = S(1 + p + p^2)$ | $S(1 + p + p^2)p$ |
| 4 | Apr | $S + S(1 + p + p^2)p = S(1 + p + p^2 + p^3)$ | |

➤  Clearly, the balance in the account at the start of each month is a geometric series with $a = S$ and $r = p$.

➤  At the start of the $n$th month the balance will be $S_n = \dfrac{S(p^n - 1)}{p - 1}$.

➤  When $S = 1000$ and $x = \frac{1}{2}$, $p = 1.005$ so $S_n = \dfrac{1000(1.005^n - 1)}{0.005}$

➤  When $S_n = 100\ 000$, $0.5 = (1.005^n - 1)$, so $1.005^n = 1 + 0.5 = 1.5$, giving $n = \dfrac{\log 1.5}{\log(1.005)} = 81.3$

Michelle will have $100 000 in her account at the start of the 82nd month, i.e. after 6 years and 10 months.

 This example started with letters and introduced numbers later. The use of letters made the pattern clear. If a problem works with numbers from the start it is important not to oversimplify the numbers, or the pattern may be lost.

## Example 19.23

A small ball is dropped from a height of 1 m on to a horizontal floor. Each time the ball strikes the ground it rebounds to 3/5 of the height it has just fallen.

(a)   Show that, when the ball strikes the floor for the third time, it has travelled a distance 2.92 m.

(b)   Show that the total distance travelled by the ball cannot exceed 4 m.

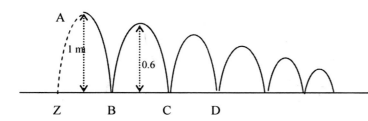

(The diagram shows the ball travelling horizontally as well as vertically, for the sake of clarity. In practice it is bouncing on the spot)

(a)   ➤  1st strike   (AB)    1
       ➤  2nd strike   (BC)    $2 \times 0{\cdot}6$
       ➤  3rd strike   (CD)    $2 \times 0.6^2$
       ➤  Distance travelled to 3rd strike $= 1 + 2 \times 0{\cdot}6 + 2 \times 0{\cdot}6^2 = 2.92$

(b)   ➤      Notice that after the ball has struck at B the distances between bounces form a GP. To find the total distance travelled, imagine that the ball started from Z, and then subtract the extra 1 m.

       ➤  GP for bounces from Z:  $2[1 + 1(0.6) + 1(0.6)^2 + 1(0.6)^3 + \ldots]$   $a = 1$, $r = 0.6$

       ➤  Theoretically the ball will bounce indefinitely, so find $S_\infty$.  $S_\infty = 2 \times \dfrac{1}{1 - 0.6} = 5$

       ➤  Removing the extra 1 m: Total distance travelled cannot exceed  4 m.


## EXERCISE 19F (There are more practical series problems in Exercise 14B)

1.    Uncle Benjamin decides to give some pocket money to his nephews Darren and Ricardo. He offers two possible schemes, both with a first payment of $5. Under scheme A the money is paid weekly and is increased by $1 each week. Under scheme B the money is paid monthly and is increased by 50% each month. Darren opts for scheme A and Ricardo opts for scheme B. Both boys save the money.

      Which boy has most money (a) after 1 year?   (b) after 2 years?

2.    Osmond is given an interest-free loan to buy a second-hand car. He repays the loan in monthly instalments. He repays $20 the first month, $22 the second month and the repayments continue to rise by $2 per month until the loan is repaid.

      Given that the final monthly repayment is $114,

      (a)  show that the number of months it will take Osmond to repay the loan is 48,

      (b)  find the amount, in dollars, of the loan.

3. Write the recurring decimal 0.333…as a geometric series, and hence write the decimal as a single fraction.

4. Island Produce offers Mustafa a ten-year contract. This gives a starting salary of $30 000 a year with an annual increase of 8% of the previous year's salary.

   (a) Show that the amounts of annual salary form a geometric sequence and write down its common ratio.

   (b) How much does Mustafa expect to earn in the tenth year?

   (c) Show that the total amount Mustafa would expect to earn over the ten years is nearly $435 000.

   After considering the offer, Mustafa asks for a different scheme of payment. This has the same starting salary of $30 000 but with a fixed annual pay rise $d.

   (d) Find $d$ if the total amount paid out over 10 years is to be the same under the two schemes.

5. Fay borrows $50 000 towards the cost of her house. Compound interest is charged on the loan at 12% per annum. She agrees to pay back the loan in 25 equal instalments, at yearly intervals, the first repayment being made exactly 1 year after the loan is taken out. Calculate the value of each instalment.

6. When a child's ball is dropped from a height $h$ metres onto a hard, flat floor, it rebounds to a height of $\frac{3}{4}h$ metres. The ball is dropped initially from a height of 1.2 m.

   (a) Find the maximum height to which the ball rises after two bounces.

   (b) Find the total distance that the ball has travelled when it hits the floor for the tenth time.

   (c) Assuming that the ball continues to bounce in the same way indefinitely, find the total distance that the ball travels.

7. A savings scheme pays 5% per annum compound interest. A deposit of $500 is invested in this scheme at the start of each year.

   (a) Show that at the start of the third year after the annual deposit has been made, the amount in the scheme is $1576.25.

   (b) Find the amount in the scheme at the start of the 15th year after the scheme has been started.

   (c) How long will it take for the amount in the scheme to reach $25 000 ?

8. When a new stretch of highway was opened, the peak rate at which traffic flowed was found to be 5000 vehicles per hour,

   (a) One mathematical model for future rates of flow was that the peak rate would increase by 6% each year. Using this model

      (i) express the peak rate of flow after $n$ years, in terms of $n$.

      (ii) find the integer value of $n$ for which the peak rate of flow first exceeds 20 000 vehicles per hour.

   (b) An alternative model stated that the peak rate of flow after $n$ years denoted by $f_n$ vehicles per hour would be given by $f_n = 5000\{6 - 5 \times (0.96)^n\}$

      Using this model

      (i) find the predicted percentage increase in the peak rate of flow over the first year,

      (ii) find the integer value of $n$ for which the peak rate of flow first exceeds 20 000 vehicles per hour.

   (c) Describe briefly how the predictions of the models in (a) and (b) differ as $n$ becomes large.

# SUMS OF SERIES

In unit 1 chapter 1 we met the rules for combining sums. They are:

$$\sum_{r=1}^{n} a_r + \sum_{r=1}^{n} b_r = \sum_{r=1}^{n} (a_r + b_r) \quad (1) \qquad k\sum_{r=1}^{n} a_r = \sum_{r=1}^{n} ka_r \qquad (2)$$

$$\sum_{r=1}^{n} k = nk \qquad (3) \qquad \sum_{r=1}^{p} a_r + \sum_{r=p+1}^{n} a_r = \sum_{r=1}^{n} a_r \qquad (4)$$

These can be used in conjunction with the sums of the first $n$ integers, squares and cubes, which we have proved by induction in example 19.3 and exercise 19C. You should memorise the formulae for these sums and you may use them without proof.

$$\sum_{r=1}^{n} r = \tfrac{1}{2}n(n+1) \qquad (5)$$

$$\sum_{r=1}^{n} r^2 = \tfrac{1}{6}n(n+1)(2n+1) \qquad (6)$$

$$\sum_{r=1}^{n} r^3 = \tfrac{1}{4}n^2(n+1)^2 \qquad (7)$$

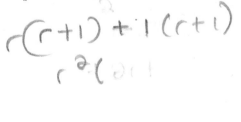

## Example 19.24

The $k$th term of a series is given by the formula $u_k = 4k - 3$.

(a) Show that the $k$th partial sum, $S_k$, of the series is $S_k = 2k^2 - k$.

(b) Hence find $S = \sum_{k=1}^{n} S_k$

(a) $S_k = \sum_{r=1}^{k} (4r - 3) = 4\sum_{r=1}^{k} r - 3\sum_{r=1}^{k} 1$

$$= 4\cdot\tfrac{1}{2}k(k+1) - 3k$$
$$= 2k^2 - k.$$

(b) $S = \sum_{k=1}^{n} (2k^2 - k) = 2\sum_{k=1}^{n} k^2 - \sum_{k=1}^{n} k$

$$= 2\cdot\tfrac{1}{6}n(n+1)(2n+1) - \tfrac{1}{2}n(n+1)$$
$$= \tfrac{1}{6}n(n+1)(4n-1)$$

## EXERCISE 19 G (Mixed Questions)

1. In a G.P. the fifth term is 100 and the seventh term is 400. Find the first term.

2. (a) Given that the first and second terms of an A.P. are 12 and 6 respectively, find the sum of the first hundred terms.

   (b) Given that the first and second terms of a G.P. are 12 and 6 respectively, show that the sum of the first ten terms is $\dfrac{3069}{128}$.

3. (a) Prove, by induction or otherwise, that

$$\sum_{r=1}^{n} r^2 = \tfrac{1}{6}n(n+1)(2n+1).$$

(b) An arithmetic series has first term 1 and common difference 2, the $n$th partial sum is denoted by $A_n$.

Write down and simplify an expression in terms of $n$ for $A_n$, and hence find $\displaystyle\sum_{n=1}^{N} A_n$.

4. A geometric series has first term 3 and common ratio 2, and the $n$th partial sum is denoted by $G_n$. Write down an expression in terms of $n$ for $G_n$.

$$3\left(2^n - 1\right)$$

5. A progression has a first term of 12 and a fifth term of 18.
   (a) Find the sum of the first 25 terms if the progression is arithmetic.
   (b) Find the 25th term if the progression is geometric.

$a = 0.99$
$r = 0.99$

6. Evaluate, correct to the nearest whole number, $\displaystyle\sum_{r=1}^{100} (0.99)^r$.

#7. An A.P. has a first term $a$ and common difference $-1$. The sum of the first $n$ terms is equal to the sum of the first $3n$ terms. Express $a$ in terms of $n$.

8. The $r$th term of an A.P. is $(1 + 4r)$. Find, in terms of $n$, the sum of the first $n$ terms of the progression.

9. (a) The first four terms $u_1, u_2, u_3, u_4$ of an arithmetic progression are such that
   $$u_4 - u_2 = 15 \quad \text{and} \quad 4u_3 = 9u_1.$$
   Find the value of $u_1$.

   (b) The first four terms $v_1, v_2, v_3, v_4$ of a geometric progression are such that
   $$v_4 - v_2 = 15 \quad \text{and} \quad 4v_3 = 9v_1.$$
   Find the value of $v_1$.

10. (i) Find the sum of the first $n$ terms of the sequence $u_1, u_2, u_3, u_4, \ldots, u_r, \ldots$ in each of the following cases.
    (a) $u_r = 3$.
    (b) $u_r = (-0.9)^r$.
    #(c) $u_r = 1 + (-1)^r$.
    (d) $u_r = 3r + 4$.

    (ii) Identify one sequence from (b) to (d) above which is convergent, and identify one which is periodic.

#11. The sum of the first twenty terms of an A.P. is 50, and the sum of the next twenty terms is $-50$. Find the sum of the first hundred terms of the progression.

12. The positive integers, starting at 1, are grouped into sets containing 1, 2, 4, 8, ... integers, as indicated below, so that the number of integers in each set after the first is twice the number of integers in the previous set.

$$\{1\}, \ \{2, 3\}, \ \{4, 5, 6, 7\}, \ \{8, 9, 10, 11, 12, 13, 14, 15\}, \ldots$$

(i) Write down expressions, in terms of $r$, for

$\dfrac{n}{2}(2+2n-2)$

    (a)   the number of integers in the $r$th set,

    (b)   the first integer in the $r$th set,

    (c)   the last integer in the $r$th set.

  (ii)  Given that the integer 1 000 000 occurs in the $r$th set, find the value of $r$.

13.  A geometric series has first term 1 and common ratio $r$. Given that the sum to infinity of the series is 5, find the value of $r$.

Find the least value of $n$ for which the sum of the first $n$ terms of the series exceeds 4.9.

14.  (a)  Find the sum of all the even numbers between 99 and 999.

    (b)  A G.P. has a common ratio of $-\frac{1}{2}$, and the sum of its first 3 terms is 18. Find the sum to infinity.

15.  A G.P. has a non-zero first term $a$ and common ratio $r$, where $0 < r < 1$. Given that the sum of the first 8 terms of the progression is equal to half the sum to infinity, find the value of $r$, correct to 3 d.p.

Given also that the 17th term of the progression is 10, find $a$.

16.  A sequence of positive integers $u_1, u_2, u_3,\ldots$ is defined by $u_1 = 1$ and $u_{n+1} = 3u_n + 2$ for $n \geq 1$, thus the first few terms of the sequence are 1, 5, 17, 53,...

Prove by induction that $u_n = 2(3^{n-1}) - 1$.

$a=0.99$
$r=0.99 \quad r<1$
$S=\dfrac{0.99(1-0.99^{100})}{1-0.99}$

17.  (a)  Find the $k$th term of the series $1(4) + 2(7) + 3(10) + 4(13) + \ldots$

    (b)  Show, by mathematical induction or otherwise, that $S_k$, the sum to $k$ terms of this series, is $k(k+1)^2$.

    (c)  Find $\displaystyle\sum_{k=1}^{n} S_k$ .

18.  The common ratio, $r$, of a geometric series is given by $r = \dfrac{6x}{x^2 + 5}$.

Find all the values of $x$ for which the geometric series converges.

19.  $12, p, q, 96$ are four consecutive terms of a progression.

    (a)  If the progression is an A.P. find the common difference, and hence find $p$ and $q$.

    (b)  If the progression is a G. P. find the common ratio and hence find $p$ and $q$.

20.  The first 4 terms of an A.P. are $8, 14, 3x + 4y$ and $4x + 5y$ respectively, where $x$ and $y$ are constants. Find the values of $x$ and $y$.

# EXERCISE 19 X

1.  (a)  Write down the first four terms of the series $\displaystyle\sum_{r=1}^{20}(-1)^{r+1}r^2$.    [3]

    (b)  Express the infinite series $3 + 5 + 7 + 9 + \ldots$ in sigma ($\Sigma$) notation.    [3]

2. Find the sum of the convergent geometric series $S = 2 + \frac{1}{2} + \frac{1}{8} + \frac{1}{32} + \dots$ [6]

3. An arithmetic progression, A.P., is such that its first, third and ninth terms are respectively equal to the first third and fifth terms of a geometric progression, G.P.. If the first term of the A.P. is 1, find
   (a) the common difference for the A.P. [3]
   (b) the common ratio for the G.P.. [3]

4. The sum $S_n$, of the first $n$ terms of a series is given by $S_n = n(3n - 4)$
   Show that the terms of the series are in arithmetic progression with common difference 6. [7]

5. A geometric progression with first term $h$ and common ratio $k$ has sum to infinity equal to $2h$.
   Find the common ratio $k$. [4]

6. An arithmetic progression has first term 3 and common difference $d$.
   The sum of the first six terms is 48 and the sum of the first $n$ terms is 168. Find
   (a) the value of $d$ [3]
   (b) the value of $n$. [5]

7. (a) Prove by mathematical induction that $\sum_{r=1}^{n} r^3 = \frac{1}{4} n^2 (n+1)^2$ for all positive integers $n$. [7]
   (b) Hence, or otherwise, find in terms of $n$ the sum of the cubes of the first $n$ even positive integers. [5]

8. (a) An arithmetic series has $n$th term $a + (n - 1)d$, where $a$ and $d$ are real constants. Prove that the $n$th partial sum, $S_n$, of the series is given by $S_n = \frac{n}{2}\{2a + (n-1)d\}$. [3]
   (b) Hence, express $\ln(2 \times 2^2 \times 2^3 \times \dots \times 2^{48} \times 2^{49})$ in the form $k \ln 2$, where $k$ is an integer. [5]

9. (a) Given that $S_r$ denotes the sum of the first $r$ positive integers, prove by mathematical induction that $S_r = \frac{1}{2}r(r + 1)$. [6]
   (b) Hence, find the sum $S = \sum_{r=1}^{n} S_r$ in terms of $n$. [9]

10. The common ratio, $r$, of a geometric series is given by $r = \dfrac{5x}{4 + x^2}$.
    Find all the values of $x$ for which the geometric series converges. [10]

11. (a) Prove by mathematical induction that $\sum_{r=1}^{n} r^3 = \frac{1}{4} n^2 (n+1)^2$. [8]
    (b) Deduce that $\sum_{r=1}^{n} (2r)^3 = 2n^2 (n+1)^2$ [2]

12. Given that the sum of the first $n$ terms of a series, S, is $9 - 3^{2-n}$
    (a) find the $n$th term of S [3]
    (b) show that S is a geometric progression [3]
    (c) find the sum to infinity of S. [2]

13. (a) Use the fact that $\dfrac{1}{k} - \dfrac{1}{k+1} = \dfrac{1}{k(k+1)}$ to show that

$$\sum_{r=1}^{n}\left(\frac{1}{k(k+1)}\right) = 1 - \frac{1}{n+1}.$$  [4]

   (b) Deduce that $\displaystyle\sum_{r=1}^{\infty}\left(\frac{1}{k(k+1)}\right) = 1.$  [3]

14. Given the series $\dfrac{1}{1.4} + \dfrac{1}{4.7} + \dfrac{1}{7.10} + \ldots$

   (a) obtain the $n$th term of the series  [3]

   (b) find the sum of the first $n$ terms of the series  [8]

   (c) find the sum to infinity if it exists.  [2]

15. The sum to infinity of a GP is 10 times the first term.

   Find the common ratio of the GP.  [3]

16. (a) Show that the series $\log_a b + \log_a(bc) + \log_a(bc^2) + \ldots + \log_a(bc^{n-1})$, where $a, b, c > 0$, $n \geq 1$
   is an arithmetic progression whose sum $S_n$ to $n$ terms is $\frac{1}{2}n\log_a b^2 c^{n-1}$.  [6]

   (b) Find $S_n$ when $n = 6$ and $a = b = c = 5$.  [2]

17. Given the series $\dfrac{1}{2} + \dfrac{1}{2^4} + \dfrac{1}{2^7} + \dfrac{1}{2^{10}} + \ldots$

   (a) show that the series is geometric  [3]

   (b) find the sum of the series to $n$ terms  [3]

   (c) show that as $n$ approaches infinity, the sum of the series approaches $\dfrac{4}{7}$.  [2]

18. The $n$th term of a sequence $\{a_n\}$ is defined by

$$a_n = \sum_{k=1}^{n}\frac{1}{k(k+1)}.$$

   (a) Express $a_n$ in terms of $n$ for $n \geq 1$.  [2]

   (b) Determine whether the sequence converges.  [1]

19. Using the fact that $\displaystyle\sum_{r=1}^{n} kr = k\sum_{r=1}^{n} r = k\left[\frac{n(n+1)}{2}\right]$, where $k$ is a constant, and that

$\displaystyle\sum_{r=1}^{n} k = k\sum_{r=1}^{n} 1 = kn$, find, **in simplest form**, the value of $\displaystyle\sum_{r=1}^{n}(6r + 5)$.  [3]

20.  (a)  Given that the sum of the first $n$ terms of the series $\sum_{r=1}^{n} (6r + 5)$ is $3n^2 + 8n$, calculate

the first five partial sums of the series.  [2]

(b)  Prove by mathematical induction that $\sum_{r=1}^{n} (6r + 5) = n(3n + 8)$.  [9]

21.  Prove by mathematical induction that $\sum_{r=1}^{n} r(r + 1) = \frac{1}{3}n(n + 1)(n + 2)$, for all $n \in \mathbb{N}$  [9]

22.  The first four terms of an A.P. are $2, 5, 2x + y + 7$ and $2x - 3y$ respectively, where $x$ and $y$ are constants.  Find the values of $x$ and $y$.  [8]

23.  (a)  Find the sum to $n$ terms of the geometric series $4 + 2 + 1 + \frac{1}{2} + \dots$  [5]

(b)  Deduce the sum to infinity of the series.  [2]

24.  Prove by mathematical induction that

$$\sum_{r=1}^{n} r(r + 1) = \frac{1}{3}n(n + 1)(n + 2), \text{ for all } n \in \mathbb{N}.$$  [9]

25.  Find the sum of the arithmetic progression    $72, 69, 66, \dots -24, -27$.  [5]

26.  The sum, $S_n$, of the first $n$ terms of a series is given by $S_n = 2n(n - 2)$.
(a)  Find, in terms of $n$, the $n$th term of the series.  [4]
(b)  Hence, show that the series is an AP.  [2]
(c)  For the AP in (b) above, identify
(i)  the first term  [1]
(ii)  the common difference  [1]

27.  (a)  Use the fact that $\dfrac{1}{r} - \dfrac{1}{r+1} = \dfrac{1}{r(r+1)}$  to show that $S_n = \sum_{r=1}^{n} \left( \dfrac{1}{r(r+1)} \right) = 1 - \dfrac{1}{n+1}$  [5]

(b)  Deduce, that as $n \to \infty$, $S_n \to 1$.  [1]

28.  Verify that the sum, $S_n$, of the series $\dfrac{1}{2} + \dfrac{1}{2^3} + \dfrac{1}{2^5} + \dots$, to $n$ terms, is $S_n = \dfrac{2}{3}\left(1 - \dfrac{1}{2^{2n}}\right)$.  [4]

29.  Three consecutive terms, $x - d, x$ and $x + d, d > 0$, of an arithmetic series have a sum 21 and product 315.  Find the value of
(a)  $x$  [2]
(b)  the common difference $d$.  [4]

30. (a) Show that the terms of $\displaystyle\sum_{r=1}^{m} \ln 3^r$ are in arithmetic progression. [3]

    (b) Find the sum of the first 20 terms of this series. [4]

    (c) Hence, show that $\displaystyle\sum_{r=1}^{2m} \ln 3^r = (2m^2 + m)\ln 3$. [3]

31. A plot of land is rented on the understanding that the rent for the first year will be $\$\,64$ and in subsequent years will always be $\frac{7}{8}$ of what it was the year before. Calculate, **to the nearest dollar**, the TOTAL amount of rent paid for the first 15 years. [5]

32. An athlete plans a training schedule which involves running 20 km in the first week of training. In each subsequent week, the distance is to be increased by 10 % over the previous week.

    (a) Write down an expression for the distance to be covered in the $n$th week. [3]

    (b) Find the week in which the athlete would first cover more than 100 km. [4]

33. Once a person ingests medication, the human body eliminates the medication from the body in such a way that a fixed percent of the amount is removed daily. A person had a daily dose of Q units of a certain medication and his body eliminated 60% of the medication daily.

    (a) Write down, in terms of Q, the amount of the medication **not** eliminated

        (i) at the end of the first day [1]

        (ii) at the end of the second day [2]

    (b) What is the value of Q, if the amount of medication in the body at the end of the second day is 56 units. [1]

34. At the start of a month, a customer owes a credit card company $1000.00. In the middle of the month, the customer pays $\$P$ to the company, where $P < 1000.00$, and at the end of the month, the company adds interest at the rate, R, of 3% of the amount still owing. This process is repeated in each subsequent month.

    (a) Find the value of P for which the customer owes $1000.00 at the start of every month. [6]

    (b) Find the value of P for which the whole amount is paid off after the second payment. [6]

    (c) Assuming that the debt has not been paid off after 4 payments, show that the amount still owing at the beginning of the 5th month can be expressed as

$$\$\left[ 1000R^4 - \frac{PR(R^4 - 1)}{R - 1} \right], \quad R = 1.03$$ [6]

    (d) Show that the value of P for which the whole amount owing is exactly paid off after the $n$th payment is given by

$$P = \frac{1000R^{n-1}(R - 1)}{R^n - 1}.$$ [7]

35. Ms Janis Smith takes out an endowment policy with an insurance company which involves making a fixed payment of $\$P$ each year. At the end of $n$ years, Janis expects to receive a payout of a sum of money which is equal to her total payments together with interest added at the rate of $\alpha$ % per annum of the total sum in the fund.

    (a) Show that the total sum in the fund at the end of the second year is $\$P(R + R^2)$, where

$$R = \left( 1 + \frac{\alpha}{100} \right).$$ [7]

(b)  Show, by mathematical induction, or otherwise, that the total sum in the fund at the end

of the $n$th year is $ \$\dfrac{P\,R(R^{n}-1)}{R-1}$.  [12]

(c)  Find the value of P, to the NEAREST dollar, when $n = 10$, $\alpha = 8$ and the payout is \$100,000.00.  [6]

36.  Twelve hundred people visited an exhibition on its opening day.  Thereafter, the attendance fell each day by 4% of the number on the previous day.

(a)  Obtain an expression for the number of visitors on the $n$th day  [2]

(b)  Find the total number of visitors for the first $n$ days.  [3]

(c)  The exhibition closed after 10 days.  Determine how many people visited during the period for which it was opened.  [3]

(d)  If the exhibition had been kept open indefinitely, what would be the maximum number of visitors?  [2]

# A FOOTNOTE ON THE CONVERGENCE OF SEQUENCES AND SERIES

If a sequence converges to a limit other than 0  (i.e. A = $\{a_n\}$ and $\lim\limits_{n\to\infty} a_n = c, c \neq 0$), then the related series $S = \sum a_n$ does not converge.

E.g.  if $a_n = \dfrac{2n+1}{n}$,  then $\lim\limits_{n\to\infty} a_n = 2$  and  $S_n = \dfrac{3}{1} + \dfrac{5}{2} + \dfrac{7}{3} + \dfrac{9}{4} + \cdots$  We can see that $2 < a_n \leq 3$ for all $n$, so $S_n > 2n$.  As $n \to \infty$, $S_n \to \infty$, so S does not converge.

It is safe to say that for a series $S = \sum a_n$ to converge $\lim\limits_{n\to\infty} a_n$ must be 0.  However the reverse is not true.  The harmonic series is an example of a series where $\lim\limits_{n\to\infty} a_n = 0$, but  S does not converge.

$H = \{\frac{1}{n}\}$  so  $S = 1 + \frac{1}{2} + \frac{1}{3} + \frac{1}{4} + \frac{1}{5} + \dots + \frac{1}{n} + \dots$

Since $\lim\limits_{n\to\infty}\frac{1}{n} = 0$, you might expect the series to converge just as the geometric series $\sum(\frac{1}{2})^r$ did; but its non-convergence can be shown as follows:

Group together all the terms of S so that each group, after the initial 1, contains $2^r$ terms, $r = 0, 1, 2, \dots$

then $S = 1 + \frac{1}{2} + (\frac{1}{3} + \frac{1}{4}) + (\frac{1}{5} + \frac{1}{6} + \frac{1}{7} + \frac{1}{8}) + (\frac{1}{9} + \frac{1}{10} + \dots + \frac{1}{16}) + ( \dots + \frac{1}{32}) + \dots$

Since each number in each bracket is greater than the last number in the bracket, the sum of the first bracket is greater than $2 \times \frac{1}{4}$, of the second bracket is greater than $4 \times \frac{1}{8}$, of the third bracket is greater than $8 \times \frac{1}{16}$; so we have an infinite number of brackets each of which has a sum is greater than $\frac{1}{2}$.  So S must diverge.

This can be summarised: $\lim\limits_{n\to\infty} a_n = 0$ is a necessary but not a sufficient condition for $S = \sum a_n$ to converge.

# What you have learnt in Chapter 19

*Sequences and Series:*

A **sequence** is a set or *succession* of terms, written either as $\{a_r\}$ or $\{a_1, a_2, a_3, a_4, a_5, \dots\}$

A **series** is a *sum* of terms of a sequence, written either as $\displaystyle\sum_{r=1}^{n} a_r$ or $a_1 + a_2 + a_3 + a_4 + a_5 + \dots + a_n$

The **sum of the series** is the result of adding the terms of the series, and is either a formula involving $n$, the number of terms in the series, or, where $n$ is known, a single number.

The **$n$th partial sum** of a series is the sum of the first $n$ terms of the series.

*Arithmetic Progressions and Geometric Progressions (A.P.s and G.P.s):*

For an A.P.   the general term is $u_n = a + (n-1)d$

the sum of the first $n$ terms is $S_n = \frac{1}{2} n [2a + (n-1)d]$

or $S_n = \frac{1}{2} n [a + l]$, where $l$ is the last term of the series

For a G.P.   the general term is $u_n = ar^{n-1}$

the sum of the first $n$ terms is $S_n = \dfrac{a(1-r^n)}{1-r}$, $r < 1$ or $S_n = \dfrac{a(r^n - 1)}{r-1}$, $r > 1$

*Convergence:*

All A.P.s are divergent

A G.P. is convergent iff $|r| < 1$. In that case $S_\infty = \dfrac{a}{1-r}$

*The Maclaurin Series:*

Many functions can be written as power series using Maclaurin's Theorem

$$f(x) = f(0) + f'(0)x + \frac{f''(0)}{2!} x^2 + \frac{f'''(x)}{3!} x^3 + \frac{f'''(0)}{4!} x^4 + \dots$$

provided that

❖ it is possible to find all the derivatives of $f(x)$

❖ all the derivatives are defined when $x = 0$

❖ the series is convergent – this may mean placing certain restrictions on the range of $x$.

*Proof by Induction:*

1. Show that the proposition $P_n$ is true for $n = 1$

2. Assume that $P_n$ is true for $n = k$

3. Show that this assumption leads to the fact that $P_n$ is true for $n = k + 1$

4. Combine steps 1 and 3 to conclude that $P_n$ is true for all $n$.

# CHAPTER 20
# 2.2c  The Binomial Theorem

This chapter requires a knowledge of the principles of counting, and in particular of selections (combinations).

> **!**  Students are advised to study at least the counting section of chapter 23 before embarking on this chapter.

> **?**  *Investigation*

Expand the bracket $(a + b)^n$ for values of $n$ from 0 to 5.

Can you deduce the expansion of $(a + b)^6$ without having to multiply it out?

What patterns can you see (i) in the indices and (ii) in the coefficients, in your results?

You should have discovered the following things:

When $(a + b)^n$ is expanded for integral values of $n$ the first few results are:

| $n$ | $(a + b)^n$ |
|---|---|
| 0 | $1$ |
| 1 | $a + b$ |
| 2 | $a^2 + 2ab + b^2$ |
| 3 | $a^3 + 3a^2b + 3ab^2 + b^3$ |
| 4 | $a^4 + 4a^3b + 6a^2b^2 + 4ab^3 + b^4$ |
| 5 | $a^5 + 5a^4b + 10a^3b^2 + 10a^2b^3 + 5ab^4 + b^5$ |
| 6 | $a^6 + 6a^5b + 15a^4b^2 + 20a^3b^3 + 15a^2b^4 + 6ab^5 + b^6$ |

Fig. 20.1

The coefficients can be summarised by the following table

```
                    1
                 1     1
              1     2     1
           1     3     3     1
        1     4     6     4     1
     1     5    10    10     5     1
  1     6    15    20    15     6     1
```

Fig. 20.2

This is known as Pascal's triangle, and in it each row can be deduced from the row above by adding the two numbers diagonally above.

❖ In Pascal's triangle the row starting 1, $n$, … gives the coefficients for the expansion of $(a + b)^n$

❖ In the expansion of $(a + b)^n$ the powers of $a$ decrease by 1 for each successive term, while

the powers of $b$ increase by 1 for each successive term

the sum of the powers of $a$ and $b$ is the same ($n$) for every term.

❖ The expansion of $(a + b)^n$ has $n + 1$ terms

Pascal's triangle is a useful way of finding the coefficients when $n$ is small, but for larger $n$ a general formula would be more useful.

Consider the expansion $(a + b)^6 = (a + b)(a + b)(a + b)(a + b)(a + b)(a + b)$

For the first term, $a^6$, the $a$ must be chosen from all 6 brackets, so there is just one way of getting $a^6$.

What will the coefficient of $a^5b$ be? If the brackets are expanded in full, $a^5b$ will occur 6 times, once for each of the $b$s from the 6 brackets – in other words the number of ways in which one $b$ can be chosen from six brackets, i.e. $^6C_1$ or $\begin{pmatrix} 6 \\ 1 \end{pmatrix}$ ways, so the coefficient of $a^5b$ is $^6C_1$.

Similarly, the third term, in $a^4b^2$, needs $b$s to be chosen from exactly two of the six brackets, and $a$s from each of the other 4, so there are $^6C_2$ terms containing $a^4b^2$. The coefficient of $a^4b^2$ is $^6C_2$.

In the same way the coefficient of $a^3b^3$ is $^6C_3$ etc.

*(handwritten) $(a+b)^n = a^n + {}^n_1C\, a^{n-1}b + {}^n_2C\, a^{n-2}b^2 \cdots$*

The expansion of $(a + b)^6$ is therefore

$$(a + b)^6 = a^6 + {}^6C_1 a^5b + {}^6C_2 a^4b^2 + {}^6C_3 a^3b^3 + {}^6C_4 a^2b^4 + {}^6C_5 ab^5 + b^6$$

Generalising this result, the expansion of $(a + b)^n$, when $n$ is a positive integer, is

$$(a + b)^n = a^n + {}^nC_1 a^{n-1}b + {}^nC_2 a^{n-2}b^2 + \ldots + {}^nC_r a^{n-r}b^r + \ldots + b^n,$$

*(handwritten) $1 + nx + \dfrac{n(n-1)x^2}{2!} + n(n-1)(x)\cdots$*

which can also be written in $\Sigma$ notation

$$(a + b)^n = \sum_{r=0}^{n} {}^nC_r\, a^{n-r}b^r$$

*(handwritten) $1 + nx + \dfrac{n(n-1)x^2}{2!} + \dfrac{n(n-1)(n-2)x^3}{3!}$*

and is known as the **Binomial Theorem**.

The binomial coefficient $^nC_r$ represents the number of ways of choosing $r$ objects from $n$ and therefore requires that $n, r \in \mathbb{W}$ and $r \leq n$.

Putting $a = 1$, $b = x$, and writing the coefficients in expanded form gives an alternative form of the binomial theorem:

$$(1 + x)^n = 1 + nx + \frac{n(n-1)}{2!}x^2 + \frac{n(n-1)(n-2)}{3!}x^3 + \ldots + x^n$$

It is sometimes useful to remember the theorem in this form, and as we shall see later, in this form it can sometimes be used when $n$ is any real number and not just a positive integer.

---

The Binomial theorem states that

$$(a + b)^n = a^n + {}^nC_1 a^{n-1}b + {}^nC_2 a^{n-2}b^2 + \ldots + {}^nC_r a^{n-r}b^r + \ldots + b^n \quad \text{for } n, r \in \mathbb{W}$$

$$= \sum_{r=0}^{n} {}^nC_r\, a^{n-r}b^r$$

$$(1 + x)^n = 1 + nx + \frac{n(n-1)}{2!}x^2 + \frac{n(n-1)(n-2)}{3!}x^3 + \ldots + x^n$$

## Example 20.1

Write out the expansion of (i) $(3 + x)^5$  (ii) $(2 - 3x)^4$  (iii)  $(1 - 2x)^5$

(i) ➤ Putting $a = 3, b = x, n = 5$:  $(3 + x)^5 = 3^5 + {}^5C_1 3^{5-1} x + {}^5C_2 3^{5-2} x^2 + + {}^5C_3 3^{5-3} x^3 + {}^5C_4 3^{5-4} x^4 + x^5$

$$= 243 + 5 \times 81x + 10 \times 27x^2 + 10 \times 9x^3 + 5 \times 3x^4 + x^5$$

$$= 243 + 405x + 270x^2 + 90x^3 + 15x^4 + x^5$$

(ii) ➤ Putting $a = 2, b = -3x, n = 4$:

$$(2 - 3x)^4 = 2^4 + {}^4C_1 2^{4-1}(-3x) + {}^4C_2 2^{4-2}(-3x)^2 + {}^4C_3 2^{4-3}(-3x)^3 + (-3x)^4$$

$$= 16 + 4 \times 8 \times -3x + 6 \times 4 \times 9x^2 + 4 \times 2 \times -27x^3 + 81x^4$$

$$= 16 - 96x + 216x^2 - 216x^3 + 81x^4$$

(iii) ➤ Using the alternative version with $-2x$ for $x$ and $n = 5$:

$$(1 - 2x)^5 = 1 + 5(-2x) + \frac{5.4}{2}(-2x)^2 + \frac{5.4.3}{3.2}(-2x)^3 + \frac{5.4.3.2}{4.3.2}(-2x)^4 + (-2x)^5$$

$$= 1 - 10x + 40x^2 - 80x^3 + 80x^4 - 32x^5$$

## Example 20.2

Find the coefficient of $x^3$ in the expansion of (i) $(2 + x)^7$    (ii) $(4 - 5x)^8$

➤ For the term in $x^3$, $r = 3$

(i)  $n = 7$, ∴ term in $x^3$ is ${}^7C_3 2^{7-3} x^3 = 35 \times 16 \times x^3$    ∴ coefficient of $x^3$ is 560

(ii)  $n = 8$, ∴ term in $x^3$ is ${}^8C_3 4^{8-3}(-5x)^3 = 56 \times 1024 \times (-5x)^3$   ∴ coefficient of $x^3$ is $-7168000$

## Example 20.3

Find the coefficient of $x^3$ in the expansion of  $(2 - 3x)(1 + 2x)^5$

In a question like this, the coefficient will be made up of 2 parts. This can be seen clearly if the original expression is first expanded by multiplying out the first bracket.

$$(2 - 3x)(1 + 2x)^5 = 2(1 + 2x)^5 - 3x(1 + 2x)^5$$

The term in $x^3$ can now be seen to be made up of 2 times the $x^3$ term of $(1 + 2x)^5$ minus $3x$ times the $x^2$ term of $(1 + 2x)^5$

Term in $x^3$ is ∴  $2 \times {}^5C_3 1^{5-3}(2x)^3 - 3x {}^5C_2 1^{5-2}(2x)^2 = 10 \times 8 x^3 - 3 \times 10 \times 4x^3$

Coefficient of $x^3 = 80 - 120 = -40$

## Example 20.4

Find the coefficient of $x^2$ in the expansion of  $\left(x + \dfrac{1}{x}\right)^6$

The term in $x^2$ will be the term which contains $x^4$ and $\left(\dfrac{1}{x}\right)^2$ (notice that the indices 4 and 2 add to 6)

For this term $n = 6, r = 2$ ∴ coefficient $= \begin{pmatrix} 6 \\ 2 \end{pmatrix} = 15$

# EXERCISE 20 A

Remember always to simplify the coefficients – they should be given as single number.

1. Find the first four terms in the expansion in ascending powers of $x$ of the following
   (a) $(1 + x)^{13}$
   (b) $(1 - x)^{15}$
   (c) $(1 + 3x)^{10}$
   (d) $(2 - 5x)^7$

2. Find the coefficient of $x^5$ in the expansion of each of the following
   (a) $(2 + x)^7$
   (b) $(3 - x)^8$
   (c) $(a - 3x)^9$
   (d) $(\frac{3}{2} + 2x)^9$

3. The coefficient of $x^3$ in the expansion of $(1 + px)^5$ is 80. Find $p$.

4. In the expansion of $(1 - ax)^{13}$, where $a$ is a positive constant, the coefficient of $x^2$ is 702.
   (i) Find the value of $a$.
   (ii) Evaluate the coefficient of $x^3$.

5. If the ratio of the coefficients of $x^6$ and $x^7$ in the expansion of $(2 + ax)^{11}$ is $14 : 25$ find the value of $a$. ✱

6. Given that the expansion of $(1 + ax)^n$ begins $1 + 36x + 576x^2$, find the values of $a$ and $n$.

7. Find the coefficient of $a^3b^5$ in the expansion of
   (i) $(3a - 2b)^8$
   (ii) $(5a + \frac{1}{2}b)^8$

8. Find the coefficient of $x^2$ in the expansion of $\left( x^4 + \dfrac{4}{x} \right)^3$

9. Find the term independent of $x$ in the expansion of   (i) $\left( 2x + \dfrac{5}{x} \right)^6$   (ii) $\left( x + \dfrac{1}{2x} \right)^8$.

10. In the expansion of $(1 - 2x)^{11}$ the coefficient of $x^3$ is $k$ times the coefficient of $x^2$.
    Evaluate $k$.

11. Find the coefficient of $a^4b^4$ in the expansion of $\left( a + \dfrac{b}{2} \right)^8$.

12. Evaluate the coefficient of $x^5$ in the expansion of $\left( x^2 - \dfrac{2}{x} \right)^7$.

13. Find the coefficient of $x^5$ in the expansion of $(1 - x)(1 + 2x)^8$

14. Find the term in $x^6$ in the expansion of $(1 + 2x)(2 + 3x)^6$

15. Evaluate the coefficient of $x^9$ in the expansion of $(1 + 2x)(3 + x)^{11}$

16. Find the coefficient of $x^3$ in the expansion of $(9 + 8x)\left(1 - \dfrac{x}{3}\right)^9$.

17. Given that the coefficient of $x$ in the expansion of $(1 + ax)(1 + 5x)^{40}$ is 207, determine the value of $a$.

18. If the coefficient of $x^3$ in the expansion of $(1 + ax)(1 + 2x)^6$ is $-20$, find the value of $a$.

19. In the expansion of $(a + bx)(1 - x)^4$ the first two terms are $2 + 3x$. Find $a$ and $b$.

20. In the expansion of the expression $(2 - x)(a + bx)^{10}$ the coefficient of $x^7$ is zero. Find, in its simplest form, the ratio $a : b$.

21. Given that the coefficient of $x^2$ in the expansion of $(4 + kx)(2 - x)^6$ is zero, find the value of $k$.

22. The coefficient of $x^3$ in the expansion of $(2 + ax)(1 - 3x)^6$ is 405, find the value of $a$.

23. The coefficient of $x^2$ in the expansion of $(2 + x)(1 - ax)^5$ is 0. Find the positive value of $a$.

24. Find the coefficient of $x^3$ in the expansion of $(10 - 7x)\left(1 + \dfrac{x}{5}\right)^8$.

25. $f(x) \equiv (1 + x)^4 - (1 - x)^4$.
    Expand $f(x)$ as a series in ascending powers of $x$.

#26. Expand $(2\sqrt{2} + \sqrt{3})^4$ in the form $a + b\sqrt{6}$, where $a$ and $b$ are integers.

## Example 20.5

Write out the first four terms in the expansion of $(1 + 2x)^8$ in ascending powers of $x$. Hence find the value of $(1.002)^8$ correct to 5 decimal places

$$(1 + 2x)^8 = 1 + 8(2x) + \tfrac{1}{2}{\cdot}8{\cdot}7\,(2x)^2 + \tfrac{1}{6}{\cdot}8{\cdot}7{\cdot}6\,(2x)^3 + \ldots$$

$$= 1 + 16x + 112x^2 + 448x^3 + \ldots$$

putting $x = 0.001$ gives

$$(1.002)^8 = (1 + 2(0.001))^8 = 1 + 16(0.001) + 112(0.001)^2 + 448(0.001)^3 + \ldots$$

$$= 1 + 0.016 + 0.000112 + 0.000000448 + \ldots$$

$$= 1.016112448 + \ldots$$

$$\approx 1.01611 \text{ correct to 5 d.p.}$$

N.B. it is clear that no further terms of the expansion are needed for this degree of accuracy, since even the term in $x^3$ does not affect the fifth decimal place.

# Example 20.6

Use the binomial theorem to find the value of $1.998^7$ correct to 5 decimal places.

➢ $1.998 = 2 - 0.002$, so find $(2 - 0.002)^7$

➢ For an approximation we only use the first few terms of an expansion. Using the first few terms of the expansion of $(2 - 0.002)^7$ as it stands will not give an accurate approximation. Why not?

*since 2 is greater than 1 sucessive terms get larger, not smaller*

➢ Rewrite the bracket so that the first number in the bracket is 1: $2^7(1 - 0.001)^7$

➢ Apply binomial theorem: $1.998^7 = 2^7\{1 + 7(-0.001) + 21(-0.001)^2 + 35(-0.001)^3 + \ldots\}$

➢ Simplify: $= 128(1 - 0.007 + 0.000021 - 0.000000035 + \ldots\} = 127.10668352$

➢ Give the answer: $1.998^7 = 127.10668$

# PROOF OF THE BINOMIAL THEOREM

We derived the binomial theorem intuitively, but obviously a rigorous proof is necessary. Since we know the result, an obvious method of proof would be induction; but for the alternative form it is simpler to use Maclaurin's theorem.

If    $f(x) = (1 + x)^n$,   then          $f(0) = 1$ and     $1 + nx + \dfrac{n(n-1)x^2}{2!}, \ldots$

$f'(x) = n(1 + x)^{n-1}$          $f'(0) = n$

$f''(x) = n(n-1)(1 + x)^{n-2}$          $f''(0) = n(n-1)$

$f'''(x) = n(n-1)(n-2)(1 + x)^{n-3}$          $f'''(0) = n(n-1)(n-2)$  etc

The first two conditions for the Maclaurin series are met – all the derivatives can be found and are defined when $x = 0$.

I  When $n \in \mathbb{W}$ the series is finite so convergence is not an issue.

substituting for $f(0)$, $f'(0)$ etc in Maclaurin's theorem when $n \in \mathbb{W}$ gives

$$(1 + x)^n = 1 + nx + \frac{n(n-1)}{2!}x^2 + \frac{n(n-1)(n-2)}{3!}x^3 + \ldots + x^n$$

or    $(1 + x)^n = 1 + {}^nC_1 x + {}^nC_2 x^2 + \ldots + {}^nC_r x^r + \ldots + x^n$   which is the binomial theorem.

II  When $n \in \mathbb{R}$ but is not a positive integer the coefficients in the series are not integers and so the series is not finite. We can no longer use the combinations notation, but defining the coefficient of the general term as

$\dbinom{n}{r}$ where $\dbinom{n}{r} = \dfrac{n(n-1)(n-2)\ldots(n-r-1)}{r!}$ we get

$$(1 + x)^n = 1 + \binom{n}{1}x + \binom{n}{2}x^2 + \binom{n}{3}x^3 + \ldots + \binom{n}{r}x^r \ldots + \ldots$$

or    $(1 + x)^n = 1 + nx + \dfrac{n(n-1)}{2!}x^2 + \dfrac{n(n-1)(n-2)}{3!}x^3 + \ldots$

This series must be convergent, and this is only the case when $|x| < 1$

$\boxed{!}$    When $n$ is not a positive integer, but is a negative number or fraction, the binomial expansion can only be used in its alternative form, with 1 at the start of the bracket, and the modulus of the second term in the bracket less than 1.

## Example 20.7

Use the binomial theorem to expand $(1-x)^{-1}$, $|x| < 1$, as a series of powers of $x$ up to the term in $x^4$.

Use this result to estimate $0.99^{-1}$ to 8 d.p.

$$(1-x)^{-1} = 1 + (-1)(-x) + \frac{(-1)(-2)}{2!}(-x)^2 + \frac{(-1)(-2)(-3)}{3!}(-x)^3 + \frac{(-1)(-2)(-3)(-4)}{4!}(-x)^4 + \ldots$$

$$= 1 + x + x^2 + x^3 + x^4 + \ldots$$

When $x = 0.01$, $0.99^{-1} = 1 + .01 + .0001 + .000001 + .00000001 + \ldots = 1.01010101$

## Example 20.8

Use the binomial theorem to write out the first 3 terms in the power expansion of $\sqrt{4-x}$, stating the values of $x$ for which the expansion is valid.

Hence estimate the value of $\sqrt{3}$ to 3 s.f.

➢  $\sqrt{4-x} = (4-x)^{\frac{1}{2}}$

➢  There is a problem here: since the power of the bracket is fractional, the number in the bracket must be 1. This means that the 4 must be factored out:

$$\sqrt{4-x} = \sqrt{4(1-\tfrac{1}{4}x)} = 2\sqrt{1-\tfrac{1}{4}x} = 2(1-\tfrac{1}{4}x)^{\frac{1}{2}}$$

$$= 2\left\{1 + \left(\frac{1}{2}\right)\left(-\frac{x}{4}\right) + \left(\frac{1}{2}\right)\left(-\frac{1}{2}\right)\left(\frac{-x}{4}\right)^2\left(\frac{1}{2!}\right) + \ldots\right\}$$

$$= 2\left\{1 - \left(\frac{x}{8}\right) - \left(\frac{x^2}{128}\right) + \ldots\right\} = 2 - \frac{1}{4}x - \frac{1}{64}x^2 + \ldots$$

➢  The expansion is valid for $|\tfrac{1}{4}x| < 1$, i.e. $|x| < 4$

➢  Putting $x = 1$ gives $\sqrt{3} = 1.734375 = 1.73$ to 3 s.f.

## Example 20.9

Portia has $50 000 dollars to save towards buying a house, which she hopes to do in 4 years time. She wants to be able to calculate how much her investment will be worth if she invests the money at compound interest but is not sure what the interest rate is likely to be. She therefore uses the binomial theorem, with an unknown interest rate of R% per annum, to give herself a basis for calculating the value of her investment after 4 years. What is the formula she derives?

Let $r = \tfrac{1}{100}R$, then $P = 50\,000(1 + r)^4$

$$= 50\,000(1 + {}^4C_1 r + {}^4C_2 r^2 + {}^4C_3 r^3 + r^4)$$

$$= 50\,000(1 + 4r + 6r^2 + 4r^3 + r^4)$$

# EXERCISE 20 B

1.  Use the expansion of $(1 + x)^5$ to find the **exact** value of $(1.001)^5$.

2.  (a) Expand $(2 + y)^5$ and work out the coefficients.
    (b) Deduce the value *exactly*, in decimal form, of (i) $(2.01)^5$ (ii) $(1.9)^5$.

3.  Expand $(1 - x)^{11}$ in ascending powers of $x$ up to and including the term in $x^4$. Hence evaluate $(0.98)^{11}$ correct to 3 decimal places.

4.  Use binomial expansions to evaluate the following to the stated degree of accuracy:
    (a) $(1.01)^7$   4 d.p.
    (b) $(0.99)^9$   6 d.p.
    (c) $(2.02)^6$   5 s.f.

5.  Write out the first 4 terms in the binomial expansion of
    (a) $(1 + x)^{-1}$          (b) $(1 + 2x)^{\frac{1}{2}}$          (c) $(8 - x)^{\frac{1}{3}}$

6.  Write out the first 4 terms in the binomial expansion of $(1 + x)^{-\frac{1}{2}}$.

    Hence evaluate $\dfrac{1}{\sqrt{1.2}}$ correct to 3 d.p.

7.  Write out the first 4 terms in the binomial expansion of $(4 + x)^{\frac{1}{2}}$.
    Hence evaluate $\sqrt{5}$ correct to 4 s.f

*8.  If $f(x) = \dfrac{5}{(1 + 3x)(1 - 2x)}$ express $f(x)$ in partial fractions.

    Hence, or otherwise, write out the first 4 terms of the binomial expansion of $f(x)$, stating the values of $x$ for which the expansion is valid. ✳

9.  Expand $\sqrt{\dfrac{1 - x}{1 + x}}$ up to and including the term in $x^3$, stating the values of $x$ for which the expansion is valid.

## SOME PROPERTIES OF THE BINOMIAL COEFFICIENT $_nC_R$

    Remember that $^nC_r$ is shorthand for $\dfrac{n!}{r!(n-r)!}$

Look again at the results of the investigation on page 140. It should be possible to generalise these results and prove their validity.

If Pascal's triangle is rewritten using combination notation, figure 20.2 becomes figure 20.3:

$$1$$
$$1 \quad 1$$
$$1 \quad 2 \quad 1$$
$$1 \quad 3 \quad 3 \quad 1$$
$$1 \quad 4 \quad 6 \quad 4 \quad 1$$
$$1 \quad 5 \quad 10 \quad 10 \quad 5 \quad 1$$
$$1 \quad 6 \quad 15 \quad 20 \quad 15 \quad 6 \quad 1$$

Fig. 20.2

$${}^{0}C_{0}$$
$${}^{1}C_{0} \quad {}^{1}C_{1}$$
$${}^{2}C_{0} \quad {}^{2}C_{1} \quad {}^{2}C_{2}$$
$${}^{3}C_{0} \quad {}^{3}C_{1} \quad {}^{3}C_{2} \quad {}^{3}C_{3}$$
$${}^{4}C_{0} \quad {}^{4}C_{1} \quad {}^{4}C_{2} \quad {}^{4}C_{3} \quad {}^{4}C_{4}$$
$${}^{5}C_{0} \quad {}^{5}C_{1} \quad {}^{5}C_{2} \quad {}^{5}C_{3} \quad {}^{5}C_{4} \quad {}^{5}C_{5}$$
$${}^{6}C_{0} \quad {}^{6}C_{1} \quad {}^{6}C_{2} \quad {}^{6}C_{3} \quad {}^{6}C_{4} \quad {}^{6}C_{5} \quad {}^{6}C_{6}$$

Fig. 20.3

The symmetry of the numbers in fig. 20.2 suggests that ${}^{n}C_{n-r} = {}^{n}C_{r}$. Is this always so?

Also, it appears that each number in fig. 20.2 can be obtained by adding the two numbers diagonally above it, i.e. ${}^{n}C_{r} = {}^{n-1}C_{r-1} + {}^{n-1}C_{r}$. Again, is this always true?

## Example 20.9

Prove (i) ${}^{n}C_{n-r} = {}^{n}C_{r}$   (ii) ${}^{n}C_{r} = {}^{n-1}C_{r-1} + {}^{n-1}C_{r}$

(i) $\displaystyle {}^{n}C_{n-r} = \frac{n!}{(n-r)!\{n-(n-r)\}!}$

$$= \frac{n!}{(n-r)!\,r!}$$
$$= {}^{n}C_{r}$$

(ii) $\displaystyle {}^{n-1}C_{r-1} + {}^{n-1}C_{r} = \frac{(n-1)!}{(r-1)!(n-r)!} + \frac{(n-1)!}{r!(n-r-1)!}$

$$= \frac{(n-1)!}{r!(n-r)!}\{r+(n-r)\}$$

$$= \frac{n(n-1)!}{r!(n-r)!}$$

$$= \frac{n!}{r!(n-r)!}$$

$$= {}^{n}C_{r}$$

# EXERCISE 20 C

Using factorials, and remembering that $0! = 1$

1. Show that (a) $^rC_0 = 1$,      (b) $^rC_r = 1$,      (c) $^rC_1 = r$   for all $r \in \mathbb{W}$

2. Show that $^nC_r + {}^nC_{r-1} = {}^{n+1}C_r$

3. Show that $^nC_{r+1} = {}^nC_r \left( \dfrac{n-r}{r+1} \right)$

4. Show that $^{n+1}C_r = {}^nC_r \left( \dfrac{n+1}{n-r+1} \right)$

5. Show that $^{n+1}C_{r+1} = {}^nC_r \left( \dfrac{n+1}{r+1} \right)$

# EXERCISE 20 X

1. Calculate the value of $k$ if the coefficient of $x^6$ in the expansion of $(4 + kx)^{10}$ is 840. [3]

2. Find the coefficient of $x^6$ in the expansion of $(1 - 3x)(1 + 2x)^9$ as a series of ascending powers of $x$. [6]

3. Find the term independent of $x$ in the expansion of $\left( 2x^3 - \dfrac{1}{x} \right)^8$ [5]

4. Without expanding $\left( \dfrac{3x^2}{2} - \dfrac{1}{3x} \right)^9$ completely, find

   (a) the term independent of $x$. [4]

   (b) the coefficient of $x^6$. [2]

5. The coefficient of $x^5$ in the expansion of $(1 + 5x)^8$ is equal to the coefficient of $x^4$ in the expansion of $(a + 5x)^7$. Find the value of $a$. [5]

6. (a) Write down, and simplify as far as possible, the FIRST THREE terms in the expansion of $(1 - ux)(2 + x)^4$. [7]

   (b) Given that the coefficient of $x^2$ is 0, find

      (i) the value of $u$

      (ii) the coefficient of $x$. [2]

7. Find the term independent of $x$ in the expansion of $\left(3x - \dfrac{1}{2x^2}\right)^9$. [5]

8. If $(1 + ax)^n \equiv 1 + 6x + 16x^2 + \ldots$ find the values of the constants $a$ and $n$. [6]

9. Given that the coefficient of $x^2$ is zero in the binomial expansion of $(1 - ax)(1 + 2x)^5$, find the value of $a$ and the coefficient of $x^3$. [12]

10. (a) Express the binomial coefficients $^nC_k$ and $^nC_{k-1}$, for $n > k \geq 1$, in terms of factorials. [2]

    (b) Hence, show that

       (i) $^nC_k = {}^nC_{n-k}$ [2]

       (ii) $^nC_k + {}^nC_{k-1} = {}^{n+1}C_k$. [6]

11. The coefficient of $x^7$ is sixteen times the coefficient of $x^{11}$ in the expansion of $(a + x)^{18}$.

Given that $^nC_k = {}^nC_{n-k}$, find the possible value(s) of the real constant $a$. [7]

12. By substituting suitable values of $x$ on both sides of the expansion of $(1 + x)^n = \displaystyle\sum_{r=0}^{n} {}^nC_r\, x^r$, show that

    (a) $\displaystyle\sum_{r=0}^{n} {}^nC_r = 2^n$ [2]

    (b) $\displaystyle\sum_{r=0}^{n} {}^nC_r(-1)^r = 0$ [2]

13. If $^{(x-2)}C_2 = \frac{5}{2}\left({}^4C_3\right)$, $x > 3$,

    (a) show that $x^2 - 5x - 14 = 0$ [4]

    (b) find $x$. [2]

14. (a) Expand $(1 + ux)(2 - x)^3$ in powers of $x$ up to the term in $x^2$, $u \in \mathbb{R}$. [6]

    (b) Given that the term in $x^2$ is zero, find the value of $u$. [2]

# What you have learnt in Chapter 20

## *The Binomial Coefficient*

$$^nC_r = \frac{n!}{r!(n-r)!} = \binom{n}{r}$$

$$^nC_r = {}^nC_{n-r}$$

## *The Binomial Theorem:*

The Binomial theorem states that

for $n \in \mathbb{N}$

$$(a+b)^n = a^n + {}^nC_1 a^{n-1}b + {}^nC_2 a^{n-2}b^2 + \ldots + {}^nC_r a^{n-r}b^r + \ldots + b^n$$

$$= \sum_{r=0}^{n} {}^nC_r\, a^{n-r}b^r$$

or $(1+x)^n = 1 + nx + \dfrac{n(n-1)}{2!}x^2 + \dfrac{n(n-1)(n-2)}{3!}x^3 + \ldots + x^n$

for $n \in \mathbb{R}$  $|x| < 1$

$$(1+x)^n = 1 + nx + \frac{n(n-1)}{2!}x^2 + \frac{n(n-1)(n-2)}{3!}x^3 +$$

or $(1+x)^n = 1 + \binom{n}{1}x + \binom{n}{2}x^2 + \binom{n}{3}x^3 + \ldots + \binom{n}{r}x^r + \ldots$

# CHAPTER 21
# 2.2d Errors

## ERRORS

When taking measurements, and when rounding off figures from a calculator, approximations, and therefore errors, are introduced. If further calculations are involved, the error may be compounded, so it is often useful to know the extent of the possible final error.

Now suppose a measurement is made, and there is an error of 3 cm. How serious is this? Obviously the answer depends on what the measurement is of. An error of 3 cm when marking out a 100 m race track is of little significance, while an error of 3 cm when deciding whether a piece of equipment 95 cm long will fit in a given space could be crucial. We therefore need certain:

## DEFINITIONS

If the true value of a quantity is $x$, and the approximate value is $a$, then the actual error, $e$, is $a - x$. Since it is the size, not the sign, of the error that is usually important we define

the **Absolute Error**, $\varepsilon$, is the modulus of the difference between the true value and the approximate value: $\varepsilon = |a - x|$,

the **Relative Error** is the ratio of the absolute error to the true value: relative error $= \dfrac{\varepsilon}{x}$, (if the true value is not known a similar result can be obtained by replacing the true value with the approximate value),

the **Percentage Error** is relative error $\times 100\%$.

Given that errors are involved in the numbers used for these calculations, 2 significant figures (or even 1 s.f.) is usually considered accurate enough for answers.

## Example 21.1

A garden plot measuring 12.3 m by 8.6 m was estimated by the gardener to be 12.5 m by 8.5 m. Find

(a)    the error in (i) the area and (ii) the perimeter of the plot, and

(b)    the percentage errors of each to one decimal place.

>     True area of the plot $= 12.3 \times 8.6 = 105.78$ m²,        true perimeter $= 41.8$ m

>     Estimated area of the plot $= 12.5 \times 8.5 = 106.25$ m²,    estimated perimeter $= 42$

>     Absolute error in area $= |105.78 - 106.25| = 0.47$ m²,   absolute error in perimeter $= 0.2$ m

>     % error in area $= \dfrac{0.47}{105.78} \times 100 = 0.44\% \approx 0.4\%$,

>     % error in perimeter $= \dfrac{0.2}{41.8} \times 100 = 0.478 \approx 0.5\%$

## Example 21.2

A cricket ball is to be covered with leather. The diameter is estimated to be 7 cm, but is in fact 7.2 cm.

Calculate the relative error in the estimated area of leather needed to cover the ball.

true area $= 4\pi r^2 = 51.84\pi$

estimated area $= 49\pi$

absolute error $= 2.84\pi$

relative error $= \dfrac{2.84\pi}{51.84\pi} = 0.0548 \approx 0.055$

 Notice that the $\pi$ cancels when finding relative or % errors, so should not be entered as a decimal

### *Rounding errors*

When a figure is given correct to, say, 1 decimal place, the true value may not be known, and therefore the absolute error cannot be found. In such cases we are concerned with error bounds.

A height measured as 1.26 m could in reality be anywhere between 1.255 and 1.265 m. In other words, the measurement allows a possible error of $\pm\,0.005$, i.e. a maximum absolute error of 0.005.

 The minimum and maximum possible values (1.255 and 1.265 in the above example) are known as the **error bounds.**

## Example 21.3

A wall is measured as 4.8 m long by 3.1 m high, correct to 1 d.p. Find the error bounds for the area of the wall, and the maximum percentage error in the area. Give answers correct to 1 d.p.

Length of wall $= 4.8 \pm 0.05$,  height of wall $= 3.1 \pm 0.05$

Maximum area of wall $= 4.85 \times 3.15 = 15.2775$ m²

Minimum area of wall $= 4.75 \times 3.05 = 14.4875$ m².

Error bounds are  $14.5 < A < 15.3$

'Measured area' $= 4.8 \times 3.1 = 14.88$

Maximum % error $= \dfrac{15.2775 - 14.88}{14.88} \times 100 = 2.671 \approx 2.7\%$

This example introduces the idea of ***Maximum Percentage Error***. This arises for two reasons:

(i)   the error in the measurement is not actually known – it can be anything up to $|\,0.05\,|$, so that is the **maximum** rounding error,  but also

(ii)  using the maximum rounding error, the difference between the 'measured area' ($m$) and the maximum possible area ($d_1$) and between the 'measured area' and the minimum possible area ($d_2$) are slightly different: $d_1 - m = 0.3975$, while $m - d_2 = 0.3925$. The difference occurs because in finding the area **two** measurements involving errors are being combined.

The next example shows how care must be taken when combining errors.

## Example 21.4

Wesley is making a model aeroplane; he wants to cut a strut of wood 31 cm long. He buys a 50 cm dowel, and measures 19 cm to cut off it. Find the range of possible lengths of his strut if the measurements are both correct to the nearest cm. Hence find the maximum relative error in the length of the strut.

Dowel: $49.5 \le l < 50.5$

Offcut: $18.5 \le x < 19.5$

To find the maximum length of the strut you need the *greatest* length of dowel and the *least* length of offcut M = 50.5 − 18.5 = 32

For the minimum length of the strut you need the *least* length of dowel and the *greatest* length of offcut m = 49.5 − 19.5 = 30

   30 < strut < 32

maximum absolute error = 32 − 31 = 1  (or 31 − 30)

i.e. strut could be up to 1 cm too long or too short.

maximum relative error = $\dfrac{1}{31}$ = 0.032

# EXERCISE 21 A

Unless otherwise stated, give relative and percentage errors correct to 2 significant figures.

1.  A radio component in the form of a cuboid is measured to be 1.23 cm by 0.51 cm by 0.46 cm. It's true size is 1.24 cm by 0.55 cm by 0.48 cm. Find the percentage error in the volume of the component.

2.  The title page of a book is 22.8 cm by 14.3 cm. An artist produces a design which is 23 cm by 14.5 cm. Find the relative error in the area of the artist's design.

3.  The dimensions of a playing field are given as 48 m long and 37 m wide, correct to the nearest m. Find the maximum absolute error in

    (a)  the perimeter of the field

    (b)  the area of the field

4.  A running track is constructed so that the inside of the inside lane is supposed to consist of two straight sections of length 100 m and two semicircles of radius 32 m. When the track is constructed the straight section turns out to be 100.3 m long and the radius of the semicircle 32.5 m. Find the absolute error in the perimeter of the inside of the track.

The area inside the track is to be covered with grass. The area is calculated so that the correct amount of seed can be purchased. If the calculation is based on the intended dimensions, what is the relative error in the area?

5.  The deck of a ferry is 30 m long. The ferry is to transport new cars from the manufacturer who estimates the length of a car to be 3.3 m to 1 decimal place.

    (a)  What are the error bounds for the length of one car?

    (b)  What is the maximum number of cars the ferry can be certain of fitting on to the deck?

6. Dotha measures the length of a die and finds that it is $\frac{3}{16}$ inch. She uses her calculator to write the length as a decimal *which she then corrects to 3 s.f.*

   (a) What is the percentage error in the value for the length which she records?

   (b) She uses this value to calculate the volume of the die. What is the relative error in the volume of the die?

✱ 7. Dwayne wants to add a shelf to his bookcase. The distance between the uprights is 125 cm, and he buys a shelf 125 cm long. Both measurements are correct to the nearest cm. What is the maximum amount that he might have to cut off the end of the shelf so that it will fit?

8. Marcus and Chris are asked to find the value of $0.98^4$. Chris uses his calculator, but Marcus has mislaid his, and he approximates the answer by writing down the FIRST THREE terms of the binomial expansion of $(1-x)^4$ and substituting a suitable value for $x$.

   (a) What approximate value for $0.98^4$ does Marcus get?

   (b) To how many significant figures is his answer correct?

   (c) What is his percentage error?

9. Given that, for a small angle $t$ radians, $\sin(\frac{1}{4}\pi + t) \approx \frac{1}{\sqrt{2}}(1 + t - \frac{1}{2}t^2)$, use your calculator to estimate the absolute error in using this approximation when $t = 0.1$, giving your answer to 5 decimal places.

10. (a) Expand $(1+2x)^{12}$ in ascending powers of $x$ up to and including the term in $x^3$, simplifying each coefficient.

    (b) By substituting a suitable value for $x$, which must be stated, into your answer to part (a), calculate an approximate value of $(1.02)^{12}$.

    (c) Use your calculator, writing down all the digits in your display, to find a more exact value of $(1.02)^{12}$.

    (d) Calculate, to 3 sf, the percentage error of the approximation found in part (b).

# EXERCISE 21 X

1. A craftsman estimated the side of a square tile to be 15 cm, but the actual measurement was 15.15 cm. Calculate, to two significant figures, the percentage error in the actual measurement of the area of the tile. [4]

2. The true length and breadth of a rectangular lawn are known to be 50 m and 35 m respectively. If the length is measured as 50.3 m and the breadth as 35.5 m, calculate

   (a) the error in the area of the lawn [3]

   (b) the percentage error in the area, correct to the nearest 1 decimal place. [2]

3.  A sphere of radius 10 cm is manufactured. If the manufacturing process resulted in a possible error of $\pm\,0.01$ cm for the radius, estimate, using differential calculus or otherwise

    (a)   the relative error in the radius                                                                      [4]

    (b)   the percentage error in the volume of the sphere.                                       [2]

4.  The length of a rectangle is given as 26 cm and the width is given as 15 cm, both correct to the nearest cm. Find the maximum absolute error in the area of the rectangle.       [7]

5.  A craftsman estimated the side of a square tile to be 16 cm, but the actual measurement was 16.14 cm. Calculate, to two decimal places, the percentage error in the actual measurement of the area of the tile.                                                                                      [4]

6.  The length of one side of a square tile is given as 25 cm correct to the nearest cm.

    (a)   Within what limits does the length of the tile lie?                                         [1]

    (b)   Calculate

        (i)    the SMALLEST absolute error in the area of the tile                          [3]

        (ii)   the MAXIMUM percentage error in the area of the tile.                       [3]

# What you have learnt in Chapter 21

If the true value of a quantity is $x$, and the approximate value is $a$, then the actual error, $e$, is $a - x$.

the ***Absolute Error*** $\varepsilon = |a - x|$,

the ***Relative Error*** $= \dfrac{\varepsilon}{x}$,

the ***Percentage Error*** = relative error × 100%.

the ***Error Bounds*** of an inexact quantity are the minimum and maximum possible values that the quantity can have

# CHAPTER 22
# 2.2e  Roots of Equations

Most equations of the form $f(x) = 0$ (e.g. $x \sin x - 1 = 0$) do not have exact solutions which can be determined by algebraic manipulation. The roots can of course be found approximately by accurate plotting of the graph of $y = f(x)$ for all values of $x$, but this is not usually a viable method. In this chapter we shall look at methods that enable roots of such equations to be found numerically to any required degree of accuracy. We shall see that we can sometimes know how many roots an equation has, how to determine the approximate location of the roots, and, knowing one approximation to a root, how to obtain more accurate approximations.

Suppose you climb a mountain, whose peak is 1500 m above sea level, and you start climbing from a point at a height of 550 m above sea level. Will you, at any time in your climb to the peak, be at a height of 1021 m? Obviously you must go through that height at some point – you might well do so more than once if there is a valley between you and the summit and you have to drop down into the valley before rising up again.

This is the idea embodied in

## THE INTERMEDIATE VALUE THEOREM

which states that

 If $y = f(x)$ is a continuous function in the interval $a < x < b$, and if $k$ is a number which lies between the values of $f(a)$ and $f(b)$ (either $f(a) < k < f(b)$ or $f(b) < k < f(a)$), then there exists at least one number $c$ such that $a < c < b$ and $f(c) = k$.

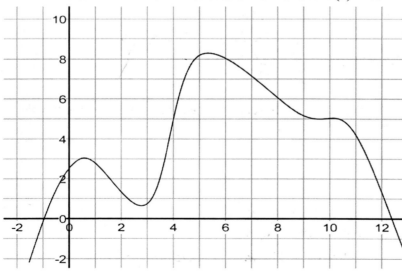

Fig. 22.1

For the graph in the diagram, fig. 22.1, $f(-1) = 0$ and $f(10) = 5$. The theorem says that if we choose any $y$-value between 0 and 5 we will be able to find at least one value for $x$ between $-1$ and 10 which gives that value.

For instance, is there a point on the graph in the range $-1 < x < 10$ whose $y$-coordinate is 4? The theorem says there is, and the graph confirms that it has an $x$-coordinate of about 3.8.

Is there a point in the range whose $y$-coordinate is 2? The theorem says there is, and the graph confirms that there is not just one, but 3, whose $x$-coordinates are approximately –0.75, 1.5, 3.5.

The Intermediate Value Theorem has a useful application for the **location of roots of an equation**. If we can show that the function $f(x)$ changes sign between two given values of $x$, then the graph of $y = f(x)$ must have crossed the $x$ axis between those two values, which means that the equation $f(x) = 0$ has at least one root between the two values.

 This is best expressed in the form: If a function $y = f(x)$ is continuous in the interval $a \leq x \leq b$, and if $f(a) . f(b) < 0$, then the equation $f(x) = 0$ has a root between $a$ and $b$.

# Example 22.1

Show that the equation $x^2 - 3x - 1 = 0$ has a root between 3 and 4.

> Let $f(x) = x^2 - 3x - 1$, then $f(x)$ is continuous on [3, 4]
> and $f(a) = f(3) = 9 - 9 - 1 = -1$ and $f(b) = f(4) = 16 - 12 - 1 = 3$
> $f(a) . f(b) = -1 \times 3 = -3 < 0$
> $\therefore x^2 - 3x - 1 = 0$ has a root between 3 and 4

In this case this can be confirmed using the quadratic formula,

which gives one root of the equation as 3.303.

It can also be illustrated on a sketch graph

Fig. 22.2

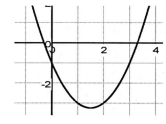

The method is particularly useful when there is no formula for solving the equation and so an exact root cannot be found.

# Example 22.2

Given that the 3 roots of the equation $12x^3 - 112x^2 + 267x - 77 = 0$ all lie in the range $0 < x < 10$, find the integral values of $x$ between which each of these roots lie.

Let $f(x) = 12x^3 - 112x^2 + 267x - 77$, which is continuous on [0, 10]

then $f(0) = -77$, $f(1) = 90$, $f(2) = 105$, $f(3) = 40$, $f(4) = -33$, $f(5) = -42$, $f(6) = 85$

from this we see that $f(0).f(1) < 0$, $f(3).f(4) < 0$ and $f(5).f(6) < 0$,

so the three roots lie in the intervals [0, 1], [3, 4] and [5, 6].

> **!** When applying the intermediate value theorem, you must always ensure and state that the function is continuous in the required interval.

# EXERCISE 22A

N.B. Questions involving trig functions should be worked in **radians**.
       All the given functions are continuous over the stated range.

In questions 1 – 4 show that the equation $f(x) = 0$ has a root between the stated values.

1. $f(x) = x^3 + x - 11$, between 2 & 3

2. $f(x) = x^2 - \cos x$, between 0 & 1

3. $f(x) = \dfrac{6}{x} - 2^x$, between 1 & 2

4. $f(x) = x - 3\sin x$ between 2 & 3.

5. By finding the value of $f(x)$ for different integral values of $x$ deduce an approximate root for the equations $f(x) = 0$ when

   (i) $f(x) = x^3 - 8x - 4$, $x > 0$          (ii) $f(x) = \dfrac{2x - 5}{x + 1}$, $x > 0$

6. Show that the equation $(x + 1)^5 - (x + 2)^3 - 4 = 0$ has a root in the interval $[0.8, 1]$.

7. Show that the equation $2\sin x + x - 3 = 0$ has a root in the interval $[1.1, 1.2]$.

## APPROXIMATE LOCATION OF ROOTS

If it is thought that there is a root of an equation within a particular interval, the intermediate value theorem can be used to check whether this is so. (However, it is possible for the intermediate value theorem not to pick up the presence of a root. If there are an *even* number of roots within the specified interval $[a, b]$ then $f(a).f(b)$ will be positive, since the graph of $y = f(x)$ will have crossed the $x$-axis an even number of times in the interval.)

Although many functions $y = f(x)$ cannot be sketched, it is often possible to rearrange the equation $f(x) = 0$ so that the roots of the equation are represented by the points of intersection of two different functions.

Consider the equation mentioned above, $x \sin x - 1 = 0$. How many roots does this equation have in the interval $0 \le x \le 2\pi$?

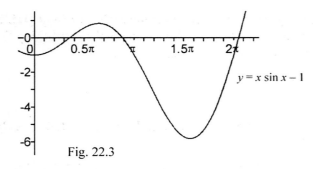

Fig. 22.3

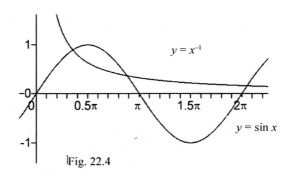

Fig. 22.4

Figure 22.3 shows the actual graph of $y = x \sin x - 1$, but plotting this would take a lot of work. The equation $x \sin x - 1 = 0$ can be rearranged to give $x \sin x = 1$, which can be further rearranged as $\sin x = x^{-1}$. Now $y = \sin x$ and $y = x^{-1}$ are both curves whose graphs can easily be sketched, and this has been done in figure 22.4. These curves intersect twice in the interval $[0, 2\pi]$. These points of intersection are the points where $\sin x = x^{-1}$, so they give the roots of the equation $x \sin x - 1 = 0$. From the sketch we can see that the roots probably lie in the intervals $[0.3\pi, 0.5\pi]$ and $[0.8\pi$ and $\pi]$. This can be verified using the intermediate value theorem. We have used a sketch to find the number of roots and their approximate locations.

A polynomial function $y = f(x)$ of $n$th degree (i.e. $x^n$ is the highest power of $x$) has a maximum number of turning points of $n - 1$. $n$ is therefore the maximum possible number of roots of the equation $f(x) = 0$. Various strategies can be used to show that under certain circumstances there are fewer roots:

(i) If $y = f(x)$ is a strictly increasing or strictly decreasing function, the graph can only cross the $x$-axis once, and there is therefore one and only one root to the equation $f(x) = 0$. Remember that for an increasing function $f'(x) > 0$, and for a decreasing function $f'(x) < 0$.

(ii) If the stationary points can be found, then a point of inflexion reduces the number of possible roots by 2; a maximum point where $f(x) < 0$ or a minimum point where $f(x) > 0$ also reduces the number of possible roots by 2.

# FINDING BETTER APPROXIMATIONS TO ROOTS OF EQUATIONS – ITERATION FORMULAE

## I    *Using the Intermediate Value Theorem*

When it is already known that an equation has a root within a given interval, that interval can be made progressively smaller by repeated use of the intermediate value theorem. Such repetitions of a process are known as iterations. This can be done in a number of ways: by interval bisection – dividing the interval in half and deciding in which half the root lies; or by linear interpolation – using a formula to divide the interval at a point which is closer to the root than the mid-point is, or by a combination of these. These methods will enable the root to be found to any degree of accuracy, but can be quite slow.

Examples 22.3 and 22.4 both solve the same problem: to find the root of the equation $x^2 + 8x - 10 = 0$ which lies between 1 and 2. The two examples arrive at the same result but 22.3 uses only the intermediate value theorem, and 22.4 uses a combination of the ivt and another method outlined below.

## Example 22.3

(a)    Show that the equation $x^2 + 8x - 10 = 0$ has a root, $\alpha$, between 1 and 2.

(b)    By repeated use of the intermediate value theorem, find the value of $\alpha$ correct to 1 d.p.

    (a)    ➤   Let $f(x) = x^2 + 8x - 10$.       $f(x)$ is continuous on [1, 2]

         ➤   Find $f(1)$ and $f(2)$:    $f(1) = -1 < 0$, $f(2) = 10 > 0$

         ➤   Apply i.v.t:    $f(1) \cdot f(2) < 0$, $\therefore$ by intermediate value theorem $f(x) = 0$ has a root in the interval [1, 2]

    (b)   The interval **bi**section method entails finding $f(1.5)$ to test in which of the intervals [1, 1.5] or [1.5, 2] $\alpha$ lies; but from the values we have already obtained for $f(1)$ and $f(2)$ we can see that $\alpha$ probably lies much closer to 1 than to 2. We can therefore save time by finding e.g. $f(1.2)$

        $f(1.2) = 1.04 > 0$. Now $f(1) = -1 < 0$, so $\alpha$ lies in [1, 1.2]

        $f(1.1) = 1.01 > 0$, so $\alpha$ lies in [1, 1.1]

        At this stage we can see that the answer to 1 d.p. must be either 1.0 or 1.1. How can we tell, quickly, which is the correct answer? By testing $f(1.05)$

        $f(1.05) = -0.4975 < 0$. $\alpha$ must therefore lie 'to the right of' 1.05 and so be closer to 1.1 than 1.0

        $\alpha = 1.1$ to 1 d.p. (the quadratic formula gives $1.0990 \approx 1.1$)

## II    *The Newton-Raphson Method*

✳      Suppose it is known that $x_1$ is an approximation to a root $\alpha$ of the equation $f(x) = 0$, where $f(x)$ is a continuous, differentiable function in the interval containing $x_1$ and $\alpha$.
The sketch, fig 22.5 shows the curve $y = f(x)$ in the vicinity of the root.

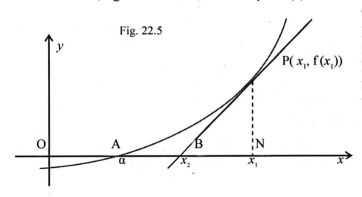

Fig. 22.5

The curve cuts the $x$-axis at A $(\alpha, 0)$
P is the point on the curve whose $x$- coordinate is $x_1$ so P is the point $(x_1, f(x_1))$
Let N $(x_1, 0)$ be the foot of the perpendicular from P to the $x$-axis, and let the tangent at P cut the $x$-axis at B $(x_2, 0)$

If the tangent at P makes an angle $\theta$ with the $x$-axis, then $\tan \theta = f'(x_1)$,

but in triangle PBN $\quad \tan \theta = \dfrac{PN}{BN} = \dfrac{f(x_1)}{x_1 - x_2}$, so $\dfrac{f(x_1)}{x_1 - x_2} = f'(x_1)$ . $\qquad ①$

Rearranging $①$ gives $\quad x_2 = x_1 - \dfrac{f(x_1)}{f'(x_1)}$

Now B is closer to A than is N, so $x_2$ is a better approximation to $\alpha$ than $x_1$.

> If $x_r$ is an approximation to a root $\alpha$ of the equation $f(x) = 0$, then
>
> $$x_{r+1} = x_r - \frac{f(x_r)}{f'(x_r)}$$
>
> gives a better approximation to $\alpha$.

## Example 22.4

(a) Show that the equation $x^2 + 8x - 10 = 0$ has a root, $\alpha$, between 0 and 1.

(b) Taking the number 1 as the first approximation for $\alpha$, use the Newton-Raphson method 3 times to find a better approximation to $\alpha$.

    (a)       see example 22.3

    (b)       $f(x) = x^2 + 8x - 10$, $f'(x) = 2x + 8$, $x_1 = 1$

$$x_2 = x_1 - \frac{f(x_1)}{f'(x_1)} = 1 - \frac{-1}{10} = 1.1$$

$$x_3 = x_2 - \frac{f(x_2)}{f'(x_2)} = 1.0990196, \quad \text{similarly,} \quad x_4 = 1.0990195$$

$\alpha = 1.09902$ to 6 s.f.

!    Notice how quickly the approximation converges with the Newton-Raphson method.

Where a function is easily differentiable the Newton-Raphson method is therefore usually the easiest and most reliable method to find approximations to roots, especially if a high degree of accuracy is required.

As long as the first approximation is close enough to the actual root, the sequence of approximations generated by the Newton-Raphson method will usually converge quickly, but under certain circumstances the sequence may diverge. Situations to beware of include:

(i) the first approximation, $x_1$, is too far from the actual root, $\alpha$, – there might be a turning-point or a discontinuity between the approximation and the root,

(ii) $f'(x_1)$ is too small or 0.

(iii) $f'(x)$ increases too rapidly.

## Example 22.5

The function $f$ is given by $f(x) = x^3 + 4x + 3$

(a) show that

    (i) $f$ is a strictly increasing function

(ii)   the equation  $f(x) = 0$ has a real root  $\alpha$  in the interval  $[-1, 0]$

(iii)   the equation  $f(x) = 0$ has no other root in the interval  $[-1, 0]$

(iv)   by choosing  $x_0 = -1$ as an initial approximation to the root  $\alpha$, that a better approximation of $\alpha$ is  $x_1 = -0.67$

(b)   Let  $x_n$  be the $n$th approximation of $\alpha$.  Find $x_{n+1}$ in terms of $x_n$.

(a)   (i)   ➤   Differentiate f:  $f'(x) = 3x^2 + 4$

   ➤   Apply principle for increasing functions:  $x^2 \geq 0$ for all $x$,  $\therefore$  $f'(x) > 0$ for all $x$,
      $\therefore$ f is a strictly increasing function

(ii)   ➤   Use i.v.t:      f is continuous on $[-1, 0]$
      $f(-1) = -2 < 0$, $f(0) = 3 > 0$  $\therefore$  $f(-1)\, f(0) < 0$
      $\therefore$  by intermediate value theorem $f(x) = 0$ has a root in the interval $[-1, 0]$

(iii)   ➤   Since f is strictly increasing it has just one root, it can therefore have no other root in the interval $[-1, 0]$ (nor anywhere else)

(iv)   ➤   Apply N-R method:   $x_2 = x_1 - \dfrac{f(x_1)}{f'(x_1)}$

$$= -1 - \frac{-2}{7} = -0.7142857$$

$$x_3 = x_2 - \frac{f(x_2)}{f'(x_2)} = -0.674222$$

$$= -0.674222$$

similarly  $x_4 = -0.6735932$

   ➤   give answer:   $\alpha$ is $-0.67$ to 2 d.p.

(b)   ➤   substitute for f and f′ in  Newton-Raphson formula:

$$x_{n+1} = x_n - \frac{f(x_n)}{f'(x_n)}$$

$$= x_n - \frac{x_n^3 + 4x_n + 3}{3x_n^2 + 4}$$

   ➤   simplify:      $x_{n+1} = \dfrac{x_n(3x_n^2 + 4) - (x_n^3 + 4x_n + 3)}{3x_n^2 + 4}$

$$= \frac{2x_n^3 - 3}{3x_n^2 + 4}$$

# Example 22.5

(a)   Show that the equation  $3x^3 = 5x^2 + 1$ has a root, $\alpha$, in the interval $[1, 2]$.

(b)   Show that if 1 is taken as the first approximation to $\alpha$ the Newton-Raphson method fails.

(c)   Taking 2 as the first approximation to $\alpha$, use the Newton-Raphson method twice to find $\alpha$ correct to 2 d.p.

(a)   ➤   rearrange the original equation:  $3x^3 - 5x^2 - 1 = 0$

- ➤ let $f(x) = 3x^3 - 5x^2 - 1$
- ➤ apply i.v.t: $f(x)$ is continuous on $[1, 2]$

$$f(1) = -3 < 0, \quad f(2) = 3 > 0$$

$$\therefore f(1) \cdot f(2) < 0$$

$\therefore$ by i.v.t $f(x) = 0$ has a root in the interval $[1, 2]$

(b) ➤ find f'(x): $f'(x) = 9x^2 - 10x$

➤ apply N-R: $x_1 = x_0 - \dfrac{f(x_0)}{f'(x_0)}$ where $x_0 = 1$

$$= 1 - \frac{-3}{-1} = -2$$

➤ apply N-R again: $x_2 = x_1 - \dfrac{f(x_1)}{f'(x_1)}$

$$= -2 - \frac{-45}{56} = -1.2$$

➤ comment: this sequence of 'approximations' is clearly not converging in the interval $[1, 2]$, so the N-R method starting with $x_0 = 1$ will not lead to a good approximation to $\alpha$.

(c) ➤ apply N-R: $\qquad x_1 = x_0 - \dfrac{f(x_0)}{f'(x_0)}$, where $x_0 = 2$

$$= 2 - \frac{3}{16} = 1.8125$$

➤ apply N-R again: $\qquad x_2 = x_1 - \dfrac{f(x_1)}{f'(x_1)}$

$$= 1.8125 - \frac{0.437}{11.44} = 1.774$$

➤ give answer: $\alpha = 1.77$ to 2 d.p.

Why did N-R not work in part (b)?

A sketch might help. Putting $f'(x) = 0$ shows that $y = f(x)$ has turning points at $x = 0$ and $x = \frac{10}{9} = 1.\dot{1}$

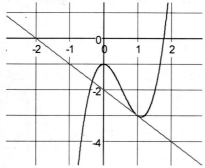

The minimum point therefore lies in the interval $[1, 2]$.

Drawing the tangent at $(1, -3)$ leads to the result found in part (b) that if $x_0 = 1$, then $x_1 = -2$. For the N-R method to work, the tangent must be drawn to the right of the minimum point. In part (c) we used $x_0 = 2$, but any value for $x_0 > 10/9$ would have been satisfactory.

Can you see why N-R would not work for $x_0 = 10/9$ ?

A note on using your calculator when using iteration formulae.

Most scientific calculators can be programmed to work iteration formulae very simply. If this method does not work on your calculator, read your instructions to find out how it can be done.

➤ Enter the value of $x_0$ followed by $=$

➤ Now enter the iteration formula using Ans in place of $x_0$ (Make sure you enter the formula as a single expression, using brackets, etc to make this possible)

➤ The first time you press $=$ you will get $x_1$ , the next time gives you $x_2$, etc.

➤ Don't forget to write down the value of each iteration before finding the next.

> ! Since the Newton-Raphson method involves differentiation, angles in any problems involving trig functions must be measured in RADIANS.

# EXERCISE 22 B

1. Show that each of the following equations has a root between $x = 0$ and $x = 1$. Find this root correct to 1 decimal place by repeated use of the intermediate value theorem.

   (a) $3x^3 - 9x + 2 = 0$  (b) $2x^3 - 1 = 0$  (c) $x^4 + 3x = 3x^3$

2. Show that each of the following equations has a root between $x = 0$ and $x = 1$. Use the Newton-Raphson method three times, with the given first approximation, to find this root correct to 2 decimal places.

   (a) $3x^3 - 9x + 2 = 0$, $x_0 = 0$  (b) $2x^3 - 1 = 0$, $x_0 = 1$

3. Use the Newton-Raphson method to find a root for each of the following equations, using the given initial value, $x_0$, and giving your answer correct to the required degree of accuracy.

   (a) $x^2 = 7$; $x_0 = 2.5$, to 3 d.p.  (b) $x^3 - 5x^2 + 3 = 0$, $x_0 = 1$, to 3 d.p

   (c) $x \ln x = 1$; $x_0 = 2$, to 4 d.p.  (d) $\sin x = x - \frac{1}{2}$; $x_0 = 1.5$, to 4 d.p.

   (e) $e^x = 3 - x$; $x_0 = 1$, to 5 d.p.

4. (a) By using sketch graphs, or otherwise, show that the equation $x = 3 \sin x$ has only one positive root, $\alpha$.

   (b) Show that $\alpha$ lies between 2.2 and 2.3.

   (c) Find a better approximation to $\alpha$ correct to 2 d.p.

5. Use the Newton-Raphson method, with initial approximation 14, to find correct to 2 d.p. the positive root of the equation $x = 5 \ln (x + 2)$

6. (a) Show that the equation $x^4 + x^2 - 80 = 0$ has a root between 2 and 3.

   (b) Taking $x = 3$ as the first approximation, use the Newton-Raphson method twice to obtain a better approximation.

7. (a) Sketch the functions $y = \ln x$ and $y = 4 - x$ on the same axes, and state why the equation $\ln x = 4 - x$ has only one real root.

   (b) Show that this root lies between 2.9 and 3.

   (c) By taking 2.9 as a first approximation to this root and applying the Newton-Raphson process once, or otherwise, find a second approximation, giving your answer to 3 s.f.

8.  Given that $x$ is measured in radians and $f(x) = \sin x - 0.4x$,

    (a) show that the equation $f(x) = 0$ has a root $\alpha$ in the interval $[2, 2.5]$

    (b) Using 2.1 as a first approximation to $\alpha$, use the Newton-Raphson method once to find a second approximation to $\alpha$, giving your answer to 2 dp

#9. (a) By drawing an appropriate sketch, find how many roots the equation $\cos x = e^{-x}$ has in the interval $[0, 2\pi]$.

    (b) Given that an approximation to one of the roots is 5, use the Newton-Raphson method 3 times to find a better approximation to this root.

#10. $f(x) = 2^x - x^3$.

    (a) Show that a root, $\alpha$, of the equation $f(x) = 0$ lies in the interval $[1.3, 1.4]$.

    (b) Taking 1.37 as the first approximation, apply the Newton-Raphson method once to $f(x)$ to find a second approximation to $\alpha$. Give your answer to 3 d.p.

11. The function f is defined by $f(x) = \ln(x^2 + 3) - \frac{3}{2}x$.

    (a) Show that the equation $f(x) = 0$ has a solution between $x = 0$ and $x = 1$.

    (b) Show that $f'(1) = -1$.

    (c) Use the Newton-Raphson method, with $x_0 = 1$ as an initial approximation to this solution, to find the iterates $x_1$ and $x_2$ correct to 5 decimal places. Hence give the solution as accurately as possible after these two iterations.

    A modified form of the Newton-Raphson iteration is $x_{n+1} = x_n - \dfrac{f(x_n)}{f'(x_0)}$.

    (d) Use this modified form, again with $x_0 = 1$, to obtain iterates $x_1$ and $x_2$ correct to 5 decimal places.

    (e) Hence identify one possible advantage of using this modified form of the Newton-Raphson iteration compared with the correct Newton-Raphson iteration.

# EXERCISE 22 X

1.  Show that the equation, $x^3 = 10 - 3x$, has a root between 1 and 2. [5]

2.  Given that $f(x) = x^3 - 5x^2 + 3x$, show that $f(x) = 0$ possesses a root in the interval $[\frac{1}{2}, 1]$.
    By considering suitable values of $x$ greater than 1,
    show that there is another root of $f(x) = 0$ greater than 1. [7]

3.  Show that the equation $x^3 = 8 + 4x$ has a root in the closed interval $[2, 3]$. [5]

4.  (a) Show that there is a root of the equation $x^3 - 5x + 1 = 0$ between 0 and 1. [2]

    (b) If $x = 0$ is a first approximation to this root, find a better approximation, correct to one significant figure, using the Newton-Raphson method. [3]

5.  (a) Show that the equation $2x^3 + 6x - 1 = 0$ has a root, $\alpha$, between 0 and 1. [3]

    (b) A first approximation to $\alpha$ is 0.5. Find a better approximation to the root, $\alpha$, to one decimal place. [3]

6. Given the equation $\sin x - x + 0.5 = 0$, consider the interval $|x - 0.5| \le 1$ and
   (a) explain, using a graphical method, why there is only one root of the equation    [4]
   (b) find an interval in which this root lies    [2]

7. (a) Show that the equation $2x^3 - 5x + 1 = 0$ has a root, $\alpha$, between $-1$ and $+1$.    [3]
   (b) A first approximation to $\alpha$ is 0.4. Find a second approximation for $\alpha$ to 1 decimal place.    [4]

8. (a) With the aid of a suitably labelled diagram, explain the Newton-Raphson method for finding successive approximations of a differentiable function.    [4]
   (b) The function f is given by $f(x) = x^3 + x - 3$
   Show that
       (i) f is a strictly increasing function    [3]
       (ii) the equation $f(x) = 0$ has a real root $\alpha$ in the interval $[1, 2]$    [3]
       (iii) the equation $f(x) = 0$ has no other root in the interval $[1, 2]$    [3]
       (iv) by choosing $x_0 = 1$ as an initial approximation to the root $\alpha$, that a better approximation of $\alpha$ is $x_1 = 1.25$.    [2]
   (c) Let $x_n$ be the $n$th approximation of $\alpha$. What is $x_{n+1}$ in terms of $x_n$?    [5]

9. The function g is given by
   $$g(x) = 6 - 4x - x^3.$$
   (a) Show that
       (i) g is everywhere strictly decreasing    [3]
       (ii) the equation $6 - 4x - x^3 = 0$ has a real root, $\beta$, in the interval $(1, 2)$    [3]
       (iii) $\beta$ is the ONLY real root of the equation.    [4]
   (b) Using $x_1 = 2$ as a first approximation to $\beta$, illustrate by means of a diagram how the Newton-Raphson method can be used to obtain a second approximation, $x_2$, to $\beta$.    [5]
   (c) Show that if $x_n$ is the $n$th approximation to $\beta$, then
   $$x_{n+1} = \frac{2x_n^3 + 6}{3x_n^2 + 4}.$$    [6]
   (d) Use the Newton-Raphson method to obtain an approximation to $\beta$, giving your answer correct to two decimal places.    [4]

10. (a) Explain carefully and in your own words the Newton-Raphson method for numerically determining the root of the equation $f(x) = 0$, where $f(x)$ is a continuous, differentiable function.    [9]
    (b) (i) Sketch the functions $y = \sin x$ and $y = x^2$ on the same axes and state why the function $f(x) = \sin x - x^2$ has exactly two real roots.    [4]
        (ii) Construct a table for the function $f(x) = \sin x - x^2$ for $x = 0, 0.25, 0.5, 0.75, 1$    [3]
        (iii) Give the interval in which the non-zero root, $\alpha$, of $f(x) = 0$ lies.    [3]
        (iv) Starting with a first approximation of $\alpha$, find a better approximation.    [3]
        (v) Calculate the value of $\alpha$ to 3 decimal places.    [3]

11. The function f is given by $f(x) = x^4 - 4x + 1$

    Show that

    (a) when $x > 1$, f if strictly increasing                                    [5]

    (b) $f(x) = 0$ has a root in each of the intervals $[0, 1]$ and $[1, 2]$        [9]

    (c) $f(x) = 0$ has no other roots in the interval $[0, 2]$                      [5]

    (d) if $x_1$ is an approximation to the root of $f(x) = 0$ in $[1, 2]$, the Newton-Raphson method

        gives a second approximation $x_2 = \dfrac{3x_1^{\,4} - 1}{4(x_1^{\,3} - 1)}$ in $[1, 2]$.        [6]

12. (a) The function $f : \mathbb{R} \to \mathbb{R}$ is given by $f(x) = x^3 + 2x - 2$.
        Show that

        (i)   f is a strictly increasing function                                 [3]

        (ii)  the equation $f(x) = 0$ has a root $\alpha$ in the interval $[0, 1]$   [3]

        (iii) the equation $f(x) = 0$ has **no other root** in the interval $[0, 1]$  [3]

    (b) By starting with $x_1 = 0.5$ as a first approximation to the root, $\alpha$, use the Newton-Raphson
        method to find a second approximation, $x_2$, to the root $\alpha$ correct to 3 decimal places.    [4]

13. The function, f, is given by $f(x) = 6 - 4x - x^3$.

    (a) Show that

        (i)   f is everywhere decreasing                                          [4]

        (ii)  the equation $f(x) = 0$ has a real root, $\alpha$, in the closed interval $[1, 2]$    [4]

        (iii) $\alpha$ is the only real root of the equation $f(x) = 0$.            [4]

    (b) If $x_n$ is the $n$th approximation to $\alpha$, use the Newton-Raphson method to show that the $(n + 1)$st

        approximation $x_{n+1}$ is given by $x_{n+1} = \dfrac{2x_n^{\,3} + 6}{3x_n^{\,2} + 4}$ .        [8]

14.

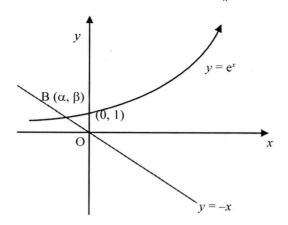

The diagram above (**not drawn to scale**) shows the graphs of the two functions $y = e^x$ and $y = -x$.

    (a) State the equation in $x$ that is satisfied at B $(\alpha, \beta)$, the point of intersection of the two graphs.   [2]

    (b) Show that $\alpha$ lies in the closed interval $[-1, 0]$                   [7]

15. (a) Sketch the functions $y = \sin x$ and $y = x^2$ on the SAME axes. [5]

(b) Deduce that the function $f(x) = \sin x - x^2$ has EXACTLY two real roots. [3]

(c) Find the interval in which the non-zero root $\alpha$ of $f(x)$ lies. [4]

(d) Starting with a first approximation of $\alpha$ at $x_1 = 0.7$, use one iteration of the Newton-Raphson method to obtain a better approximation of $\alpha$ to 3 decimal places. [8]

# What you have learnt in Chapter 22

## *Location of Roots of Equations*

Not all equations can be solved algebraically.

Although a graphical method can often be used to find good approximations, graphs are usually more useful to find the approximate location of roots and then one of the following methods used to gain better approximations.

## *Intermediate Value Theorem*

If a function $y = f(x)$ is continuous in the interval $a \leq x \leq b$, and if $f(a) . f(b) < 0$, then the equation $f(x) = 0$ has at least one root between $a$ and $b$.

The intermediate value theorem can be used repeatedly to narrow the interval in which the root lies.

## *Newton-Raphson Method*

If $x_r$ is an approximation to a root $\alpha$ of the equation $f(x) = 0$, then

$$x_{r+1} = x_r - \frac{f(x_r)}{f'(x_r)}$$

often gives a better approximation to $\alpha$.

# CHAPTER 23

# 2.3a Counting

## COUNTING

If you have to make a decision and you want to know how many choices are available to you, you can list the possible choices and <u>count</u> how many there are. When the choice is limited, this is easy to do, but when a lot of choices are involved, it would be difficult to list all the possibilities and count them, so in many circumstances it is possible to apply certain rules that enable the count to be made without long lists. These rules are the principles of counting.

For example, suppose for a meal you had the choice of two soups, followed by the choice of chicken or meat or fish, followed by the choice of two desserts. How many possible meal combinations are there?

To list the possibilities a systematic approach is necessary. One good way of doing this is to draw a tree diagram:

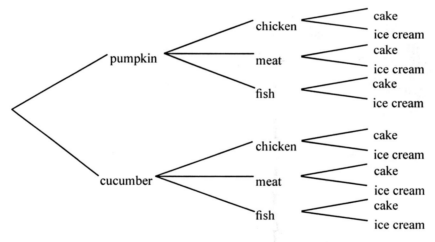

Following each of the branches in turn we can see that there are a total of 12 different meal combinations.

After the first choice between 2 soups there are 3 main courses to choose between, giving a total of 6 (= 2 × 3) possible combinations so far. When the dessert is then chosen, each of the 6 previous combinations can be followed by 1 of 2 desserts, so there are a total of 12 (6 × 2) meal combinations.

This illustrates the first principle of counting:

> **The Multiplication Principle**
>
> If $k$ items are to be chosen, one from each of $k$ sets, and the first set contains $n_1$ items, the second set contains $n_2$ items, ... the $k$th set contains $n_k$ items, then the number of different possible choices that can be made is $n_1 n_2 n_3 n_4 \dots n_k$

## Example 23.1

A company needs to appoint an accountant, a secretary and a security guard. There are 2 applicants for the accountant's job, 4 applicants for secretary and 8 applicants for security guard. In how many different ways could the three appointments be made?

the accountant could be chosen in 2 ways

the secretary could be chosen in 4 ways

the security guard could be chosen in 8 ways

∴ the 3 appointments could be made in $2 \times 4 \times 8 = 64$ different ways

## Example 23.2

8 athletes run in a 100m race. Gold, silver and bronze medals are awarded to the first three contestants to finish (in order) How many possible lists of medal winners are there?

➤ the gold medallist could be any one of the 8 runners $(n_1 = 8)$

➤ there are now 7 runners left who might finish second to get the silver medal $(n_2 = 7)$

➤ similarly, there are 6 possible candidates for the bronze medal $(n_3 = 6)$

➤ the number of possible lists of medal winners is therefore $8 \times 7 \times 6 = 336$

## Example 23.3

In a certain Caribbean island, car registration plates consist of two letters of the alphabet (excluding I and O) followed by three digits from 0 to 9. How many different registration plates are available (letters and digits may be repeated)?

➤ the first letter can be chosen from 24

➤ the second letter can be chosen from 24

➤ the first digit can be chosen from 10, so can each of the other two digits

➤ number of registration plates available = $24 \times 24 \times 10 \times 10 \times 10 = 24^2 \times 10^3 = 576\ 000$

## Example 23.4

A password must consist of 7 characters. The characters can be either numerical digits (0 to 9) or letters of the alphabet (A to Z). How many possible passwords are there?

Each character can be chosen from any of the 10 digits **or** 26 letters. There are therefore 36 characters to choose from.

Number of possible passwords = $36^7$ which is nearly eighty thousand million (80 billion)

This example illustrates the second principle of counting:

**The Addition Principle**

If an item is to be chosen from any one of $k$ disjoint (mutually exclusive) sets (i.e. there is no overlap in the elements of the sets), and the first set contains $n_1$ items, the second set contains $n_2$ items, ... the $k$th set contains $n_k$ items, then the number of different possible choices that can be made is $n_1 + n_2 + n_3 + n_4 + \ ... \ + n_k$

## Example 23.5

Sun Island College has 4 sixth-form classes. There are 55 students in 12 Arts, 49 students in 12 Science, 51 students in 13 Arts and 45 students in 13 Science.

A sixth-form student must be chosen to represent the college at a conference. In how many ways can he/she be chosen if

(a) the student must be an arts student

(b) the student must be from grade 13

(c) any sixth-former can be chosen?

    (a) There are $55 + 51 = 106$ arts students, so there are 106 possible choices

    (b) There are $51 + 45 = 96$ grade 13 students, so there are 96 possible choices

    (c) There are $55 + 49 + 51 + 45 = 200$ students altogether so there 200 possible choices

## EXERCISE 23 A

1. There are 12 runners in a race. In how many different ways can the first, second and third prizes be awarded?

2. A club needs to elect a committee of 3 officers. 1 candidate is proposed for president, 5 candidates are proposed for secretary and 3 candidates are proposed for treasurer. In how many ways could the officers be elected? (each candidate can only stand for one position)

3. How many different arrangements are there of the letters of the word FILM?

4. Judith has 5 detective novels, 3 Harry Potter books and 4 romances. Tracian wants to borrow a book from Judith. How many choices does she have if

    (a) she doesn't mind which kind of book she borrows

    (b) she wants to borrow either a Harry Potter or a romance?

5. A password is made up of 5 digits chosen from 1, 2, 3, 4 and 5.

   How many different passwords are there if

    (a) each digit can only be used once

    (b) digits can be repeated?

6. A family of 2 parents, 4 girls and 3 boys is given 2 tickets for the pantomime.

   In how many ways can the two who will go to the pantomime be chosen if

    (a) one boy and one girl must be chosen

    (b) one parent goes with one child

## PERMUTATIONS AND COMBINATIONS

All the examples so far can be solved simply by using the multiplication and addition rules, but where large numbers of choices need to be made some further rules are useful.

Consider again example 23.2. At the start of the race the runners are assigned to lanes, of which there are eight. How many different ways are there of assigning the runners to their lanes?

The first lane can be assigned to any of the 8 runners,

the second to any of the remaining 7 runners,

the third lane to any of the remaining 6,

and so on until the last runner must be placed in the last lane.

There are therefore $8 \times 7 \times 6 \times 5 \times 4 \times 3 \times 2 \times 1$ $(= 40320)$ ways of assigning the runners to lanes.

Since we meet numbers like this very frequently we have special notation for it.

$8 \times 7 \times 6 \times 5 \times 4 \times 3 \times 2 \times 1 = 8!$ which we read as eight factorial or factorial 8 (some old books use $\lfloor 8$ )

> The product of all the integers from $n$ down to 1 is known as the **factorial** of $n$.
>
> The factorial of zero is defined as 1.
>
> $$n! = n(n-1)(n-2)(n-3) \dots 3 . 2 . 1$$
>
> $$0! = 1$$

$n!$ is a special example of what is known as a **permutation**, that is an ordered arrangement of objects.

8! represented the number of ways in which 8 runners could be arranged in 8 lanes.

Similarly if a family of 5 needed to be lined up for a photograph, they could be arranged in 5! ways.

Let's now take another look at example 23.2.

This is another example of a permutation: 3 medals are awarded to the top 3 runners (in order), but this time there are 8 runners but only 3 medals.

The answer was $8 \times 7 \times 6$, which is the first 3 factors of 8!

The remaining factors $5 \times 4 \times 3 \times 2 \times 1$ are not needed because the placing of the remaining 5 runners is not important.

We see, therefore, that the arrangement of 3 runners chosen from 8 is

$$8 \times 7 \times 6 = \frac{8 \times 7 \times 6 \times (5 \times 4 \times 3 \times 2 \times 1)}{5 \times 4 \times 3 \times 2 \times 1}$$

$$= \frac{8!}{5!}$$

$$= \frac{8!}{(8-3)!}$$

Generalizing this result we get

> **The Permutations Rule** ─ ORDER MATTERS
>
> If $r$ items must be chosen from a set of $n$ different items, and arranged in order, then the number of possible arrangements, known as the **permutation** of $r$ items chosen from $n$, is
>
> $$^{n}P_{r} = \frac{n!}{(n-r)!}$$

## Example 23.6

From a club of 12 members a 5-a-side football team is to be chosen in which the 5 chosen players are assigned places on the field. How many possible teams could be picked?

> An **arrangement** of 5 players chosen from 12 is required – i.e. $^{12}P_5 = 95040$

Taking a further look at this example, the manager may wish to select his team first, and then assign the places later. How many **selections** could he make?

When he selects the names he can do so in $^{12}P_5$ ways, but each possible team list is inevitably written in some order.

Suppose 5 of the players are A, B, C, D and E, then his list of teams will include ABCDE, ABCED, ABDCE… in fact all the permutations of the 5 players A, B, C, D and E (of which there are 5!)

These 5! permutations all represent the same team.

To avoid this duplication of teams on his list when he counts the possible number of team selections, he must divide the number of team arrangements by 5!

The result is what is known as a **combination**. The number of possible teams is $\dfrac{12!}{7!5!} = 792$
Generalizing this result gives

---

**The Combinations Rule** – ORDER DOES NOT MATTER

If a sample of $r$ items must be chosen from a set of $n$ different items, so that the order in which they are chosen does not matter, then the number of possible <u>selections</u>, known as the **combination** of $r$ items chosen from $n$, is

$$^nC_r = \frac{n!}{n!(n-r)!} \quad \text{which can also be written} \quad \binom{n}{r}$$

---

## Example 23.7

There are 19 students in a class. 4 students must be chosen to represent the class in a competition. In how many ways can this be done?

> This is a **choice** or **selection** in which the order does not matter: $^{19}C_4 = 3876$

 The important thing when solving many counting problems is to decide whether it is an arrangement (permutation) or selection (combination) that is required.

# EXERCISE 23 B

State whether each question requires a permutation or a combination for its solution.
(Ask yourself whether the order of the items matters.)

Then solve the question.

1. The 7 members of the school netball team will travel with the coach and a parent in 2 cars. In how many ways can the 4 girls who will travel with the coach be chosen?

2. Three brothers visit the dentist. In how many ways could they decide who goes first, second and third?

3.   A DJ is making the programme for tonight's disco.  He has 25 CDs but only time to play 11 of them.  In how many ways can he arrange his programme?

4.   Ricardo is going on holiday.  He has 8 shirts.  In how many ways can he choose the 5 shirts to take with him?

5.   A delivery van must make 10 deliveries.  In how many ways can the driver plan his route?

6.   In how many ways could the university allocate the places?

Some problems involve a combination of 2 or more of the different rules:

## Example 23.8

A committee consists of 8 men and 4 women.  In how many ways can a sub-committee of 3 men and 1 woman be chosen?

➢   Combinations rule:  The men can be chosen in $^8C_3 = 56$ ways and the woman can be chosen in $^4C_1 = 4$ ways

➢   Multiplication rule:  so the sub-committee can be chosen in $56 \times 4 = 224$ ways

## EXERCISE 23 C

1.   How many different arrangements are there of the letters of the word TALKING?

2.   There are 11 runners in a race.  Prizes are awarded to the first 4 finishers.  In how many ways can the prizes be awarded?

3.   How many different ways are there of selecting an executive committee of three from a general committee of eighteen members?

4.   In how many ways can seven books be arranged on a shelf when the books are selected from 10 different books?

5.   A supermarket stocks 6 different varieties of biscuit.  A customer selects 3 packets of different varieties.  In how many ways could she make her choice?

6.   The university has 12 places available for students to study maths. 20 students have applied.  In how many ways could the university allocate the places? 15 students sit a scholarship examination.  Scholarships are awarded to the top 5 candidates.  In how many ways can the scholarships be awarded?

7.   Grandfather gives 5 books to his 5 grandchildren.  In how many ways can they be distributed if each child receives one book?

8.   A password must consist of 3 letters, followed by two digits, followed by another letter.  If there are 24 letters and 9 digits available, how many possible passwords are there when

(a)   the characters can be repeated?

(b)   the characters cannot be repeated?

9.   An artist who has 13 paintings available is asked to provide 3 for an exhibition.

(a)   In how many different ways could she do this?

She takes the 3 paintings to the gallery.

(b)   In how many ways can the curator arrange them?

10.   Mr and Mrs DaSilva have 5 sons and 3 daughters. They decide to take 3 of their children on holiday with them. In how many ways can they make their selection if

(a)   any child can be chosen        (b)   only boys can be chosen

(c)   only girls can be chosen       (d)   2 boys and 1 girl must be chosen?

11.   A DJ has 8 reggae tracks and 5 rap tracks. He needs to include 4 tracks on his show. In how many ways can he arrange his programme if it must include

(a)   any 4 reggae tracks?           (b)   2 reggae tracks followed by 2 rap tracks?

12.   An artist has 8 portraits and 10 landscape paintings. He wishes to display 5 of the paintings at an exhibition.

In how many ways can he select the paintings to exhibit if

(a)   any painting can be chosen

(b)   only landscapes may be chosen

(c)   there must be 3 landscapes and 2 portaits?

13.   A sixth form committee consists of 3 delegates from class 12S, 4 delegates from class 12A, and 5 delegates each from classes 13S and 13A.

If there are 25 students in 12A, 21 students in 12S, 19 students in 13A and 15 students in 13S, in how many ways could the committee be chosen?

## _Harder Problems involving Permutations and Combinations_

Many problems involving permutations (and combinations) also involve restrictions and complications. **Each problem must be treated on its own merits** – there is no rule that covers all problems.

However there are two particular types of problem which frequently occur.

The first (see examples 23.9, 23.10) is when 2 or more of the original items are the same.

The second is when a number of items of the arrangement need to be kept together (see example 23.11)

## Example 23.9

Find the number of <u>different</u> arrangements of the letters of the words (a) TABLE  and  (b) FLUFF

(a)  5 items can be arranged in 5! = 120 ways. Since all the letters of TABLE are different, there are 120 different arrangements.

(b)  suppose the 5 letters of the word FLUFF are written on 5 cards, which are drawn at random and arranged in order as they are drawn. There are 5! arrangements of the cards. However several of the arrangements will look the same. If the original cards had been written as $F_1$ L U $F_2$ and $F_3$ then with the subscripts we have 5! arrangements, but removing the subscripts $F_2LUF_3F_1$ will look exactly the same as $F_1LUF_2F_3$ – in fact 3! arrangements (the permutations of the 3 F's among themselves) will be the same. So the number of different arrangements is $\dfrac{5!}{3!}$.

Generalizing this result gives

> If $N$ items are to be arranged in order, where $n_1$ of them are identical, another $n_2$ are the same as each other but not the same as the previous $n_1$, and so on; then the number of different arrangements of the items is $\dfrac{N!}{n_1!n_2!...}$

## Example 23.10

Find the number of different arrangements of the letters of the word STATISTICS.

➢   There are 10 letters altogether:  3 Ss, 3 Ts, 1 A, 2 Is and 1 C

➢   Number of arrangements is therefore $\dfrac{10!}{3!3!2!} = 50400$

## Example 23.11

Ten books need to be arranged on a shelf.  4 of the books are by Jane Austen and the rest by different authors. In how many ways can the books be arranged if those by Austen must be kept together?

The 4 Austen books need to be kept together and treated as a unit.  We then have 7 units (6 not Austen and one set of Austen)  These can be arranged in 7! ways.  However although they must be together the order of the Austen books is not fixed – they can be arranged among themselves in 4! ways.

∴ total number of different arrangements = 7! × 4! = 120960

Now some examples which require extra thought as well as the rules:

## Example 23.12

A photograph is to be taken of the 8 members of a family.  How many arrangements of the seating are possible if

(a)   there are no restrictions

(b)   the grandparents must sit in the central 2 seats

(c)   2 brothers must not sit next to each other.

(a)   8! = 40320

(b)   Start by placing the grandparents   x x x $G_1$ $G_2$ x x x  – there are 2! ways of doing this.

the remaining family members can be arranged around them in 6! ways

$$2! \times 6! = 1440$$

(b)   the number of ways in which the 2 brothers are *not*  next to each other = the total number of arrangements minus the number of ways in which the 2 *are*  together.

Following the method of example 23.11:

Number of arrangements with brothers together = 7! × 2! = 10080

Number of arrangements with brothers separated = 40320 – 10080 = 30240

## Example 23.13

In the legend of King Arthur and his Knights, the king and his knights sat at a round table so that no member of the group had a 'top seat'.

One day 12 of them sat down to eat together.

In how many ways could they be arranged around their round table?

> Since there is no 'top seat' it doesn't matter where the first knight sits – the others arrange themselves around him – there are therefore 11! possible arrangements.

## Example 23.14

A quiz team of 6 members is selected from a group of 9 boys and 7 girls.

In how many ways can the team be chosen if it must contain

(a) at least 1 girl

(b) at least 5 boys

> (a) at least 1 girl means that there cannot be no girls.
>
> total no. of selections = $^{16}C_6$ = 8008
>
> no. of selections with no girls = $^9C_6$ = 84
>
> ∴ no. of selections with a least 1 girl = 8008 − 84 = 7924
>
> (b) if there must be at least 5 boys, then there can be either 5 boys and 1 girl or 6 boys and no girls
>
> number of selections with at least 5 boys = $^9C_5 \times {}^7C_1 + {}^9C_6$ = 966

 In these harder questions always start with the restrictions and complications and fit the straightforward parts around that.

## EXERCISE 23 D

1. Find the number of different arrangements of the letters of the word JAMAICA.

2. Find the number of different arrangements of the letters of the word TRINIDAD.

3. Find the number of different arrangements of the letters of the word PARALLEL.

4. Find the number of different arrangements of the letters of the word BARBADOS.

5. In how many ways can 6 books be arranged on a shelf if 2 particular books must be placed next to each other?

6. Six people are going to travel in a six-seater minibus but only three of them can drive. In how many different ways can they seat themselves?

7. A group of 6 boys and 5 girls are to be photographed together. The girls are to sit on 5 chairs placed in a row and the boys are to stand in a line behind them. Find the number of different possible arrangements.

8.   How many 3-digit numbers **greater than** 400 can be formed using 1, 2, 3, 4 and 5 as digits when
   (a)   EACH digit is used ONLY once
   (b)   the repetition of each digit is allowed

9.   The judges in a 'Beautiful Baby' competition have to arrange 10 babies in order of merit. In how many ways could this be done?

   Two babies are to be selected to be photographed. In how many ways can this selection be made?

10.   The number of applicants for a job is 15.
   (a)   Calculate the number of different ways in which 6 applicants can be selected for interview.
      The 6 selected applicants are interviewed on a particular day.
   (b)   Calculate the number of ways in which the order of the 6 interviews can be arranged.

11.   8 out of 13 teams will qualify for the final stages of a regional tournament.

   Calculate the number of possible combinations of the 8 qualifying teams.

12.   A man decides to plant 3 shrubs and 4 trees. In how many ways can he make his selection if he has 5 shrubs and 6 trees from which to choose?

13.   A small holiday hotel advertises for a manager and 7 other members of staff. There are 4 applicants for the position of manager and 10 other people apply for the other jobs at the hotel. Find the number of different ways of selecting a group of people for the 8 jobs.

14.   A book club offers a choice of 20 books of which a member chooses 6. Find the number of different ways in which a member may make his choice.

   Given that 12 of the 20 books on offer are novels and that the other 8 are biographies, find the number of different ways in which a member chooses 6 so that
   (a) he has 3 novels and 3 biographies
   (b) he has at least 4 biographies.

15.   Find the number of ways of choosing a team of 11 people from 15 if two particular people out of the 15 must be in the team.

16.   Each of 7 children, in turn, throws a ball once at a target. Calculate the number of ways the children can be arranged in order to take the throws.

   Given that 3 of the children are girls and 4 are boys, calculate the number of ways the children can be arranged in order that
   (a)   successive throws are made by boys and girls alternately,
   (b)   a girl takes the first throw and a boy takes the last throw.

17.   In how many ways can a committee of four men and four women be seated
   (a)   they can sit in any position in a row,
   (b)   no one in the row is seated next to a person of the same sex?
   (c)   they sit round a table instead of in a row? (no restrictions on who sits next to whom)

➤ Let  $\mathcal{E}$ = {students in the class}
        B = {students studying biology}
        S = {students studying statistics}

➤ Represent this on a Venn diagram   (remember: S′ represents the students not studying statistics)

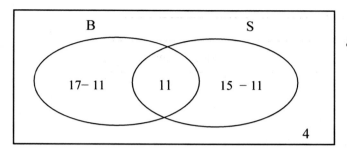

$$n(\mathcal{E}) = 25$$
$$n(B \cap S') = 17 - 11 = 6$$
$$n(B' \cap S) = 15 - 11 = 4$$
$$n(B \cup S) = 6 + 11 + 4 = 21$$
$$n[(B \cup S)'] = 25 - 21 = 4$$

➤       (i)  $P(B \cap S') = \dfrac{6}{25}$    (ii)  $P[(B \cup S)'] = \dfrac{4}{25}$

## RULES

Following from the rules that you already know for sets:  $n(A') = n(\mathcal{E}) - n(A)$  and

$n(A \cup B) = n(A) + n(B) - n(A \cap B)$  we can derive similar rules for probability:

$$P(A') = \frac{n(\mathcal{E}) - n(A)}{n(\mathcal{E})} = 1 - P(A)$$

$$P(A \cup B) = \frac{n(A \cup B)}{n(\mathcal{E})} = \frac{n(A) + n(B) - n(A \cap B)}{n(\mathcal{E})} = P(A) + P(B) - P(A \cap B)$$

If the intersection of  A and B is empty  (A ∩ B = ∅), i.e. if A and B are  **mutually exclusive** events then  $P(A \cup B) = P(A) + P(B)$

If two events are **independent** (i.e. the outcome of one does not affect the outcome of the other, e.g. the events are tossing a coin and throwing a die) then

      $P(A \cap B) = P(A) \times P(B)$

Also if the sample space is comprised of the sample points  $x_i$  ($i = 1,...n$) i.e.  $\mathcal{E} = \{x_1, x_2, x_3, ...x_n\}$  and  $P(x_i)$  is the probability of  $x_i$, then the sum of the probabilities  $\displaystyle\sum_{i=1}^{n} P(x_i)$  is 1

| | |
|---|---|
| $P(A') = 1 - P(A)$      $\displaystyle\sum_{i=1}^{n} P(x_i) = 1$ | (2) (3) |
| $P(A \cup B) = P(A) + P(B) - P(A \cap B)$ | (4) |
| for mutually exclusive events $P(A \cup B) = P(A) + P(B)$ | (5) |
| for independent events  $P(A \cap B) = P(A) \times P(B)$ | (6) |

A set of events is described as **exhaustive** if at least one of the events must occur.  Another way of saying this is that the union of exhaustive events is the entire sample space.  E.g. if the sample space is numbers on the faces of a regular 6-sided die, and events A, B and C are A = {1, 2, 4},  B = {1, 3, 5, 6}, C = {2, 4, 6}; then A and B are exhaustive (because between them they contain all the numbers 1 – 6), A and C are not exhaustive because 3 and 5 are missing, B and C are exhaustive; and so are A, B and C.

# Example 23.19

A marksman has two attempts to hit a target. The probability that his first shot hits is 0.4 and the probability that the second shot hits is 0.5. Given that the probability that he hits the target with both shots is 0.25, find the probability that he misses the target with both shots.

***Either*** : using a Venn diagram    (F is event that first shot hits the target, S that second shot hits)

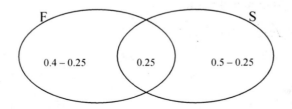

Clearly the probability that he hits at least once is $0.15 + 0.25 + 0.25 = 0.65$, so probability he misses both times is 0.35

***Or*** : using rules

$P(F) = 0.4, \quad P(S) = 0.5, \quad P(F \cap S) = 0.25$

$P(F \cup S) = P(F) + P(S) - P(F \cap S) = 0.4 + 0.5 - 0.25 = 0.65$

$P(F \cup S)' = 1 - P(F \cup S) = 1 - 0.65 = 0.35$

If you are asked to identify or determine the sample space in a particular situation you can either describe the set of all the sample points (see example 23.15) or draw a suitable sample space diagram.

# Example 23.20

A bag contains 3 green marbles and 2 red marbles. In an experiment two marbles are drawn from the bag. Identify the sample space for the experiment if (i) the first marble is replaced (ii) it is not replaced.

It is important to realise that even though 3 green marbles may look the same, they are different and must be treated individually.

Let $G_1$, $G_2$, $G_3$ represent the green marbles and $R_1$ and $R_2$ represent the red marbles.

Method 1, list all the sample points

(i)    The sample space is $\{(G_1, G_1), (G_1, G_2), (G_1, G_3), (G_1, R_1), (G_1, R_2), (G_2, G_1), (G_2, G_2), (G_2, G_3), (G_2, R_1),$
$(G_2, R_2), \quad (G_3, G_1), (G_3, G_2), (G_3, G_3), (G_3, R_1), (G_3, R_2), (R_1, G_1), (R_1, G_2), (R_1, G_3), (R_1, R_1), (R_1, R_2),$
$(R_2, G_1), (R_2, G_2), (R_2, G_3), (R_2, R_1), (R_2, R_2)\}$

(ii)    The sample space is $\{ (G_1, G_2), (G_1, G_3), (G_1, R_1), (G_1, R_2), (G_2, G_1), (G_2, G_3), (G_2, R_1), (G_2, R_2), (G_3, G_1),$
$(G_3, G_2), \quad (G_3, R_1), (G_3, R_2), (R_1, G_1), (R_1, G_2), (R_1, G_3), (R_1, R_2), (R_2, G_1), (R_2, G_2), (R_2, G_3), (R_2, R_1)\}$

Method 2, draw a sample space diagram

(i)

|   | G | G | G | R | R |
|---|---|---|---|---|---|
| G |   |   |   |   |   |
| G |   |   |   |   |   |
| G |   |   |   |   |   |
| R |   |   |   |   |   |
| R |   |   |   |   |   |

(ii)

|   | G | G | G | R | R |
|---|---|---|---|---|---|
| G | ▨ |   |   |   |   |
| G |   | ▨ |   |   |   |
| G |   |   | ▨ |   |   |
| R |   |   |   | ▨ |   |
| R |   |   |   |   | ▨ |

The sample space is represented by the unshaded squares in the tables.

## Example 23.21

A card is to be selected from a pack of cards and a die is to be thrown. Find the probability that the outcome will be (a) a heart and an odd number on the die, (b) a black card and a number greater than 4 on the die.

➢ With 52 cards in the pack, a possibility space diagram would not be a sensible method. But the throw of the die and the drawing of a card are clearly independent events so

➢ P (heart and odd number) = P(heart) × P(odd number) = $\frac{1}{4} \times \frac{1}{2} = \frac{1}{8}$

➢ P (black card and number > 4) = P(black card) × P(number > 4) = $\frac{1}{2} \times \frac{1}{3} = \frac{1}{6}$

## Example 23.22

Events A and B are such that P(A) = $\frac{3}{4}$, P (B) = $\frac{1}{2}$ and P (A ∩ B) = $\frac{3}{8}$.
State whether A and B are (a) mutually exclusive (b) independent

(a) P(A ∩ B) ≠ 0 ∴ A and B are not mutually exclusive

(b) P(A) × P(B) = $\frac{3}{4} \times \frac{1}{2} = \frac{3}{8}$ = P(A ∩ B) ∴ A and B are independent events

## Example 23.23

A race has 6 runners. Apart from the favourite all the runners are equally likely to win, while the favourite is three times as likely to win as anyone else. What is the probability that the favourite does not win the race?

➢ Let $x$ be probability of any one runner other than the favourite winning

➢ The sum of the probabilities of winning must be 1: $3x + x + x + x + x + x = 1$

$$\therefore x = \frac{1}{8} \quad \therefore P \text{ ( favourite wins)} = 3x = \frac{3}{8}$$

➢ P ( favourite does not win) = $1 - P$ ( favourite wins) = $1 - \frac{3}{8} = \frac{5}{8}$

## Example 23.24

In Sun Island High School, the sixth form comprises 4 classes each containing 25 students.

Within in each class the students are ranked by ability

Find the probability that 4 sixth form students chosen at random are

(a) all from the same class

(b) either all top of their class or all bottom of their class.

(a) ➢ The first student chosen can be from any class – call this class C

➢ the second student must then also be from C. Since one has already been chosen, there are 99 left in the sixth form to choose from, of which 24 are in C , so

$$P \text{ (second student chosen is in C)} = \frac{24}{99}$$

➢ similarly $\quad$ P (third student chosen is in C) = $\frac{23}{98}$

and $\quad$ P (fourth student chosen is in C) = $\frac{22}{97}$

➢ using P(A ∩ B) = P(A)·P(B): P (all 4 students are from same class) = $\frac{24}{99} \cdot \frac{23}{98} \cdot \frac{22}{97}$ = 0.013

(notice how closely the multiplication rule for probability mirrors the multiplication rule in counting)

(b) ➢ If T is the event that the chosen students are all top of their class and B is the event that the chosen students are all bottom of their class, then P(A) = P(B) and, since events T and B are mutually exclusive P(T ∪ B) = P(T) + P(B) = 2P(T)

- ➢ P (first student chosen is top of his/her class) = $\frac{4}{100}$

  P (second student chosen is top of his/her class) = $\frac{3}{99}$

  P (third student chosen is top of his/her class) = $\frac{2}{98}$

  P (fourth student chosen is top of his/her class) = $\frac{1}{97}$

- ➢ P (T) = $\frac{4}{100} \cdot \frac{3}{99} \cdot \frac{2}{98} \cdot \frac{1}{97} = 2.55 \times 10^{-7}$

  P(chosen students are either all top or all bottom of their class) = 2P(T) = $5.1 \times 10^{-7}$

# EXERCISE 23 E

1. A cubical die whose faces are numbered 2, 2, 4, 4, 6 and 6; and a tetrahedral die whose faces are numbered 1, 2, 3 and 4 are thrown, and the product of the numbers on the faces on which the dice are resting is calculated. Identify the sample space.

2. In an experiment the temperatures of three cooling fluids are recorded at 5 minute intervals for one hour. Determine the event space.

3. A farmer has three fields and needs to plant a field each of yams, beans and onions. Identify the sample space.

4. A coin is weighted so that the probability of the coin landing with heads up is twice the probability of it landing with tails up. What is the probability of it landing with heads up?

5. Monique, Alicia and Jean run a race. The probability that Alicia wins is $\frac{1}{2}$, and that Jean wins is $\frac{1}{6}$. What is the probability that Monique wins?

6. A security firm uses 3 breeds of guard dog. They have 4 Alsatians, 2 Rottweilers and 5 Dobermans.

   The dogs are allocated to jobs at random.

   Mr Singh wants an Alsatian and Mr Brown does not want a Rottweiler.

   Find the probability that (a) Mr Singh, (b) Mr Brown get what they want.

7. The probability that it rains today is $\frac{1}{3}$, and the probability that Sheldon's mother calls him from America is $\frac{1}{4}$. Find the probability that

   (a) it rains and his mother doesn't call

   (b) it doesn't rain but his mother does call.

8. The cards for a particular card game consist of a pack of 32 cards. There are cards numbered 1, 2, 3, 4, 5, 6, 7 and 8 in each of four colours, red, blue, green and yellow. Find the probability that a card drawn at random from the pack

   (a) is a yellow card with the number 8

   (b) is an even numbered card in green

(c) is an odd numbered card, given that it is red

(d) is not blue

(e) has a number on it less than 7.

9. Roshan and Sharon are playing a game in which a coin and a fair die, numbered 0, 1, 2, 3, 4, 5 are thrown simultaneously. If the coin lands heads up (H) the player scores the number on the die, and if it lands tails up (T) the player scores twice the score on the die.

   (a) Represent this situation in a sample space diagram, and hence find the probability of scoring

   (i) 4,     (ii) 6     (iii) 7.     $(i)\ \frac{1}{2} \times \left( \frac{1}{6} + \frac{1}{6} \right)$

   (b) If Roshan scores 4, what is the probability that Sharon will get a higher score?

10. Two tetrahedral dice, whose faces are numbered 1, 2, 3 and 4, are thrown. Calculate the probability that the PRODUCT of the scores on the base faces is

   (a) exactly 4     (b) less than 6     (c) at least 8     (d) 5, 6, or 7

11. A researcher for a TV company compiled a list of 25 local authors for a potential feature interview. On the list there were 12 authors who wrote only fiction, 5 who wrote only biography, and 7 who wrote neither fiction nor biography. The producer chose an author at random from the list. Find the probability that the chosen author wrote

   (a) both fiction and biography     (b) biography, given that he wrote fiction

12. In a class of 25 students, 7 are birdwatchers who collect stamps, 4 others collect stamps but are not birdwatchers, and 8 students neither collect stamps nor watch birds. Find the probability that a student chosen at random is a birdwatcher who does not collect stamps?

13. On the staff of a certain Caribbean school there were 12 teachers qualified to teach science, 20 qualified to teach humanities and 7 staff members without teaching qualifications. If there was a total of 36 on the staff and one staff member was chosen at random for further training, what is the probability that the chosen member of staff was already qualified to teach both science and humanities?

14. A forestry officer measures the heights of the trees in a certain area of woodland, numbers the trees for future reference and records the results according to height and type of tree in the table below.

| | Height of Tree in metres | | |
|---|---|---|---|
| | $h < 3$ | $3 \leq h \leq 8$ | $h > 8$ |
| deciduous | 3 | 36 | 21 |
| evergreen | 18 | 27 | 45 |

An official from the ministry of agriculture chooses a tree at random for monitoring by picking a number. Determine the probability that the chosen tree is

   (a) evergreen     (b) over 8 m high     (c) deciduous and less than 8 m high

   (d) over 3 m high, given that it is deciduous

15. Kingston Kings basketball club played 50 games last season. The results were classified according to whether the fixture was at home or away and whether the team won, drew or lost as shown in the table:

| Fixtures | Result | | |
|---|---|---|---|
| | Won | Drawn | Lost |
| Home | 17 | 2 | 7 |
| Away | 13 | 5 | 6 |

Before the season began the national coach chose a fixture at random to visit the club. Find the probability that the fixture he attended was:

(a) an away win      (b) a drawn match

(c) a home game      (d) a win, given that it was a home game

(e) a match where the Kings did not lose.

16. A record is kept of books borrowed from a lending library. The table below shows the number of overdue books outstanding at the end of the week

| | Adult | | Child | |
|---|---|---|---|---|
| | Man | Woman | Boy | Girl |
| **Fiction** | 17 | 10 | 14 | 20 |
| **Non-Fiction** | 9 | 14 | 5 | 11 |

One of these books has been requested by another reader. Determine the probability that

(a) it is non-fiction, borrowed by a boy

(b) it is fiction, borrowed by a female

(c) it was borrowed by a man, given that it is fiction

(d) it was borrowed by an adult or a boy.

17. Events A and B are such that $P(A) = \frac{1}{2}$, $P(B) = \frac{3}{4}$ and $P[(A \cap B)] = \frac{3}{8}$.
State whether A and B are (a) mutually exclusive (b) independent

18. If $P(A) = \frac{1}{5}$ and $P(B) = \frac{2}{3}$ find
(a) $P(A \cap B)$ and $P(A \cup B)$ if A and B are mutually exclusive events,
(b) $P(A \cap B)$ if A and B are independent events.

19. If $P(A) = \frac{3}{8}$ and $P(B) = \frac{1}{4}$, find
(a) $P(A \cup B)$ if A and B are mutually exclusive events,
(b) $P(A \cap B)$ if A and B are independent events.

20. Events C and D are such that $P(C) = \frac{1}{3}$ and $P(C \cap D) = \frac{1}{18}$
If C and D are independent events, calculate
(a) $P(D)$      (b) $P(C \cup D)$

21. If $P(A) = \frac{2}{3}$, $P(B) = \frac{1}{4}$ and $P(A \cup B) = \frac{3}{4}$, evaluate $P(A \cap B)$ and hence determine whether or not A and B are independent events.

22. Events A and C are independent. Probabilities relating to events A, B and C are
$$P(A) = \frac{1}{3}, \quad P(B) = \frac{1}{4}, \quad P(A \cap C) = \frac{1}{15} \quad \text{and} \quad P(B \cup C) = \frac{2}{5}$$
(a) Evaluate $P(C)$
(b) Show that the events B and C are independent

Often there are a number of different equally valid methods that can be used to solve a particular probability question. Questions that can be solved using rules, especially the multiplication rule can often be solved using a

# TREE DIAGRAM

Consider a situation where there are a number of coloured balls in a bag, let us say 5 red balls and 3 blue balls. If two balls are withdrawn from the bag, one after the other, what is the probability that the second ball to be withdrawn is red? Now, clearly, the situation will be different if the first ball is replaced before the second is removed. In that case, the result of the second draw is independent of the result of the first draw, namely $\frac{5}{8}$. On the other hand, if the first ball is not replaced there will only be 7 balls left in the bag to choose from, and if the first ball had been red then the probability of red second would be 4 out of 7, while if the first one had been blue, the probability of red second would be 5 out of 7.

This latter situation is a case of what we call **conditional probability**. The probability of red second, $P(R_2)$ is conditional on the outcome of the first draw. This is expressed mathematically as

$$P(R_2 \mid R_1) = \tfrac{4}{7}, \text{ and } P(R_2 \mid B_1) = \tfrac{5}{7},$$

which reads 'the probability that event $R_2$ occurs, given that event $R_1$ has occurred, is 4 out of 7, and the probability that event $R_2$ occurs, given that event $B_1$ has occurred, is 5 out of 7.' In short, 'the probability of $R_2$ given $R_1$ is four sevenths'.

We will look at the rules relating to conditional probability later, but for the moment we shall see how situations like this one can be helpfully represented pictorially by the use of a tree diagram.

First let us consider the situation when the first ball is replaced before the second is withdrawn. A case of 'with replacement'.

The two colours R and B are each represented by a 'branch' of the tree, representing the first withdrawal of a ball, and their probabilities noted along the branches:

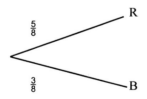

Notice that the sum of the probabilities adds to 1, since the ball must be either red or blue.

Now, since each colour can be followed by either red or blue, add two 'branches' at the end of each of the existing branches. Since the first ball has been replaced, there are still 5 red and 3 blue balls in the bag, and the probability of choosing either of these colours the second time remains the same; and since the two draws are independent, the probability of obtaining, for instance, Red followed by Blue will be $\frac{5}{8} \times \frac{3}{8} = \frac{15}{64}$.

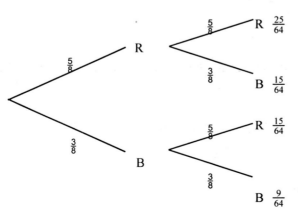

The numbers at the ends of the branches are the products of the probabilities along the branches, and represent the probabilities of the combined events, e.g. R and B. So

$$P(R_1 \cap B_2) = \tfrac{5}{8} \times \tfrac{3}{8} = \tfrac{15}{64}$$

191

Once the tree diagram has been completed it can be used to read off a number of probabilities.

For instance:  (i)      what is the probability that both balls withdrawn are blue?

(ii)     what is the probability that one ball is red and the other blue?

(iii)    what is the probability that at least one ball is red?

(iv)    what is the probability that neither ball is red?

(i)    two blue balls, $B_1 \cap B_2$, only occurs once in the tree, so $P(B_1 \cap B_2) = \frac{9}{64}$

(ii)   one red and one blue ball occur twice in the tree $R_1 \cap B_2$ and $B_1 \cap R_2$. Since the event of one ball of each of these colours is satisfied by *either* $R_1 \cap B_2$ *or* $B_1 \cap R_2$, we *add* the probabilities.

$P$ (one red and one blue ball) $= \frac{15}{64} + \frac{15}{64} = \frac{30}{64} = \frac{15}{32}$

(iii)  at least one red ball can be found by using the whole of the top branch ($\frac{5}{8}$) and the other branch containing red:

$P$(at least one red) $= \frac{5}{8} + \frac{15}{64} = \frac{55}{64}$

(iv)  no red ball can be found from answer (iii): $P$(no red ball) $= 1 - \frac{55}{64} = \frac{9}{64}$     or direct from the tree:

$P$(no red ball) $= P$(two blue balls) $= \frac{9}{64}$

!   Notice that the sum of the probabilities in the final column is 1, since the events there are mutually exclusive and exhaustive.

Let us now consider the other situation: 'without replacement'. The tree diagram will look similar but with different probabilities along the second set of branches and at the ends. These new probabilities are in fact conditional probabilities $P(R_2 \mid R_1) = \frac{4}{7}$, and $P(R_2 \mid B_1) = \frac{5}{7}$ since after a red ball has been removed there are a total of 7 balls left in the bag of which 4 are red.

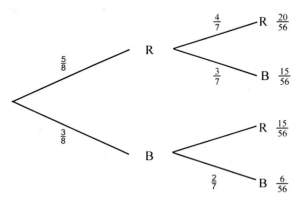

Looking at the same questions we answered in the 'with replacement' situation:

(i)    what is the probability that both balls withdrawn are blue?

(ii)   what is the probability that one ball is red and the other green?

(iii)  what is the probability that at least one ball is red?

(iv)  what is the probability that neither ball is red?

(i)    $P(B_1 \cap B_2) = \frac{3}{8} \times \frac{2}{7} = \frac{3}{28}$  $[= P(B_1) \times P(B_2 \mid B_1)]$

(ii)   $P$ (one green and one red ball) $= P(R_1 \cap G_2) + P(G_1 \cap R_2) = \frac{15}{56} + \frac{15}{56} = \frac{15}{28}$

(iii)   P (at least one red) = $\frac{5}{8} + \frac{15}{56} = \frac{50}{56} = \frac{25}{28}$

(iv)   P (no red ball) = $\frac{3}{28}$

# Example 23.25

Alan, Ben and Carl fire one shot each at a target.

Their probabilities of hitting the target are $\frac{1}{5}$, $\frac{1}{4}$ and $\frac{1}{3}$ respectively.

If they fire together, find the probability that

(a)   all three shots hit the target,

(b)   C's shot only hits the target,

(c)   at least one shot hits the target,

(d)   exactly two shots hit the target.

➤   The statement that they fire together informs you that the events are independent. You can still draw the tree diagram with the three shots separate.  H stands for hit, M for miss:

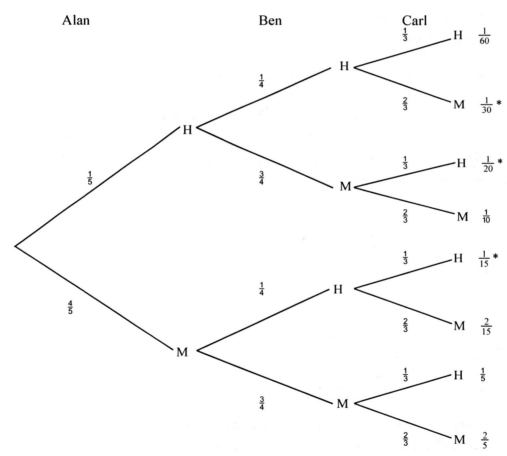

(a)   P (all hit) = $\frac{1}{60}$

(b)   P (only C hits) = P (M, M, H) = $\frac{1}{5}$

(c)   P (at least 1 hit) = P (not all misses) = $1 - \frac{2}{5} = \frac{3}{5}$

(d)   P (2 H and 1 M) = $\frac{1}{30} + \frac{1}{20} + \frac{1}{15} = \frac{3}{20}$  (see * on tree diagram)

## EXERCISE 23F

 When probabilities are given as percentages, it easier to work with either decimals or fractions

1.  A bag contains 5 blue discs, 2 yellow discs and 3 red discs. One disc is withdrawn, the colour noted and then the disc is replaced. A second disc is then withdrawn.

    (a)  Draw a tree diagram to represe nt the different ways in which the two discs can be drawn from the bag, with branches showing the probabilities of drawing the blue, yellow or red discs.

    (b)  Find the probability of drawing a red disc both times.

    (c)  Find the probability that neither disc drawn is yellow.


2.  The test centre in Maryton released figures concerning those who passed their driving test in 2004. 40% of candidates passed at the first attempt, 55% of those who took the test a second time passed, and 12% of those taking it a third time passed. Find the probability of

    (a)  a candidate passing after no more than 2 attempts

    (b)  a candidate not passing after 3 attempts.


3.  A teacher notices that in class 12S, 65% of the students complete their homework regularly. Of those who complete their homework 80% do well in tests, while only 45% of those who do not complete their homework do well in tests.

    (a)  Represent this information on a tree diagram.

    (b)  Hence find the probability that a student chosen at random did well on the most recent test.


4.  Tracey always goes to school by bus or taxi. If one day she goes by bus, then the probability that she goes by bus the next day is $\frac{3}{4}$. If one day she goes by taxi, then the probability that she goes by taxi the next day is $\frac{1}{3}$. Given that she goes to school by taxi on a particular Monday, draw a tree diagram and hence find the probability that she will go to school by bus on the Wednesday of that week.

5.  The probability that Anthony goes to school by bus is $\frac{1}{2}$, by taxi is $\frac{1}{3}$, otherwise he walks. If he walks he will certainly be late. If he catches the bus the probability that he is late is $\frac{1}{3}$, if he catches a taxi the probability that he is late is $\frac{1}{4}$. By drawing a tree diagram, or otherwise, find the probability he arrives at school on time on any particular day.

## CONDITIONAL PROBABILITY

### Example 23.26

Raymond either walks to college, or goes by bus. Which he does depends a lot on the weather. If it rains, the probability that he takes the bus is $\frac{4}{5}$ whereas if it's fine the probability that he takes the bus is $\frac{1}{4}$. The probability of rain on any particular day is $\frac{1}{6}$. What is the probability that on a day chosen at random he takes the bus.

This example can be solved using a tree diagram. Let R be the event that it rains, and B be the event that Raymond goes to college by bus. Then $P(R) = \frac{1}{6}$, $P(R') = \frac{5}{6}$, $P(B \mid R) = \frac{4}{5}$, etc. where $P(B \mid R)$ stands for the probability that he travels by bus given that it rains – or, in short, the probability of B given R.

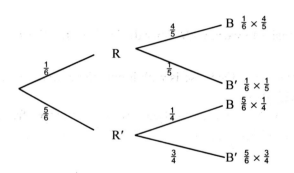

Looking at the diagram we can see that
$$P(R \cap B) = \tfrac{1}{6} \times \tfrac{4}{5}$$

But $\tfrac{1}{6} \times \tfrac{4}{5} = P(R) \times P(B \mid R)$

So $P(R \cap B) = P(R) \times P(B \mid R)$ ⟶ (7)

Rearranging this gives $P(B \mid R) = \dfrac{P(B \cap R)}{P(R)}$

Also $P(B) = P(R \cap B) + P(R' \cap B) = P(R) \times P(B \mid R) + P(R') \times P(B \mid R')$ ⟶ (8)

These are the two main rules for conditional probability

Also it follows from (7) that since for independent events A and B $P(A \cap B) = P(A) \cdot P(B)$

$P(B \mid A) = P(B)$ implies that A and B are independent

$$P(A \cap B) = P(A) \cdot P(B \mid A) = P(B) \cdot P(A \mid B) \quad (7)$$

which can also be written as $P(A \mid B) = \dfrac{P(A \cap B)}{P(B)}$ ⟶ (7)

$P(B) = P(B \mid A) \cdot P(A) + P(B \mid A') \cdot P(A')$ ⟶ (8)

if $P(B \mid A) = P(B)$ then A and B are independent ⟶ (9)

*(handwritten)* $P(A \cap B) = P(A) \cdot P(B \mid A)$

*(handwritten)* $P(A \cap B') = P(B') \cdot P(A \mid B')$

## Example 23.27

The events A and B are such that $P(A') = \tfrac{3}{4}$, $P(A \mid B) = \tfrac{1}{3}$, $P(A \cup B) = \tfrac{2}{3}$. Find

(a) $P(A)$  (b) $P(A \cap B)$

(c) $P(B)$  (d) $P(A \mid B')$

(e) state whether or not events A and B are independent

(a) ➤ $P(A) = 1 - P(A') = \tfrac{1}{4}$

(b) ➤ this is tricky since we do not have a formula connecting $P(A \cap B)$ directly to the probabilities we already know.

➤ However, we do have 2 formulae that connect what we know to $P(A \cap B)$ and $P(B)$:

$P(A \cap B) = P(B) \cdot P(A \mid B)$ (7) and $P(A \cap B) = P(A) + P(B) - P(A \cup B)$ (4)

➤ letting $P(A \cap B) = x$ and $P(B) = y$:   $x = \tfrac{1}{3}y$ and $x = \tfrac{1}{4} + y - \tfrac{2}{3}$

➤ solving for $x$:   $x = P(A \cap B) = \tfrac{5}{24}$

(c) ➤ solving for $y$:   $y = P(B) = \tfrac{5}{8}$

(d) ➤ first find $P(B')$ $P(B') = 1 - P(B) = \tfrac{3}{8}$

➤ exchanging A and B in (8) gives $P(A) = P(A \mid B) \cdot P(B) + P(A \mid B') \cdot P(B')$

so, $\tfrac{1}{4} = \tfrac{1}{3} \cdot \tfrac{5}{8} + P(A \mid B') \cdot \tfrac{3}{8}$

giving $P(A \mid B') = \tfrac{1}{9}$

(e) ➤ $P(A \mid B) = \tfrac{1}{3}$ and $P(A) = \tfrac{1}{4}$, which are not =, so A and B are not independent

## Example 23.28

A maths teacher gave her class two tests. 42% of the class passed the first test while 25% passed both tests. What percentage of those who passed the first test also passed the second test?

> ➤    let F represent the event 'passing the first test' and S 'passing the second test'

> ➤    then $P(F) = 0.42$, $P(F \cap S) = 0.25$ (remember to work in decimals, not %)

> ➤    find $P(S \mid F)$: $\quad P(S \mid F) = \dfrac{P(S \cap F)}{P(F)} = \dfrac{0.25}{0.42} = 0.595$

> ➤    give answer as %: approx 60%

(Example 23.28, like many problems involving conditional probability, could be solved using a Venn or some other diagram, but example 23.27 has to be solved using rules. Whereas when a diagram is appropriate it is often easier, rules (correctly applied) will always lead to a solution provided enough information has been given in the question.)

## EXERCISE 23G (conditional probability and miscellaneous)

1.    If A and B are dependent events such that $P(A) = \frac{5}{8}$ and $P(B \mid A) = \frac{3}{4}$, find $P(A \cap B)$.

2.    If A and B are dependent events such that $P(A) = \frac{2}{3}$ and $P(A \cap B) = \frac{1}{4}$, find $P(B \mid A)$.

3.    (a)   If $P(C) = \frac{1}{4}$ and $P(C \cap D) = \frac{1}{6}$, find $P(D \mid C)$
      (b)   If $P(Y) = \frac{1}{3}$ and $P(X \mid Y) = \frac{1}{2}$, find $P(X \cap Y)$

4.    A and B are independent events.
      Given that $P(A) = \frac{1}{2}$ and $P(B \mid A) = \frac{1}{3}$, find   (a) $P(B)$   (b) $P(B \cap A)$

5.    Given that $P(A) = \frac{1}{4}$, $\;P(B) = \frac{2}{3}$, $\;P(C) = \frac{1}{2}$, $\;P(A \mid B) = \frac{1}{4}$, $\;P(C \mid B) = \frac{2}{3}$ and $P(C \mid A) = \frac{1}{2}$, state which, if any, of the events A, B and C are independent.

6.    Given that $P(B) = \frac{1}{4}$, $P(A \mid B) = \frac{1}{3}$, $P(A \mid B') = \frac{1}{2}$ find
      (a)   $P(A)$,       (b)   $P(A \cup B)$

7.    In a large garden there are 7 fruit trees and 13 other types of tree. Six of the trees have birds nesting in them but only 2 of these are fruit trees.

      (a)   Copy and complete the table below to illustrate the information.

|  | Fruit tree | Other tree | Total |
|---|---|---|---|
| Birds nest | 2 |  | 6 |
| No nest |  |  |  |
| Total | 7 | 13 |  |

The owner of the garden has given permission for Mario to play in the garden but has instructed him not to climb any fruit trees or trees that have birds nesting in them. Mario selects a tree at random to climb.

(b)   Find the probability that Mario will obey the owner's instructions.

Given that Mario climbs a fruit tree,

(c)   find the probability that the tree has birds nesting in it.

8. There are 60 students in the sixth form of Palm Island High School. Mathematics is studied by 27 of them, sociology by 20 and 22 students study neither mathematics nor sociology.

   (a) Find the probability that a randomly selected student studies both mathematics and sociology.

   (b) Find the probability that a randomly selected mathematics student does not study sociology.

   #(c) Determine whether the event "studying mathematics" is statistically independent of the event "not studying sociology".

9. During 2003 a vet saw 125 dogs, each suspected of having a particular disease. Of the 125 dogs, 60 were female of whom 25 actually had the disease and 35 did not. Only 20 of the males had the disease, the rest did not. The case history of each dog was documented on a separate record card.

   A record card from 2003 is selected at random.

   By drawing a suitable sample space diagram, or otherwise, find the probability that the dog on the record card was

   (a) female
   (b) female and suffering from the disease
   (c) female or suffering from the disease
   (d) female, given that she was suffering from the disease

10. A golfer has 5 different clubs in a golf bag, only 1 of which is correct for the shot about to be played. The probability that the golfer plays a good shot if the correct club is chosen is $\frac{1}{3}$ and the probability of a good shot with an incorrect club is $\frac{1}{4}$.

    The golfer chooses a club at random and plays the shot.

    Find the probability that a good shot is made.

11. It may be assumed that each birth in a family is equally likely to result in a boy or a girl.

    There are 4 children in a family.

    Given that at least one child is a boy, find the probability that there are exactly 2 boys.

12. Three cards are to be drawn at random from a pack of ten cards. Six of the cards in the pack are red and numbered from 1 through 6, respectively, while the other four cards are blue and numbered from 1 through 4, respectively.

    Calculate the probabilities that

    (a) exactly two red cards will be drawn,

    (b) exactly one 2 will be drawn.

13. The Kingston Kings have a probability of $\frac{2}{3}$ of winning whenever they play. Given that the team plays 4 games, find the probability that they win more than half of the games.

14. A pencil case contains 4 red crayons, 3 blue crayons and 2 yellow crayons. Damian wants a red crayon. He picks out a crayon at random. If it is not red he picks out another (without replacement), and so on until he gets a red one.

    (a) What is the probability the first red crayon he picks is on his second attempt?

    (b) How many crayons must he remove to be sure of getting a red one?

15. Of the households in Jonesville, 35% have a refrigerator, 60% have a TV set. Given that 25% of the households have both a refrigerator and a TV, calculate the probability that a household has either a refrigerator or a TV set but not both.

# EXERCISE 23 X

1. Calculate the different number of ways a radio announcer can choose **two** calypso compact discs from **ten** available calypso compact discs and **three** reggae cassette tapes from **eight** available reggae cassette tapes. [2]

2. How many 4-digit numbers **greater than** 3000 can be formed using 1, 2, 3, 4 and 5 as digits when

    (a) EACH digit is used ONLY once [3]

    (b) the repetition of each digit is allowed [3]

3. In a group of 14 persons there are 9 males and 5 females.

    A team of 7 persons is to be selected to include 3 males and 4 females. In how many ways can the team be selected? [6]

4. (a) A singer is scheduled to sing 6 particular songs without repetition at a cultural show.

    In how many ways can his schedule be arranged? [2]

    (b) If the singer has altogether 13 suitable songs, in how many ways can he prepare his schedule of 6 songs? [4]

5. If repetitions are not permitted, using the digits 2, 3, 5, 6, 7 and 9

    (a) how many three digit numbers can be formed? [2]

    (b) how many 3 digit numbers are multiples of 5? [2]

6. A class of 13 students has 8 boys, including a set of twins, and 5 girls. How many ways can a committee of 3 be formed if one of the twins must be on the committee? [4]

7. A machine can produce a maximum of 5 items in a one-minute period. In an experiment to rate the actual output of the machine, the number of items produced in a one-hour period is recorded. Identify the sample space. [2]

8. At a football match there are four ticket offices and each can accommodate 75 people standing in line. An official observes the number of people standing in the lines at random times. Determine the sample space for this experiment. [2]

9. One morning there were 39 cars and 15 trucks lined up on a certain highway at a police spot-check. What is the probability that the police will search a car and then a truck? [4]

10. A committee of 4 people is to be selected from a group consisting of 8 males and 4 females. Determine the number of ways in which the committee may be formed if it is to contain

    (a) NO females [2]

(b) EXACTLY one female [3]

(c) AT LEAST one female. [4]

11. The following data show the reactions to a certain drug of a group of persons who participated in a large-scale experiment.

**Reactions to Drug**

| Sex | Severe Reaction | Mild Reaction | No Reaction |
|---|---|---|---|
| Male | 520 | 380 | 350 |
| Female | 330 | 480 | 300 |

TOTAL: 2,360

A person is randomly selected from the group. What is the probability that the person selected,

(a) has no reaction to the drug? [2]

(b) is female and has a mild reaction to the drug? [2]

(c) is male or has a severe reaction to the drug? [4]

(d) does not have a severe reaction to the drug, given that the person is female? [4]

12. A recent study of how a representative sample of West Indians travel to work revealed the information shown in the table below.

| Type of Transportation | Type of Worker | |
|---|---|---|
| | Urban | Rural |
| Private | 400 | 200 |
| Public | 50 | 20 |

If a worker is selected at random from this sample, what is the probability that the worker

(a) is a rural worker? [2]

(b) uses public transportation? [2]

(c) uses private transportation? [2]

(d) is a rural worker or uses public transportation? [5]

(e) is a rural worker, given he/she uses public transportation? [5]

(f) uses public transportation, given he/she is an urban worker? [5]

(g) uses public transportation, given he/she is a rural worker? [5]

13. A group of football fans travelling to a match consists of 16 men, 12 women, 8 boys and 4 girls. A person is chosen at random from the group to receive a free ticket for the match.

Find the probability that

(a) the person is a woman or girl [4]

(b) the person is either a man, woman or girl [2]

14. A bag contains 2 red balls, $R_1$ and $R_2$, 1 green ball, $G$, and 2 black balls, $B_1$ and $B_2$. Randomly, two balls are drawn together from the bag.

    (a) Describe the sample space     [2]

    (b) Determine the probability that

        (i) BOTH balls are the same colour     [2]

        (ii) AT LEAST ONE ball is black     [3]

15. A die, numbered 1 to 6 is biased in such a way that a 6 is twice as likely as any other score. Find $P(6)$, the probability that a six shows when this die is thrown.     [3]

16. (a) The letters H, R, D, S and T are consonants. In how many ways can the letters of the word HARDEST be arranged so that

        (i) the first letter is a consonant?     [3]

        (ii) the first and last letters are consonants?     [3]

    (b) Find the probability that the event in (a) (i) above occurs.     [2]

17. Only three horses, $A$, $B$ and $C$ are in a race. The probability that $A$ wins the race is twice the probability that $B$ wins. The probability that $B$ wins the race is twice the probability that $C$ wins. Find the probability of winning for EACH of the horses.     [5]

18. (a) How many four-digit numbers can be formed from the digits 1, 2, 3, 4, 5 if ALL digits can be repeated?     [2]

    (b) Determine

        (i) how many of the four-digit numbers in (a) above are even     [4]

        (ii) the probability that a four-digit number in (a) above is odd.     [2]

19. A fair coin is tossed three times.

    (a) Determine the sample space.     [3]

    (b) Calculate

        (i) the probability that two heads and one tail appear     [2]

        (ii) the probability that AT LEAST two heads appear.     [2]

20. On a particular day, a certain fuel service station offered 100 customers who purchased premium or regular gasoline, a free check of the engine oil or brake fluid in their vehicles. The services required by these customers were as follows:

15% of the customers purchased premium gasoline, the others purchased regular gasoline.

20% of the customers who purchased premium gasoline requested a check for brake fluid, the others requested a check for engine oil.

51 of the customers who purchased regular gasoline requested a check for engine oil, the others requested a check for brake fluid.

(a) Copy and complete the diagram below to represent the event space.

**Brake fluid**  **Engine oil**

**Premium gasoline**

**Regular gasoline**

51

(b) Find the probability that a customer chosen at random
  (i) who had purchased premium gasoline requested a check for engine oil
  (ii) who had requested a check of the brake fluid purchased regular gasoline
  (iii) who had requested a check of the engine oil purchased regular gasoline. [6]

21. A bag contains 12 red balls, 8 blue balls and 4 white balls. Three balls are drawn from the bag at random **without replacement**.

    Calculate
    (a) the total number of ways of choosing the three balls [3]
    (b) the probability that ONE ball of EACH colour is drawn [3]
    (c) the probability that ALL THREE balls drawn are of the SAME COLOUR. [5]

22. (a) How many numbers lying between 3 000 and 6 000 can be formed from the digits 1, 2, 3, 4, 5, 6, if no digit is used more than once in forming the number? [5]
    (b) Determine the probability that a number in 22 (a) above is even. [5]

23. In an experiment, $p$ is the probability of success and $q$ is the probability of failure in a single trial. For $n$ trials, the probability of $x$ successes and $(n-x)$ failures is represented by $^nC_x\, p^x q^{n-x}$, $n > 0$. Apply this model to the following problem.

    The probability that John will hit the target at a firing practice is $\frac{5}{6}$. He fires 9 shots. Calculate the probability that he will hit the target
    (a) AT LEAST 8 times [7]
    (b) NO MORE than seven times. [3]

# What you have learnt in Chapter 23

## COUNTING

### *Some Definitions*

The product of all the integers from $n$ down to 1 is known as the **factorial** of $n$.

The factorial of zero is defined as 1.

$$n! = n(n-1)(n-2)(n-3) \ldots 3 . 2 . 1$$

$$0! = 1$$

### *The Principles of Counting*

#### The Multiplication Principle

If $k$ items are to be chosen, one from each of $k$ sets, and the first set contains $n_1$ items, the second set contains $n_2$ items, … the $k$th set contains $n_k$ items, then the number of different possible choices that can be made is $n_1 n_2 n_3 n_4 \ldots n_k$

#### The Addition Principle

If an item is to be chosen from any one of $k$ disjoint (mutually exclusive) sets (i.e. there is no overlap in the elements of the sets), and the first set contains $n_1$ items, the second set contains $n_2$ items, … the $k$th set contains $n_k$ items, then the number of different possible choices that can be made is $n_1 + n_2 + n_3 + n_4 + \ldots + n_k$

#### The Permutations Rule

If a sample of $r$ items must be chosen from a set of $n$ different items, and arranged in order, then the number of possible <u>arrangements,</u> known as the permutation of $r$ items chosen from $n$, is

$$^n P_r = \frac{n!}{(n-r)!}$$

#### The Combinations Rule

If a sample of $r$ items must be chosen from a set of $n$ different items, so that the order in which they are chosen does not matter, then the number of possible <u>selections,</u> known as the combination of $r$ items chosen from $n$, is

$$^n C_r = \frac{n!}{n!(n-r)!} \quad \text{which can also be written} \quad \binom{n}{r}$$

#### The Repetition Rule

If $N$ items are to be arranged in order, where $n_1$ of them are identical, another $n_2$ are the same as each other but not the same as the previous $n_1$, and so on; then the number of different arrangements of the items is

$$\frac{N!}{n_1! n_2! \ldots}$$

# PROBABILITY

## *Some Definitions*

In an **experiment** or **trial**, any possible result is known as an **outcome, simple event** or **sample point**.

All the possible outcomes make up the **sample space** or **event space**.

An **event** is a set of one or more possible events.

When the outcomes of a trial are **equally likely**, the **probability** that event A occurs is its **relative frequency**

i.e. $P(A) = \dfrac{\text{the number of successful outcomes}}{\text{the total number of possible outcomes}}$

## *Some Rules*

$0 \le P(A) \le 1$

$P(A') = 1 - P(A)$ $\qquad\qquad \sum\limits_{i=1}^{n} P(x_i) = 1$

$P(A \cup B) = P(A) + P(B) - P(A \cap B)$

for mutually exclusive events $P(A \cup B) = P(A) + P(B)$

for independent events $P(A \cap B) = P(A) \cdot P(B)$

conditional probability $P(A \cap B) = P(A) \cdot P(B \mid A) = P(B) \cdot P(A \mid B)$

which can also be written as $P(A \mid B) = \dfrac{P(A \cap B)}{P(B)}$

$P(B) = P(A) \cdot P(B \mid A) + P(A') \cdot P(B \mid A')$

if $P(B \mid A) = P(B)$ then A and B are independent

## *Some Sample Space Diagrams* include

tables

Venn diagrams

tree diagrams

# CHAPTER 24
# 2.3b  Matrices and Systems of Linear Equations

## 📖 DEFINITIONS

A **matrix** (plural: matrices) is a rectangular array of numbers enclosed in round brackets, following certain rules of association, used for storing data. A matrix with $m$ rows and $n$ columns is said to be of **order** $m \times n$. Each number in the matrix is an **element** of the matrix. If $m = n$ the matrix is a **square matrix**; if $m = 1$ it is a **row matrix**; if $n = 1$ it is a **column matrix**.

In a square matrix the diagonal from top left to bottom right is known as the **leading diagonal**. A matrix all of whose elements are zero is known as a **zero matrix**, and a square matrix with every element of its leading diagonal 1 and all other elements 0 is known as a **unit matrix**. Square matrices have a number associated with them, known as the **determinant**. When the determinant is not zero the matrix has an **inverse** such that the product of the matrix and its inverse is equal to the unit matrix. A matrix whose determinant is zero does not have an inverse and is known as a **singular** matrix. If an $m \times n$ matrix is rewritten with the rows and columns interchanged the resultant $n \times m$ matrix is the **transpose** of the original matrix.

## NOTATION

Matrices are usually represented by a bold capital letter or the full array.

$$A = \begin{pmatrix} a & b & c & d \\ e & f & g & h \\ i & j & k & l \end{pmatrix}$$

A is a $3 \times 4$ matrix.

The transpose of **A** is represented by $A^T$ (or $A_r$)
In this case it is a $4 \times 3$ matrix and

$$A^T = \begin{pmatrix} a & e & i \\ b & f & j \\ c & g & k \\ d & h & l \end{pmatrix}$$

**O** is the zero matrix (the identity under matrix addition), **I** is the unit matrix (the identity under matrix multiplication), and $A^{-1}$ is the inverse of **A**. The determinant of **A** can be written det **A** or $|A|$ or $\Delta$. For example the $2 \times 3$ zero matrix **O** and the $3 \times 3$ unit matrix **I** are given by

$$O = \begin{pmatrix} 0 & 0 & 0 \\ 0 & 0 & 0 \end{pmatrix} \qquad I = \begin{pmatrix} 1 & 0 & 0 \\ 0 & 1 & 0 \\ 0 & 0 & 1 \end{pmatrix}$$

## LAWS OF ASSOCIATION

***Matrix addition and subtraction***: Matrices of the same order can be added or subtracted to form another matrix of the same order, by adding/subtracting corresponding elements

$$\begin{pmatrix} a & b & c \\ d & e & f \end{pmatrix} + \begin{pmatrix} g & h & i \\ j & k & l \end{pmatrix} = \begin{pmatrix} a+g & b+h & c+i \\ d+j & e+k & f+l \end{pmatrix}$$

(Matrix addition is associative, commutative and distributive)

***Multiplication of a matrix by a scalar***: A matrix can be multiplied by a scalar by multiplying every element of the matrix by that scalar

$$\lambda \begin{pmatrix} a & b & c \\ d & e & f \end{pmatrix} = \begin{pmatrix} \lambda a & \lambda b & \lambda c \\ \lambda d & \lambda e & \lambda f \end{pmatrix}$$

***Matrix multiplication***: Matrices can be multiplied together if the number of columns in the first matrix is the same as the number of rows in the second matrix.

So, if **A** is a $2 \times 3$ matrix and **B** is a $3 \times 4$ matrix, **AB** exists, but **BA** does not.

The product of an $m \times n$ matrix and an $n \times p$ matrix is an $m \times p$ matrix.

Matrices that can be added/multiplied are said to be **conformable** under matrix addition/multiplication.

(notice that the requirements for matrices to conform are different for addition and multiplication)

The rule for finding the elements of the product is that the element in the $r^{th}$ row and $s^{th}$ column of **AB** is formed by adding the products of the elements of the $r^{th}$ row of **A** and the $s^{th}$ column of **B**.

e.g. 
$$\begin{pmatrix} a & b \\ c & d \\ e & f \end{pmatrix}_{3 \times 2} \begin{pmatrix} p & q \\ r & s \end{pmatrix}_{2 \times 2} = \begin{pmatrix} ap + br & aq + bs \\ cp + dr & cq + ds \\ ep + fr & eq + fs \end{pmatrix}$$

**Example 24.1** Write the following as a single matrix

(a) $\begin{pmatrix} 1 & 3 \\ 5 & -3 \\ -7 & 4 \end{pmatrix} + \begin{pmatrix} 5 & -1 \\ 8 & 7 \\ 2 & -4 \end{pmatrix}$

(b) $8 \begin{pmatrix} 2 & 4 & 1 \\ 7 & -3 & 2 \end{pmatrix} - 2 \begin{pmatrix} 5 & -2 & 4 \\ 2 & 3 & -1 \end{pmatrix}$

➢ (a) $\begin{pmatrix} 1 & 3 \\ 5 & -3 \\ -7 & 4 \end{pmatrix} + \begin{pmatrix} 5 & -1 \\ 8 & 7 \\ 2 & -4 \end{pmatrix} = \begin{pmatrix} 6 & 2 \\ 13 & 4 \\ -5 & 0 \end{pmatrix}$

➢ (b) $8 \begin{pmatrix} 2 & 4 & 1 \\ 7 & -3 & 2 \end{pmatrix} - 2 \begin{pmatrix} 5 & -2 & 4 \\ 2 & 3 & -1 \end{pmatrix} = \begin{pmatrix} 16 & 32 & 8 \\ 56 & -24 & 16 \end{pmatrix} - \begin{pmatrix} 10 & -4 & 8 \\ 4 & 6 & -2 \end{pmatrix}$

$$= \begin{pmatrix} 6 & 36 & 0 \\ 52 & -30 & 18 \end{pmatrix}$$

**Example 24.2** If

$$A = \begin{pmatrix} 1 & 2 \\ 4 & 3 \end{pmatrix} \qquad B = \begin{pmatrix} 1 & 3 \\ 2 & 0 \\ 5 & 4 \end{pmatrix} \qquad C = \begin{pmatrix} 3 & -1 & 4 \\ -2 & 3 & 1 \end{pmatrix}$$

determine which of the following multiplications are not possible, and calculate those that are possible:

(a)  **A²** (= **A** × **A**)    (b)  **AB**    (c)  **BA**    (d)  **B²**    (e)  **C²**    (f)  **AC**
(g)  **CA**    (h)  **BC**    (i)  **CB**    (j)  **CCᵀ**    (k)  **CᵀC**

➤ Since the number of columns in the first matrix must be the same as the number of rows in the second matrix, the matrices in (b), (d), (e) and (g) are not conformable and their multiplications are **not** possible:

➤ (a) $\mathbf{A}^2 = \begin{pmatrix} 1 & 2 \\ 4 & 3 \end{pmatrix} \times \begin{pmatrix} 1 & 2 \\ 4 & 3 \end{pmatrix}$

$$= \begin{pmatrix} 1\times1+2\times4 & 1\times2+2\times3 \\ 4\times1+3\times4 & 4\times2+3\times3 \end{pmatrix} = \begin{pmatrix} 9 & 8 \\ 16 & 17 \end{pmatrix}$$

➤ (c) $\mathbf{BA} = \begin{pmatrix} 1 & 3 \\ 2 & 0 \\ 5 & 4 \end{pmatrix} \times \begin{pmatrix} 1 & 2 \\ 4 & 3 \end{pmatrix}$

$$= \begin{pmatrix} 1\times1+3\times4 & 1\times2+3\times3 \\ 2\times1+0\times4 & 2\times2+0\times3 \\ 5\times1+4\times4 & 5\times2+4\times3 \end{pmatrix} = \begin{pmatrix} 13 & 11 \\ 2 & 4 \\ 21 & 22 \end{pmatrix}$$

➤ (f) $\mathbf{AC} = \begin{pmatrix} 1 & 2 \\ 4 & 3 \end{pmatrix} \times \begin{pmatrix} 3 & -1 & 4 \\ -2 & 3 & 1 \end{pmatrix}$

$$= \begin{pmatrix} 1\times3+2\times(-2) & 1\times(-1)+2\times3 & 1\times4+2\times1 \\ 4\times3+3\times(-2) & 4\times(-1)+3\times3 & 4\times4+3\times1 \end{pmatrix} = \begin{pmatrix} -1 & 5 & 6 \\ 6 & 5 & 19 \end{pmatrix}$$

➤ (h) $\mathbf{BC} = \begin{pmatrix} 1 & 3 \\ 2 & 0 \\ 5 & 4 \end{pmatrix} \times \begin{pmatrix} 3 & -1 & 4 \\ -2 & 3 & 1 \end{pmatrix}$

$$= \begin{pmatrix} 1\times3+3\times(-2) & 1\times(-1)+3\times3 & 1\times4+3\times1 \\ 2\times3+0\times(-2) & 2\times(-1)+0\times3 & 2\times4+0\times1 \\ 5\times3+4\times(-2) & 5\times(-1)+4\times3 & 5\times4+4\times1 \end{pmatrix} = \begin{pmatrix} -3 & 8 & 7 \\ 6 & -2 & 8 \\ 7 & 7 & 24 \end{pmatrix}$$

➤ (i) $\mathbf{CB} = \begin{pmatrix} 3 & -1 & 4 \\ -2 & 3 & 1 \end{pmatrix} \times \begin{pmatrix} 1 & 3 \\ 2 & 0 \\ 5 & 4 \end{pmatrix}$

$$= \begin{pmatrix} 3\times1+(-1)\times2+4\times5 & 3\times3+(-1)\times0+4\times4 \\ (-2)\times1+3\times2+1\times5 & (-2)\times3+3\times0+1\times4 \end{pmatrix} = \begin{pmatrix} 21 & 25 \\ 9 & -2 \end{pmatrix}$$

➤ (j) $\mathbf{CC}^{\mathrm{T}} = \begin{pmatrix} 3 & -1 & 4 \\ -2 & 3 & 1 \end{pmatrix} \times \begin{pmatrix} 3 & -2 \\ -1 & 3 \\ 4 & 1 \end{pmatrix} = \begin{pmatrix} 26 & -5 \\ -5 & 14 \end{pmatrix}$

➤ (k) $\mathbf{C}^{\mathrm{T}}\mathbf{C} = \begin{pmatrix} 3 & -2 \\ -1 & 3 \\ 4 & 1 \end{pmatrix} \times \begin{pmatrix} 3 & -1 & 4 \\ -2 & 3 & 1 \end{pmatrix} = \begin{pmatrix} 13 & -9 & 10 \\ -9 & 10 & -1 \\ 10 & -1 & 17 \end{pmatrix}$

Notice that $\mathbf{BC} \neq \mathbf{CB}$, and that whereas $\mathbf{BC}$ is a $3 \times 3$ matrix, $\mathbf{CB}$ is a $2 \times 2$ matrix.

Similarly with $\mathbf{CC^T}$ and $\mathbf{C^TC}$

Matrix multiplication is **not** commutative, though it **is** associative and distributive.

## EXERCISE 24 A

If $\quad \mathbf{A} = \begin{pmatrix} 5 \\ 1 \end{pmatrix}$ $\qquad \mathbf{B} = (-1 \quad 4 \quad 3)$ $\qquad \mathbf{C} = \begin{pmatrix} -2 & 5 \\ 3 & 2 \end{pmatrix}$ $\qquad \mathbf{D} = \begin{pmatrix} 4 & 0 & -2 \\ -2 & 1 & 1 \\ 1 & 3 & 4 \end{pmatrix}$

$\quad \mathbf{E} = \begin{pmatrix} 3 & 3 & 2 \\ 5 & 0 & -1 \end{pmatrix}$ $\qquad \mathbf{F} = (3 \quad -2)$ $\qquad \mathbf{G} = \begin{pmatrix} 4 & -1 & 6 \\ -2 & 5 & 0 \end{pmatrix}$ $\qquad \mathbf{H} = \begin{pmatrix} 1 & -1 \\ 2 & 3 \end{pmatrix}$

1. Perform the following calculations, where possible.

   (a) $\mathbf{A} + \mathbf{F}$ 
   (b) $\mathbf{C} + \mathbf{H}$ 
   (c) $\mathbf{E} + \mathbf{H}$ 
   (d) $2\mathbf{E} + 2\mathbf{G}$

   (e) $2\mathbf{C} - \mathbf{H}$ 
   (f) $\mathbf{B} + 2\mathbf{G}$ 
   (g) $\mathbf{A} + \mathbf{C^T}$ 
   (h) $\mathbf{A} - \mathbf{F^T}$

   (i) $\mathbf{AF}$ 
   (j) $\mathbf{FA}$ 
   (k) $\mathbf{DB}$ 
   (l) $\mathbf{BD}$

   (m) $\mathbf{C}^2$ 
   (n) $\mathbf{E}^2$ 
   (o) $\mathbf{D}^2$ 
   (p) $\mathbf{CH}$

   (q) $\mathbf{CH^T}$ 
   (r) $\mathbf{H^TH}$ 
   (s) $\mathbf{C^TA}$ 
   (t) $\mathbf{A^TC}$

2. Find $\mathbf{O} + \mathbf{A}$ and $\mathbf{A} + \mathbf{O}$, also $\mathbf{O} + \mathbf{D}$ and $\mathbf{D} + \mathbf{O}$,
   What does this tell you about $\mathbf{O}$?

3. Find $\mathbf{IH}$ and $\mathbf{HI}$, also $\mathbf{ID}$ and $\mathbf{DI}$.
   What does this tell you about $\mathbf{I}$?

## INVERSE MATRICES

The multiplicative inverse of the square matrix $\mathbf{A}$ is $\mathbf{A^{-1}}$ where $\mathbf{AA^{-1}} = \mathbf{A^{-1}A} = \mathbf{I}$ .

Finding the inverse of a $2 \times 2$ matrix is straightforward, and here is the proof of the rule:

If $\mathbf{A}$ is the matrix $\begin{pmatrix} a & b \\ c & d \end{pmatrix}$, suppose its inverse $\mathbf{A^{-1}}$ is $\begin{pmatrix} e & f \\ g & h \end{pmatrix}$, then from the definition $\mathbf{AA^{-1}} = \mathbf{I}$

$$\begin{pmatrix} a & b \\ c & d \end{pmatrix} \times \begin{pmatrix} e & f \\ g & h \end{pmatrix} = \begin{pmatrix} 1 & 0 \\ 0 & 1 \end{pmatrix}$$

$$\therefore \quad \begin{pmatrix} ae+bg & af+bh \\ ce+dg & cf+dh \end{pmatrix} = \begin{pmatrix} 1 & 0 \\ 0 & 1 \end{pmatrix}$$

**Two matrices are equal if their corresponding elements are equal,** so

$ae + bg = 1$ ① $\qquad af + bh = 0$ ②

$ce + dg = 0$ ③ $\qquad cf + dh = 1$ ④

Solving the simultaneous equations ① and ② gives $\quad e = \dfrac{d}{ad-bc}\quad$ and $\quad g = \dfrac{-c}{ad-bc}$.

similarly solving ③ and ④ gives $\quad h = \dfrac{a}{ad-bc}\quad$ and $\quad f = \dfrac{-b}{ad-bc}$

so $\qquad \mathbf{A}^{-1} = \begin{pmatrix} \dfrac{d}{ad-bc} & \dfrac{-b}{ad-bc} \\[2mm] \dfrac{-c}{ad-bc} & \dfrac{a}{ad-bc} \end{pmatrix}$

Notice that the expression $ad - bc$ occurs in the denominator of each element. This expression is known as the **determinant** of the matrix (written $|\mathbf{A}|$ or sometimes det $\mathbf{A}$ or $\Delta$). It is the difference between the products of the elements of the diagonals of the matrix. We can take it out as a common factor and write

$$\mathbf{A}^{-1} = \frac{1}{(ad-bc)}\begin{pmatrix} d & -b \\ -c & a \end{pmatrix} \quad \text{or} \quad \mathbf{A}^{-1} = \frac{1}{|\mathbf{A}|}\begin{pmatrix} d & -b \\ -c & a \end{pmatrix} \quad \text{where } |\mathbf{A}| = ad - bc$$

An easy way to remember this formula is: to find the inverse of a $2 \times 2$ matrix interchange the elements of the leading diagonal, and change the signs of the elements of the other diagonal, divide this new matrix by the determinant.

It is sometimes useful, for instance when we do not wish to use a letter to denote a particular matrix, to use the notation $\begin{vmatrix} a & b \\ c & d \end{vmatrix}$ for the determinant of the matrix $\begin{pmatrix} a & b \\ c & d \end{pmatrix}$.

## Example 24.3

Find $\mathbf{A}^{-1}$ if (a) $\quad \mathbf{A} = \begin{pmatrix} 4 & 1 \\ 3 & 2 \end{pmatrix}$ $\qquad$ (b) $\quad \mathbf{B} = \begin{pmatrix} 5 & 10 \\ 2 & 4 \end{pmatrix}$

(a) $\quad |\mathbf{A}| = 4 \times 2 - 1 \times 3 = 8 - 3 = 5$ $\qquad$ (b) $\quad |\mathbf{B}| = 5 \times 4 - 10 \times 2 = 20 - 20 = 0$

$\qquad \mathbf{A}^{-1} = \dfrac{1}{5}\begin{pmatrix} 2 & -1 \\ -3 & 4 \end{pmatrix}$ $\qquad\qquad$ $\mathbf{B}$ is a singular matrix, and has no inverse.

## Example 24.4

$\mathbf{A} = \begin{pmatrix} 1 & -1 \\ 2 & 3 \end{pmatrix}$ and $\mathbf{B} = \begin{pmatrix} -2 & 2 \\ 16 & 9 \end{pmatrix}$. Find the matrix $\mathbf{X}$ such that $\mathbf{AX} = \mathbf{B}$.

$\underline{\text{Method 1}}$, using simultaneous equations: Let $\mathbf{X} = \begin{pmatrix} a & b \\ c & d \end{pmatrix}$

Then $\mathbf{AX} = \begin{pmatrix} 1 & -1 \\ 2 & 3 \end{pmatrix}\begin{pmatrix} a & b \\ c & d \end{pmatrix}$

$\qquad = \begin{pmatrix} a-c & b-d \\ 2a+3c & 2b+3d \end{pmatrix}$

Equating corresponding elements of **AX** and **B** gives the simultaneous equations $a - c = -2$ and $2a + 3c = 16$, so that $a = 2$ and $c = 4$ and $b - d = 2$ and $2b + 3d = 9$, giving $b = 3$, and $d = 1$

so $\mathbf{X} = \begin{pmatrix} 2 & 3 \\ 4 & 1 \end{pmatrix}$

<u>Method 2</u>, using inverse matrices: pre-multiply $\mathbf{AX = B}$ by $\mathbf{A^{-1}}$ then $\mathbf{A^{-1}AX = X = A^{-1}B}$

$|\mathbf{A}| = 3 + 2 = 5$, so $\mathbf{A^{-1}} = \dfrac{1}{5} \begin{pmatrix} 3 & 1 \\ -2 & 1 \end{pmatrix}$

and $\mathbf{X = A^{-1}B} = \dfrac{1}{5} \begin{pmatrix} 3 & 1 \\ -2 & 1 \end{pmatrix} \begin{pmatrix} -2 & 2 \\ 16 & 9 \end{pmatrix}$

$$= \dfrac{1}{5} \begin{pmatrix} 10 & 15 \\ 20 & 5 \end{pmatrix} = \begin{pmatrix} 2 & 3 \\ 4 & 1 \end{pmatrix}$$

# EXERCISE 24 B

1. Find, where they exist, the inverses of the following matrices

(a) $\begin{pmatrix} 1 & 3 \\ 2 & 5 \end{pmatrix}$   (b) $\begin{pmatrix} 8 & 2 \\ 5 & 3 \end{pmatrix}$   (c) $\begin{pmatrix} 4 & 2 \\ 2 & 1 \end{pmatrix}$   (d) $\begin{pmatrix} -3 & 7 \\ 1 & 2 \end{pmatrix}$

(e) $\begin{pmatrix} 3 & -5 \\ 2 & 2 \end{pmatrix}$   (f) $\begin{pmatrix} 1 & -1 \\ 4 & -2 \end{pmatrix}$   (g) $\begin{pmatrix} 6 & 4 \\ -3 & -2 \end{pmatrix}$   (h) $\begin{pmatrix} -5 & 2 \\ 1 & 2 \end{pmatrix}$

(i) $\begin{pmatrix} 7 & 4 \\ 2 & 1 \end{pmatrix}$   (j) $\begin{pmatrix} 4 & 3 \\ 8 & 5 \end{pmatrix}$

2. If $\mathbf{A} = \begin{pmatrix} 2 & 1 \\ 5 & 3 \end{pmatrix}$ and $\mathbf{B} = \begin{pmatrix} 2 & 4 \\ 1 & 3 \end{pmatrix}$   find **X** and **Y** such that $\mathbf{AX = B}$  and $\mathbf{YA = B}$

3. If $\mathbf{C} = \begin{pmatrix} -3 & -4 \\ -1 & 1 \end{pmatrix}$ and $\mathbf{D} = \begin{pmatrix} 2 & 3 \\ -2 & 1 \end{pmatrix}$   find **P** and **Q** such that $\mathbf{CP = D}$  and $\mathbf{QC = D}$

## SIMULTANEOUS EQUATIONS

Consider the matrix equation $\begin{pmatrix} a & b \\ c & d \end{pmatrix} \begin{pmatrix} x \\ y \end{pmatrix} = \begin{pmatrix} r \\ s \end{pmatrix}$

Equating elements on left- and right-hand sides gives the pair of simultaneous equations

$ax + by = r$ ①
$cx + dy = s$

This shows that a set of simultaneous equations can be written as a matrix equation.

The simultaneous equations can therefore be solved by solving the matrix equation.

Let $\begin{pmatrix} a & b \\ c & d \end{pmatrix} = \mathbf{A}$. Then if both sides of this matrix equation are pre-multiplied by $\mathbf{A}^{-1}$ we get

$$\mathbf{A}^{-1}\mathbf{A}\begin{pmatrix} x \\ y \end{pmatrix} = \mathbf{A}^{-1}\begin{pmatrix} r \\ s \end{pmatrix}$$

but $\mathbf{A}^{-1}\mathbf{A} = \mathbf{I}$, so $\begin{pmatrix} x \\ y \end{pmatrix} = \mathbf{A}^{-1}\begin{pmatrix} r \\ s \end{pmatrix}$

## Example 24.5

Using matrices, solve the simultaneous equations $\quad 3x + y = 9$

$\qquad\qquad\qquad\qquad\qquad\qquad\qquad\qquad 5x = 7 + y \qquad\qquad ②$

➤ equation ② must be rewritten in form ①: $\quad 5x - y = 7$

➤ now write equations in matrix form: $\quad \begin{pmatrix} 3 & 1 \\ 5 & -1 \end{pmatrix}\begin{pmatrix} x \\ y \end{pmatrix} = \begin{pmatrix} 9 \\ 7 \end{pmatrix} \quad ③$

➤ Let $\mathbf{A} = \begin{pmatrix} 3 & 1 \\ 5 & -1 \end{pmatrix}$; find $\mathbf{A}^{-1}$: $\quad \mathbf{A}^{-1} = -\frac{1}{8}\begin{pmatrix} -1 & -1 \\ -5 & 3 \end{pmatrix}$

➤ premultiply ③ by $\mathbf{A}^{-1}$: $\quad \begin{pmatrix} x \\ y \end{pmatrix} = \mathbf{A}^{-1}\begin{pmatrix} 9 \\ 7 \end{pmatrix} = -\frac{1}{8}\begin{pmatrix} -1 & -1 \\ -5 & 3 \end{pmatrix}\begin{pmatrix} 9 \\ 7 \end{pmatrix}$

$$= -\frac{1}{8}\begin{pmatrix} -16 \\ -24 \end{pmatrix} = \begin{pmatrix} 2 \\ 3 \end{pmatrix}$$

➤ equating elements: $\quad x = 2$ and $y = 3$

## EXERCISE 24 C

1. Use matrices to solve the following pairs of simultaneous equations:

(a) $x + y = 6$
$\quad\;\; x - y = 2$

(b) $3x + 2y = 22$
$\quad\;\; x + y = 9$

(c) $5x - 3y = 1$
$\quad\;\; 3x + 4y = 18$

(d) $x + y = 5$
$\quad\;\; 3x - 2y + 5 = 0$

(e) $4x + y = 14$
$\quad\;\; 2x - y = 1$

(f) $2x = 3y - 19$
$\quad\;\; 3x + 2y = 17$

(g) $8y = 3 - x$
$\quad\;\; 3x - 2y = 9$

(h) $2x - 5y + 11 = 0$
$\quad\;\; 3x + 4y = 18$

# 3 × 3 MATRICES

When we worked with 2 × 2 matrices, we used the letters $a$, $b$, $c$ and $d$ for the four elements of a general matrix.

When working with matrices with more than 2 rows and columns, rather than giving a different letter to every element, it is easier to have a systematic way of representing the elements. One method is to use a single letter with two suffixes for each element; the first suffix denotes the row and the second suffix denotes the column. Another method, which we shall adopt here, is to use a different letter for each column, and numerical suffixes for the rows. A general $3 \times 3$ matrix, $\mathbf{M}$, will then be

either $\mathbf{M} = \begin{pmatrix} a_{11} & a_{12} & a_{13} \\ a_{21} & a_{22} & a_{23} \\ a_{31} & a_{32} & a_{33} \end{pmatrix}$ or $\mathbf{M} = \begin{pmatrix} a_1 & b_1 & c_1 \\ a_2 & b_2 & c_2 \\ a_3 & b_3 & c_3 \end{pmatrix}$

The definition of the determinant of the $3 \times 3$ matrix $\mathbf{M} = \begin{pmatrix} a_1 & b_1 & c_1 \\ a_2 & b_2 & c_2 \\ a_3 & b_3 & c_3 \end{pmatrix}$ is

$$\det \mathbf{M} = a_1 \begin{vmatrix} b_2 & c_2 \\ b_3 & c_3 \end{vmatrix} - b_1 \begin{vmatrix} a_2 & c_2 \\ a_3 & c_3 \end{vmatrix} + c_1 \begin{vmatrix} a_2 & b_2 \\ a_3 & b_3 \end{vmatrix}$$

This is easier to remember than might at first appear: the elements of the smaller determinants (sometimes known as **minors**) are found by covering up the row and column containing the letter that precedes it and using the remaining elements.

e.g. to find the minor of $a_2$ remove elements in row and column containing $a_2$: $\begin{pmatrix} \cancel{a_1} & b_1 & c_1 \\ \cancel{a_2} & \cancel{b_2} & \cancel{c_2} \\ \cancel{a_3} & b_3 & c_3 \end{pmatrix}$

leaving $\begin{vmatrix} b_1 & c_1 \\ b_3 & c_3 \end{vmatrix}$ as the minor $a_2$

# Example 24.6

Find the determinant of the matrix $\mathbf{B} = \begin{pmatrix} 1 & 3 & 7 \\ 2 & -1 & 0 \\ 5 & 0 & 4 \end{pmatrix}$

➢   $\det \mathbf{B} = 1 \begin{vmatrix} -1 & 0 \\ 0 & 4 \end{vmatrix} - 3 \begin{vmatrix} 2 & 0 \\ 5 & 4 \end{vmatrix} + 7 \begin{vmatrix} 2 & -1 \\ 5 & 0 \end{vmatrix}$

$= (-4 - 0) - 3(8 - 0) + 7(0 + 5) = -4 - 24 + 35 = 7$

# Example 24.7

Solve the determinant equation $\begin{vmatrix} x & 2 & 3 \\ 2 & x+3 & 6 \\ 3 & 4 & x+6 \end{vmatrix} = 0$

➢   Break down the determinant: $x \begin{vmatrix} x+3 & 6 \\ 4 & x+6 \end{vmatrix} - 2 \begin{vmatrix} 2 & 6 \\ 3 & x+6 \end{vmatrix} + 3 \begin{vmatrix} 2 & x+3 \\ 3 & 4 \end{vmatrix} = 0$

➢   Multiply out: $x[(x+3)(x+6) - 6 \times 4] - 2[2(x+6) - 6 \times 3] + 3[2 \times 4 - 3(x+3)] = 0$

$$x^3 + 9x^2 + 18x - 24x - 4x - 24 + 36 + 24 - 9x - 27 = 0$$

➤ Simplify: $\qquad\qquad x^3 + 9x^2 - 19x + 9 = 0$

➤ Use factor theorem to find a factor: $\qquad x - 1$

➤ Factorise: $\qquad\qquad\qquad (x - 1)(x^2 + 10x - 9) = 0$

➤ Find solutions using quadratic formula: $x = 1, -5 \pm \sqrt{34}$

## EXERCISE 24 D

1. Evaluate the following determinants

(a) $\begin{vmatrix} 2 & -1 & 1 \\ 4 & 2 & 0 \\ 0 & 1 & 1 \end{vmatrix}$
(b) $\begin{vmatrix} 0 & 1 & 2 \\ 2 & 1 & 0 \\ 3 & 0 & 1 \end{vmatrix}$
(c) $\begin{vmatrix} 2 & 1 & -2 \\ 1 & 3 & 4 \\ 0 & 1 & 2 \end{vmatrix}$
(d) $\begin{vmatrix} -1 & 3 & 0 \\ 5 & 2 & 1 \\ -2 & 1 & -1 \end{vmatrix}$

2. Solve the determinant equations

(a) $\begin{vmatrix} x+1 & x & -2 \\ 1 & 3 & 4 \\ 0 & x & 2 \end{vmatrix} = 0$
(b) $\begin{vmatrix} 2 & 1 & x-1 \\ x+2 & 3 & 4 \\ x+1 & 1 & 2 \end{vmatrix} = 0$
(c) $\begin{vmatrix} x & 1 & -2 \\ -2 & 1 & x \\ 1 & x-3 & 1 \end{vmatrix} = 0$

(d) $\begin{vmatrix} x-3 & 1 & -1 \\ -7 & x+5 & -1 \\ -6 & 6 & x-2 \end{vmatrix} = 0$
(e) $\begin{vmatrix} x & 1 & 2 \\ 1 & x & 2 \\ 1 & 2 & x \end{vmatrix} = 0$
(f) $\begin{vmatrix} x & 2 & 3 \\ 2 & x+3 & 6 \\ 3 & 4 & x+6 \end{vmatrix} = 0$

3. Factorise the following determinants

(a) $\begin{vmatrix} a & x & x+a \\ x+a & a & x \\ x & x+a & a \end{vmatrix}$
(b) $\begin{vmatrix} p & p & p \\ p & q & q \\ p & q & r \end{vmatrix}$

## _Cofactors_

When we learnt the rule for finding the determinant of a $3 \times 3$ matrix, we met the idea of the minor of an element of the matrix. The minor of an element $e$ is the determinant whose elements consist of the elements of the original matrix without the elements of the row and column which contain $e$. The minor of $a_1$ was therefore $\begin{vmatrix} b_2 & c_2 \\ b_3 & c_3 \end{vmatrix}$. Similarly the minor of $b_2$ must be $\begin{vmatrix} a_1 & c_1 \\ a_3 & c_3 \end{vmatrix}$ etc.

When the appropriate $+$ or $-$ sign is attached to a minor, it becomes a **cofactor**. The appropriate signs are

$\begin{vmatrix} + & - & + \\ - & + & - \\ + & - & + \end{vmatrix}$. From this we see that the cofactors of the matrix $\mathbf{M} = \begin{pmatrix} a_1 & b_1 & c_1 \\ a_2 & b_2 & c_2 \\ a_3 & b_3 & c_3 \end{pmatrix}$ are

$$A_1 = \begin{vmatrix} b_2 & c_2 \\ b_3 & c_3 \end{vmatrix} \quad B_1 = -\begin{vmatrix} a_2 & c_2 \\ a_3 & c_3 \end{vmatrix} \quad C_1 = \begin{vmatrix} a_2 & b_2 \\ a_3 & b_3 \end{vmatrix}$$

$$A_2 = -\begin{vmatrix} b_1 & c_1 \\ b_3 & c_3 \end{vmatrix} \quad B_2 = \begin{vmatrix} a_1 & c_1 \\ a_3 & c_3 \end{vmatrix} \quad C_2 = -\begin{vmatrix} a_1 & b_1 \\ a_3 & b_3 \end{vmatrix}$$

$$A_3 = \begin{vmatrix} b_1 & c_1 \\ b_2 & c_2 \end{vmatrix} \quad B_3 = -\begin{vmatrix} a_1 & c_1 \\ a_2 & c_2 \end{vmatrix} \quad C_3 = \begin{vmatrix} a_1 & b_1 \\ a_2 & b_2 \end{vmatrix}.$$

Det $\mathbf{M}$ can therefore be written as $a_1 A_1 + a_2 A_2 + a_3 A_3$

When this is multiplied out and rearranged it is $a_1 b_2 c_3 - a_1 b_3 c_2 + a_2 b_3 c_1 - a_2 b_1 c_3 + a_3 b_1 c_2 - a_3 b_2 c_1$.

Now it can be shown that this same expression will be obtained if *any* row or *any* column is used with its corresponding cofactors.

So det $\mathbf{M}$ $= a_1 A_1 + a_2 A_2 + a_3 A_3$
$= b_1 B_1 + b_2 B_2 + b_3 B_3$
$= a_1 A_1 + b_1 B_1 + c_1 C_1$ etc.

This can simplify the working when finding the determinant of a matrix which contains several zeros.

## Example 24.8

Find the determinant of the matrix $\mathbf{N} = \begin{pmatrix} 1 & 3 & 7 \\ 2 & 0 & 3 \\ 5 & 0 & 4 \end{pmatrix}$

> Column 2 contains 2 zeros, so expand the determinant using column 2 with its cofactors

Det $\mathbf{N} = a_2 A_2 + b_2 B_2 + c_2 C_2$

$$= -3 \begin{vmatrix} 2 & 3 \\ 5 & 4 \end{vmatrix} + 0 - 0$$

$$= -3(8 - 15) = 21$$

!  Take care when evaluating the cofactor determinant to add the appropriate + or − sign

The matrix whose elements are the cofactors of the matrix $\mathbf{M}$ is known as the **cofactor matrix of M**. The *transpose* of the cofactor matrix is known as the **adjoint** of the matrix $\mathbf{M}$, written adj $\mathbf{M}$.

So the cofactor matrix of $\mathbf{M}$ is $\begin{pmatrix} A_1 & B_1 & C_1 \\ A_2 & B_2 & C_2 \\ A_3 & B_3 & C_3 \end{pmatrix}$ and adj $\mathbf{M} = \begin{pmatrix} A_1 & A_2 & A_3 \\ B_1 & B_2 & B_3 \\ C_1 & C_2 & C_3 \end{pmatrix}$

## *The Inverse of a 3 × 3 Matrix*

It can be shown that the inverse of a matrix is its adjoint divided by its determinant. i.e. $\mathbf{M}^{-1} = \dfrac{\text{adj}\,\mathbf{M}}{\det \mathbf{M}}$.

We will not give the proof here, but we will verify that in a particular case $\mathbf{M}\,\mathbf{M}^{-1} = \mathbf{M}^{-1}\,\mathbf{M} = \mathbf{I}$

## Example 24.9

Find the inverse, $\mathbf{M}^{-1}$, of the matrix $\mathbf{M} = \begin{pmatrix} 1 & -3 & 0 \\ 2 & 0 & 1 \\ 4 & 1 & 3 \end{pmatrix}$

Verify that $\mathbf{M}^{-1}$ is the inverse of $\mathbf{M}$ (i.e. show that $\mathbf{M}\,\mathbf{M}^{-1} = \mathbf{M}^{-1}\,\mathbf{M} = \mathbf{I}$)

➤ Find det $\mathbf{M}$:  det $\mathbf{M} = 1(0 \times 3 - 1 \times 1) + 3(2 \times 3 - 1 \times 4) + 0 = -1 + 6 = 5$

➤ Find the cofactors of the elements of $\mathbf{M}$:

$A_1 = 0 \times 3 - 1 \times 1 = -1,$   $B_1 = -(2 \times 3 - 1 \times 4) = -2,$   $C_1 = 2 - 0 = 2;$

$A_2 = -(-9) = 10;$   $B_2 = +(3 - 0) = 3;$   $C_2 = -(1 + 12) = -13$

$A_3 = +(-3 -; 0) = -3;$   $B_3 = -(1 - 0) = -1;$   $C_3 = +(0 + 6) = 6$

➤ Write down cofactor matrix of $\mathbf{M}$ and adj $\mathbf{M}$:

$$\text{cofactor matrix} = \begin{pmatrix} -1 & -2 & 2 \\ 9 & 3 & -13 \\ -3 & -1 & 6 \end{pmatrix} \quad \text{adj } \mathbf{M} = \begin{pmatrix} -1 & 9 & -3 \\ -2 & 3 & -1 \\ 2 & -13 & 6 \end{pmatrix}$$

➤ Calculate $\mathbf{M}^{-1}$:   $\mathbf{M}^{-1} = \dfrac{\text{adj } \mathbf{M}}{\det \mathbf{M}} = \dfrac{1}{5}\begin{pmatrix} -1 & 9 & -3 \\ -2 & 3 & -1 \\ 2 & -13 & 6 \end{pmatrix}$

$$= \begin{pmatrix} -\frac{1}{5} & \frac{9}{5} & -\frac{3}{5} \\ -\frac{2}{5} & \frac{3}{5} & -\frac{1}{5} \\ \frac{2}{5} & -\frac{13}{5} & \frac{16}{5} \end{pmatrix}$$

➤ Check: $\mathbf{M}\,\mathbf{M}^{-1} = \mathbf{M}\,\frac{1}{5}(\text{adj } \mathbf{M}) = \dfrac{1}{5}\begin{pmatrix} 1 & -3 & 0 \\ 2 & 0 & 1 \\ 4 & 1 & 3 \end{pmatrix}\begin{pmatrix} -1 & 9 & -3 \\ -2 & 3 & -1 \\ 2 & -13 & 6 \end{pmatrix}$

$$= \frac{1}{5}\begin{pmatrix} 5 & 0 & 0 \\ 0 & 5 & 0 \\ 0 & 0 & 5 \end{pmatrix}$$

$$= \begin{pmatrix} 1 & 0 & 0 \\ 0 & 1 & 0 \\ 0 & 0 & 1 \end{pmatrix} = \mathbf{I}$$

and $\mathbf{M}^{-1}\,\mathbf{M} = \frac{1}{5}(\text{adj } \mathbf{M})\,\mathbf{M} = \dfrac{1}{5}\begin{pmatrix} -1 & 9 & -3 \\ -2 & 3 & -1 \\ 2 & -13 & 6 \end{pmatrix}\begin{pmatrix} 1 & -3 & 0 \\ 2 & 0 & 1 \\ 4 & 1 & 3 \end{pmatrix}$

$$= \frac{1}{5} \begin{pmatrix} 5 & 0 & 0 \\ 0 & 5 & 0 \\ 0 & 0 & 5 \end{pmatrix}$$

$$= \begin{pmatrix} 1 & 0 & 0 \\ 0 & 1 & 0 \\ 0 & 0 & 1 \end{pmatrix} = \mathbf{I}$$

## Example 24.10

Find, if it exists, the inverse, $\mathbf{M}^{-1}$, of the matrix $\mathbf{M} = \begin{pmatrix} 2 & 1 & -2 \\ 1 & 3 & 4 \\ 0 & 1 & 2 \end{pmatrix}$

➤ Find $\det \mathbf{M}$: $\det \mathbf{M} = 2(6-4) - (2-0) - 2(1-0) = 0$

➤ since $\Delta = 0$, $\mathbf{M}$ is a singular matrix and does not have an inverse.

 It is easy to make arithmetical mistakes when finding the adjoint matrix. This problem can be overcome by checking that either $\mathbf{M}\,\mathbf{M}^{-1}$ or $\mathbf{M}^{-1}\,\mathbf{M}$ equals $\mathbf{I}$. Alternatively the method of example 24.11 incorporates multiplying by the adjoint and so will reveal if mistakes have been made.

## Example 24.11

Find the inverse $\mathbf{P}^{-1}$ of the matrix $\mathbf{P} = \begin{pmatrix} 1 & -1 & 2 \\ 1 & 3 & 1 \\ 2 & 1 & 4 \end{pmatrix}$

➤ Find the cofactors of the elements of $\mathbf{P}$:

$A_1 = +(12-1) = 11,$     $B_1 = -(4-2) = -2,$     $C_1 = +(1-6) = -5;$
$A_2 = -(-4-2) = 6;$     $B_2 = +(4-4) = 0;$     $C_2 = -(1+2) = -3;$
$A_3 = +(-1-6) = -7;$     $B_3 = -(1-2) = 1;$     $C_3 = +(3+1) = 4$

➤ $\text{adj } \mathbf{P} = \begin{pmatrix} 11 & 6 & -7 \\ -2 & 0 & 1 \\ -5 & -3 & 4 \end{pmatrix}$

➤ $\mathbf{P}\,(\text{adj } \mathbf{P}) = \begin{pmatrix} 1 & -1 & 2 \\ 1 & 3 & 1 \\ 2 & 1 & 4 \end{pmatrix} \begin{pmatrix} 11 & 6 & -7 \\ -2 & 0 & 1 \\ -5 & -3 & 4 \end{pmatrix}$

$$= \begin{pmatrix} 3 & 0 & 0 \\ 0 & 3 & 0 \\ 0 & 0 & 3 \end{pmatrix} = 3\mathbf{I}$$

➤ this shows that $\Delta = \det \mathbf{P} = 3$

so $\mathbf{P}^{-1} = \dfrac{\text{adj } \mathbf{P}}{\Delta} = \dfrac{1}{3} \begin{pmatrix} 11 & 6 & -7 \\ -2 & 0 & 1 \\ -5 & -3 & 4 \end{pmatrix}$

## EXERCISE 24 E

Find the inverses, if they exist, of the following matrices

1. $P = \begin{pmatrix} 1 & 0 & 0 \\ -1 & 1 & 0 \\ 3 & 2 & 1 \end{pmatrix}$

2. $Q = \begin{pmatrix} 3 & 1 & 1 \\ 1 & 0 & -2 \\ 10 & 3 & 1 \end{pmatrix}$

3. $R = \begin{pmatrix} 2 & 0 & 1 \\ 4 & -1 & 2 \\ 3 & 1 & 1 \end{pmatrix}$

4. $S = \begin{pmatrix} 2 & 3 & 2 \\ 1 & 5 & 0 \\ 4 & 1 & 6 \end{pmatrix}$

5. $T = \begin{pmatrix} 3 & 2 & 4 \\ 1 & 5 & 3 \\ -1 & 8 & 2 \end{pmatrix}$

6. $U = \begin{pmatrix} 1 & 0 & 1 \\ -1 & 3 & 2 \\ 2 & -2 & 2 \end{pmatrix}$

# 3 × 3 SYSTEMS OF LINEAR EQUATIONS

We are familiar with the idea that lines in 2 dimensions have linear equations in 2 variables. In 3 dimensions a linear equation (in 3 variables) represents a plane. A system of three such equations will normally therefore give the point of intersection of three planes.

Just as we saw that 2 simultaneous equations in 2 variables can be written as a matrix equation involving a square 2 × 2 matrix and two column matrices, so 3 simultaneous equations in 3 variables can be written as a matrix equation:

The system $\left. \begin{array}{l} a_1x + b_1y + c_1z = d_1 \\ a_2x + b_2y + c_2z = d_2 \\ a_3x + b_3y + c_1z = d_3 \end{array} \right\}$ becomes $\begin{pmatrix} a_1 & b_1 & c_1 \\ a_2 & b_2 & c_2 \\ a_3 & b_3 & c_3 \end{pmatrix}\begin{pmatrix} x \\ y \\ z \end{pmatrix} = \begin{pmatrix} d_1 \\ d_2 \\ d_3 \end{pmatrix}$

The simultaneous equations can therefore be solved by finding $M^{-1}$ and premultiplying the matrix equation by $M^{-1}$, so that

$$M^{-1}M\begin{pmatrix} x \\ y \\ z \end{pmatrix} = M^{-1}\begin{pmatrix} d_1 \\ d_2 \\ d_3 \end{pmatrix} \quad \text{or} \quad \begin{pmatrix} x \\ y \\ z \end{pmatrix} = M^{-1}\begin{pmatrix} d_1 \\ d_2 \\ d_3 \end{pmatrix}$$

[If $\Delta = 0$, the equations cannot be solved using the inverse matrix and another method is needed to determine whether or not the equations are consistent – see below.]

## Example 24.12

Solve the simultaneous equations 
$$2x + y + z = 5$$
$$x - 2y + z = 4$$
$$x - y + 2z = 7$$

➢ Write in matrix form: $\begin{pmatrix} 2 & 1 & 1 \\ 1 & -2 & 1 \\ 1 & -1 & 2 \end{pmatrix}\begin{pmatrix} x \\ y \\ z \end{pmatrix} = \begin{pmatrix} 5 \\ 4 \\ 7 \end{pmatrix}$ ①

➢ Find adj $M$ and $\Delta$:

$$\text{adj } M = \begin{pmatrix} -3 & -3 & 3 \\ -1 & 3 & -1 \\ 1 & 3 & -5 \end{pmatrix}$$

$$\Delta = 2(-4+1) - 1(2-1) + 1(-1+2) = -6 - 1 + 1 = -6$$

➤ Solve by premultiplying ① by $\mathbf{M}^{-1}$ (= adj $\mathbf{M}/\Delta$):

➤ $$\begin{pmatrix} x \\ y \\ z \end{pmatrix} = -\frac{1}{6} \begin{pmatrix} -3 & -3 & 3 \\ -1 & 3 & -1 \\ 1 & 3 & -5 \end{pmatrix} \begin{pmatrix} 5 \\ 4 \\ 7 \end{pmatrix}$$

$$= -\frac{1}{6} \begin{pmatrix} -15-12+21 \\ -5+12-7 \\ 5+12-35 \end{pmatrix}$$

$$= -\frac{1}{6} \begin{pmatrix} -6 \\ 0 \\ -18 \end{pmatrix} = \begin{pmatrix} 1 \\ 0 \\ 3 \end{pmatrix}$$

➤ Give the answer: $x = 1, y = 0, z = 3$

## EXERCISE 24 F

Solve the following simultaneous equations using inverse matrices

1.  $3x - y + z = 1$
    $x + y - z = 7$
    $2x + y + 3z = 5$

2.  $2x + y + 2z = 1$
    $x + 4z = 2$
    $y - 2z = 1$

3.  $x - y - z = 3$
    $2x + y - z = 10$
    $x - 2y + 3z = 6$

## 📖 SOME MORE DEFINITIONS

A matrix in which all the elements below (or above) the leading diagonal are 0 is called a **triangular** matrix

e.g. $\begin{pmatrix} a_1 & b_1 & c_1 \\ 0 & b_2 & c_2 \\ 0 & 0 & c_3 \end{pmatrix}$ and $\begin{pmatrix} a_1 & 0 & 0 \\ a_2 & b_2 & 0 \\ a_3 & b_3 & c_3 \end{pmatrix}$ are triangular matrices.

When a matrix has been manipulated so that it becomes a triangular matrix, it is said to be **reduced to echelon form**.

The system of equations $\left.\begin{array}{l} a_1 x + b_1 y + c_1 z = d_1 \\ a_2 x + b_2 y + c_2 z = d_2 \\ a_3 x + b_3 y + c_1 z = d_3 \end{array}\right\}$ can be written in matrix form $\begin{pmatrix} a_1 & b_1 & c_1 \\ a_2 & b_2 & c_2 \\ a_3 & b_3 & c_3 \end{pmatrix} \begin{pmatrix} x \\ y \\ z \end{pmatrix} = \begin{pmatrix} d_1 \\ d_2 \\ d_3 \end{pmatrix}$

If the constants are included in the $3 \times 3$ matrix, the resulting matrix $\left(\begin{array}{ccc|c} a_1 & b_1 & c_1 & d_1 \\ a_2 & b_2 & c_2 & d_2 \\ a_3 & b_3 & c_3 & d_3 \end{array}\right)$ is called the

**augmented matrix** for the system.

## SOLVING A SYSTEM OF SIMULTANEOUS EQUATIONS BY ROW REDUCTION

One common method for solving 2 simultaneous equations in 2 unknowns is by elimination. This method can be extended to larger systems, by eliminating one unknown at a time until an equation in only one unknown is obtained. Applied to example 24.12 it would look like this

Solve the simultaneous equations

$$2x + y + z = 5 \quad ①$$
$$x - 2y + z = 4 \quad ②$$
$$x - y + 2z = 7 \quad ③$$

First eliminate $x$ from ① and ② by subtracting $2 \times$ ② from ①: $5y - z = -3$ ④

Next eliminate $x$ from ① and ③ by subtracting $2 \times$ ③ from ①: $3y - 3z = -9$ ⑤

We now have 2 equations in 2 unknowns, so we can use elimination again to eliminate $y$. Normally we would divide ⑤ through by 3 and then eliminate $z$, but since we want to generalise the method we will leave it as it stands and eliminate $y$.

$3 \times$ ④ $- 5 \times$ ⑤: $\qquad 12z = 36 \quad \therefore z = 3$

Putting $z = 3$ in ④ gives $y = 0$ and putting $z = 3, y = 0$ in ① gives $x = 1$
Answer: $x = 1, \; y = 0, \; z = 3$

This process can be systematised using matrices, by reducing the augmented matrix for the system to echelon form. The process is often known as **row reduction**. Repeating the same example will show how it works.

## Example 24.13

Solve the simultaneous equations

$$2x + y + z = 5$$
$$x - 2y + z = 4$$
$$x - y + 2z = 7$$

Write in matrix form:
$$\begin{pmatrix} 2 & 1 & 1 \\ 1 & -2 & 1 \\ 1 & -1 & 2 \end{pmatrix} \begin{pmatrix} x \\ y \\ z \end{pmatrix} = \begin{pmatrix} 5 \\ 4 \\ 7 \end{pmatrix}$$

➤ Write out the augmented matrix:
$$\begin{matrix} r_1 \\ r_2 \\ r_3 \end{matrix} \left( \begin{array}{ccc|c} 2 & 1 & 1 & 5 \\ 1 & -2 & 1 & 4 \\ 1 & -1 & 2 & 7 \end{array} \right)$$

➤ Eliminate $x$:
$$\begin{matrix} r_1 = s_1 \\ r_1 - 2r_2 = s_2 \\ r_1 - 2r_3 = s_3 \end{matrix} \left( \begin{array}{ccc|c} 2 & 1 & 1 & 5 \\ 0 & 5 & -1 & -3 \\ 0 & 3 & -3 & -9 \end{array} \right)$$

➤ Eliminate $y$:
$$\begin{matrix} s_1 \\ s_2 \\ 3s_2 - 5s_3 \end{matrix} \left( \begin{array}{ccc|c} 2 & 1 & 1 & 5 \\ 0 & 5 & -1 & -3 \\ 0 & 0 & 12 & 36 \end{array} \right)$$

➤ Rewrite the equations from the new augmented matrix:
$$2x + y + z = 5 \quad ①$$
$$5y - z = -3 \quad ②$$
$$12z = 36 \quad ③$$

➤ From ③, $z = 3$;

substituting in ② gives $y = 0$;

substituting in ① gives $x = 1$

If the row reduction leads to a 0 appearing on the leading diagonal, the rows of the augmented matrix should be rearranged to avoid the 0.

## Example 24.14

Solve the simultaneous equations

$$x + y + z = 5$$
$$2x + 2y - z = 7$$
$$x - y + 2z = 2$$

➢ Write out the augmented matrix:

$$\begin{array}{c} r_1 \\ r_2 \\ r_3 \end{array} \left( \begin{array}{ccc|c} 1 & 1 & 1 & 5 \\ 2 & 2 & -1 & 7 \\ 1 & -1 & 2 & 2 \end{array} \right)$$

➢ Eliminate $x$:

$$\begin{array}{c} r_1 = s_1 \\ 2r_1 - r_2 = s_2 \\ r_1 - r_3 = s_3 \end{array} \left( \begin{array}{ccc|c} 1 & 1 & 1 & 5 \\ 0 & 0 & 3 & 3 \\ 0 & 2 & -1 & 3 \end{array} \right)$$

➢ because there is a 0 on the leading diagonal, interchange rows $s_2$ and $s_3$

$$\begin{array}{c} s_1 \\ s_3 \\ s_2 \end{array} \left( \begin{array}{ccc|c} 1 & 1 & 1 & 5 \\ 0 & 2 & -1 & 3 \\ 0 & 0 & 3 & 3 \end{array} \right)$$

➢ this is in echelon form and the equations have become
$$x + y + z = 5$$
$$2y - z = 3$$
$$3z = 3$$

➢ solution: $z = 1, y = 2, x = 2$

 just as the order of the equations, and therefore the rows of the matrix, can be interchanged, so also any row can be multiplied or divided by a non-zero constant without affecting the solution to the equations, and rows can be added to or subtracted from each other.

When the rows of the matrix, or the equations themselves are manipulated according to these rules, the consistency and solutions to the equations do not change and so the systems are said to be **equivalent**.


# EXERCISE 24 G

Solve the following systems of equations using row reduction

1.     $x + y + z = 2$
       $2x + y - 3z = 1$
       $2x - y - 5z = 3$

2.     $3x + y + z = 4$
       $x + 2y - 3z = 3$
       $2x - y + 5z = 3$

3.     $2x + y + 3z = 4$
       $y - z = 4$
       $4x - 2y + z = 1$


## _Consistency of the System and Number of Solutions_

The systems of equations we have worked with so far have unique solutions. These systems can be represented geometrically by 3 planes that intersect in a single point.

If, however, the matrix representing the equations is singular, there is no unique solution.

There will be no solutions if the equations are inconsistent (i.e. the equations cannot all be true at the same time), which will happen if either two or more of the planes are parallel but not coincident (see fig. 24.1-3) or each plane is parallel to the line of intersection of the other two (see fig. 24.4).

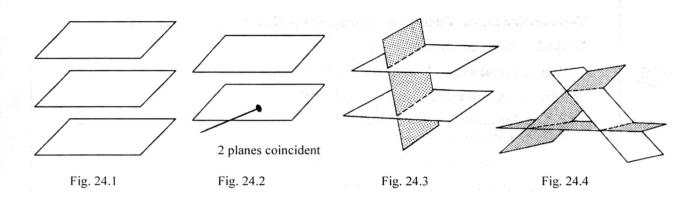

2 planes coincident

Fig. 24.1          Fig. 24.2          Fig. 24.3          Fig. 24.4

If all 3 equations represent the same plane (see fig. 24.5) then any point in that plane is a solution and there are an infinite number of solutions. If 2 planes are coincident and intersect the third plane (see fig. 24.6) or if all 3 planes have a common line (see fig. 24.7) then any point on the line of intersection is a solution and again there are an infinite number of solutions.

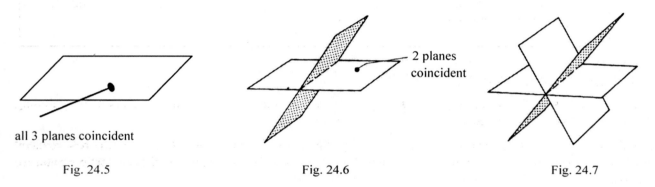

2 planes coincident

all 3 planes coincident

Fig. 24.5                    Fig. 24.6                    Fig. 24.7

## Using the augmented matrix to determine the number of solutions

If the determinant of the matrix of a system of equations is found to be non-zero, then the system is known to be consistent and have a unique solution. However, if the determinant is found to be zero, neither the consistency nor the number of solutions is determined. The augmented matrix reduced to echelon form can be used to determine these things.

Suppose that a system of equations has been written in matrix form and the augmented matrix reduced to echelon form, then the augmented matrix can be expressed as $\begin{pmatrix} a_1 & b_1 & c_1 & | & d_1 \\ 0 & b_2 & c_2 & | & d_2 \\ 0 & 0 & c_3 & | & d_3 \end{pmatrix}$

If the elements of the leading diagonal are all non-zero ($a_1 \neq 0$, $b_2 \neq 0$, $c_3 \neq 0$) then the matrix is non-singular and the system has a unique solution.

If $c_3 = 0$ and $d_3 \neq 0$ (i.e $0z = d_3$, which is impossible) the system is inconsistent and there are no solutions. The same is true if $b_2 = c_2 = 0$ and $d_2 \neq 0$.          (A)

As long as (A) is not true, then if either row 2 or row 3 or both consist entirely of zeros the system will have an infinite number of solutions.

**To determine the consistency and number of solutions of a 3 × 3 system of equations**

**Method I** using det **M**.

If det **M** ≠ 0 the system is consistent and has a unique solution

If det **M** = 0, row reduce the augmented matrix.

if any row consists entirely of zeros, the system is consistent and has an infinite number of solutions

otherwise the system is inconsistent and has no solutions.

**Method II** using only the augmented matrix

If all elements of the diagonal of the triangular matrix are non-zero, the system is consistent and has a unique solution

If either row 2 or row 3 have all zeros in the triangular matrix but a non-zero number in the augmented section ($d_2$ or $d_3$) the system is inconsistent and has no solutions (A)

If (A) is not true and any row consists entirely of zeros, the system is consistent and has an infinite number of solutions.

## Finding All the Solutions of a System

We have already seen that when $\Delta \neq 0$ the system has a unique solution which can be found either by finding the inverse matrix or by row reduction.

An infinite number of solutions occur if the three equations are the same (rows 2 and 3 in the row reduced augmented matrix consist entirely of zeros), in which case any set of values $(x, y, z)$ satisfying that equation are solutions.

Where the equations represent planes that meet in a common line (either row 2 or row 3 consists entirely of zeros) there are again an infinite number of solutions, the coordinates of all the points on that common line. The solutions can be given in terms of a parameter. See example 24.14.

## Example 24.15

For what value of $c$ does the system of equations $\begin{cases} 2x - 4y + 2z = 5 \\ 4x - y + 2z = 1 \\ 2x + cy \quad\quad = -6 \end{cases}$ not have a unique solution?

With that value of $c$, how many solutions does the system have?

➢ Solution is not unique if $\Delta = 0$

➢ Find $\Delta$: $\Delta = \begin{vmatrix} 2 & -4 & 2 \\ 4 & -1 & 2 \\ 2 & c & 0 \end{vmatrix} = 2(0 - 2c) + 4(0 - 4) + 2(4c + 2) = 4c - 12$

➢ Solution is not unique when $c = 3$

➢ When $c = 3$, augmented matrix is $\begin{pmatrix} 2 & -4 & 2 & | & 5 \\ 4 & -1 & 2 & | & 1 \\ 2 & 3 & 0 & | & -6 \end{pmatrix}$

➤ Row reduce the augmented matrix: 
$$\begin{array}{c} r_1 \\ r_2 - 2r_1 \\ r_3 - r_1 \end{array} \left( \begin{array}{ccc|c} 2 & -4 & 2 & 5 \\ 0 & 7 & -2 & -9 \\ 0 & 7 & -2 & -11 \end{array} \right)$$

$$\left( \begin{array}{ccc|c} 2 & -4 & 2 & 5 \\ 0 & 7 & -2 & -9 \\ 0 & 0 & 0 & 2 \end{array} \right)$$

➤ The last row of this augmented matrix shows that the system is inconsistent and has no solutions

## Example 24.16

Determine how many solutions each of the following systems of equations has, and if they exist, find them

(a)  $x + 2y + 3z = 4$
     $2x + y + 5z = -2$
     $4x + 5y + 11z = 6$

(b)  $2x - 3y + 5z = 1$
     $4x + z = 2$
     $8x - 12y + 20z = 4$

(a) ➤ Write out the augmented matrix: 
$$\begin{array}{c} r_1 \\ r_2 \\ r_3 \end{array} \left( \begin{array}{ccc|c} 1 & 2 & 3 & 4 \\ 2 & 1 & 5 & -2 \\ 4 & 5 & 11 & 6 \end{array} \right)$$

➤ Eliminate $x$: 
$$\begin{array}{c} r_1 \\ 2r_1 - r_2 \\ 4r_1 - r_3 \end{array} \left( \begin{array}{ccc|c} 1 & 2 & 3 & 4 \\ 0 & 3 & 1 & 10 \\ 0 & 3 & 1 & 10 \end{array} \right)$$

➤ Eliminate $y$: 
$$\left( \begin{array}{ccc|c} 1 & 2 & 3 & 4 \\ 0 & 3 & 1 & 10 \\ 0 & 0 & 0 & 0 \end{array} \right)$$

➤ Interpret:     the system is consistent and row 3 shows that there is an infinite number of solutions
➤ Find the solutions:  let one of the unknowns equal the parameter $t$, e.g. let $y = t$
➤ Find $z$:                  from row 2, $3y + z = 10$, ∴ $z = 10 - 3t$
➤ Find $x$:                  from row 1, $x + 2y + 3z = 4$, ∴ $x = 4 - 2t - 3(10 - 3t) = 7t - 26$
➤ Give the solution:   the system has an infinite number of solutions satisfying the parametric equations $x = 7t - 26$, $y = t$, $z = 10 - 3t$
➤ Geometric Interpretation: clearly none of the planes represented by the equations is coincident, so the three planes intersect in a common line.

(b) ➤ Write out the augmented matrix: 
$$\begin{array}{c} r_1 \\ r_2 \\ r_3 \end{array} \left( \begin{array}{ccc|c} 2 & -3 & 5 & 1 \\ 4 & 0 & 1 & 2 \\ 8 & -12 & 20 & 4 \end{array} \right)$$

➤ Eliminate $x$:
$$\begin{array}{c} r_1 \\ r_2 - 2r_1 \\ 4r_1 - r_3 \end{array} \left( \begin{array}{ccc|c} 2 & -3 & 5 & 1 \\ 0 & 6 & -9 & 0 \\ 0 & 0 & 0 & 0 \end{array} \right)$$

➤ Interpret:  the system is consistent and row 3 shows that there is an infinite number of solutions

➤ Find the solutions:  let one of the unknowns equal the parameter $t$, let $x = t$

➤ Find $z$:  from equation 2, $z = 2 - 4t$

➤ Find $y$:  from row 2, $2y - 3z = 0$, $\therefore y = 3 - 6t$

➤ Give the solution:  the system has an infinite number of solutions satisfying the parametric equations $x = t$, $y = 3 - 6t$, $z = 2 - 4t$

➤ Geometric Interpretation: the first and third equations represent the same plane, so the system consists of two planes intersecting in a line.

# $2 \times 2$ Systems of Linear Equations

We looked at $3 \times 3$ systems of equations before $2 \times 2$ systems because the echelon form is clearer with the larger number of equations, but simultaneous equations in two unknowns can be treated in the same way as those in three unknowns.

In example 24.5 and exercise 24C we used the inverse matrix method to solve simultaneous equations in two unknowns.  This was possible because the matrix was not singular and the equations had unique solutions. When $\Delta = 0$, row reduction can be used to determine consistency and solutions.

## Example 24.17

Use row reduction of the augmented matrix to determine how many solutions each of the following systems of equations has, and if they exist, find the solutions.

(a) $\begin{cases} y = 2x - 4 \\ x + 2y = 7 \end{cases}$  (b) $\begin{cases} 2x - y = 5 \\ 2y - 4x = 3 \end{cases}$  (c) $\begin{cases} x + 2y = 7 \\ 3x + 6y = 21 \end{cases}$

(a) ➤ rearrange equations:  $\begin{cases} -2x + y = -4 \\ x + 2y = 7 \end{cases}$

➤ write out augmented matrix:  $\begin{array}{c} r_1 \\ r_2 \end{array} \left( \begin{array}{cc|c} -2 & 1 & -4 \\ 1 & 2 & 7 \end{array} \right)$

➤ eliminate $x$:  $\begin{array}{c} r_1 \\ r_1 + 2r_2 \end{array} \left( \begin{array}{cc|c} -2 & 1 & -4 \\ 0 & 5 & 10 \end{array} \right)$

➤ find solutions:  $5y = 10$, $\therefore y = 2$, $x = 7 - 4 = 3$

➤ interpret:  equations are consistent with unique solution $x = 3$, $y = 2$.

(b) ➤ rearrange equations:  $\begin{cases} 2x - y = 5 \\ -4x + 2y = 3 \end{cases}$

➤ write out augmented matrix:  $\begin{array}{c} r_1 \\ r_2 \end{array} \left( \begin{array}{cc|c} 2 & -1 & 5 \\ -4 & 2 & 3 \end{array} \right)$

➤ eliminate $x$:

$$\begin{array}{c} r_1 \\ r_2 + 2r_1 \end{array} \left( \begin{array}{cc|c} 2 & -1 & 5 \\ 0 & 0 & 13 \end{array} \right)$$

➤ interpret:   the equations are inconsistent and have no solutions.

(c) ➤ write out augmented matrix:

$$\begin{array}{c} r_1 \\ r_2 \end{array} \left( \begin{array}{cc|c} 1 & 2 & 7 \\ 3 & 6 & 21 \end{array} \right)$$

➤ eliminate $x$:

$$\begin{array}{c} r_1 \\ 3r_1 - r_2 \end{array} \left( \begin{array}{cc|c} 1 & 2 & 7 \\ 0 & 0 & 0 \end{array} \right)$$

➤ interpret:   the equations are consistent but do not have a unique solution.

➤ Geometric interpretation:

(a)   lines intersect in a single point (3, 2)

(b)   equations can be written as $\begin{cases} 4x - 2y = 10 \\ 4x - 2y = -3 \end{cases}$

these are clearly the equations of parallel lines
the equations are therefore inconsistent, with no solution

(c)   dividing the second equation by 3 gives $\begin{cases} x + 2y = 7 \\ x + 2y = 7 \end{cases}$

so both equations represent the same line
the equations are therefore consistent
    and the solution is all points on the line $x + 2y = 7$

---

**To determine the consistency and number of solutions of a $2 \times 2$ system of equations**

If det $\mathbf{M} \neq 0$ the system is consistent and has a unique solution

If det $\mathbf{M} = 0$, row reduce the augmented matrix.

if any row consists entirely of zeros, the system is consistent and has an infinite number of solutions

otherwise the system is inconsistent and has no solutions.

---

## Example 24.18

A pair of simultaneous equations is given by

$$\left. \begin{array}{l} 5x + 6y = q \\ px + 3y = 7 \end{array} \right\} \quad \text{where } p, q \in \mathbb{R} \qquad \begin{array}{l} ① \\ ② \end{array}$$

Find the values of $p$ and $q$ for which the system of equations above has

(a)   a unique solution,

(b)   an infinite number of solutions,

(c)   no solution

➢ write in matrix augmented form:
$$\begin{matrix} r_1 \\ r_2 \end{matrix} \begin{pmatrix} 5 & 6 & | & q \\ p & 3 & | & 7 \end{pmatrix}$$

➢ eliminate $x$:
$$\begin{matrix} r_1 \\ pr_1 - 5r_2 \end{matrix} \begin{pmatrix} 5 & 6 & | & q \\ 0 & 6p-15 & | & pq-35 \end{pmatrix}$$

(a) ➢ unique solution if $6p - 15 \neq 0$ i.e. $p \neq 2\frac{1}{2}$

(b) ➢ infinite number of solutions if $6p - 15 = 0$ AND $pq - 35 = 0$
i.e. $p = 2\frac{1}{2}$ and $q = 14$

(c) ➢ no solutions if $6p - 15 = 0$ AND $pq - 35 \neq 0$
i.e. $p = 2\frac{1}{2}$ and $q \neq 14$

# EXERCISE 24 H

1. Use matrices to determine whether the following systems of equations have no solutions, one solution or an infinite number of solutions. Where solutions exist, find them.

(a) $x + 2y + 3z = 2$
$x - y + 2z = 4$
$3x + z = 4$

(b) $2x - 2y + z = 3$
$-3x + 3y + 2z = -1$
$2x - 2y + 3z = 5$

(c) $x + y + z = -1$
$2x - 2y + z = 5$
$3x - y + 2z = -1$

(d) $2x - y + 4z = 0$
$3x + 4y + 12z = 0$
$7x - 9y + 8z = 0$

(e) $2x + y + 3z = 7$
$2x - y + 4z = 9$
$-x + 2y - z = 1$

(f) $2x + y + z = 4$
$5x - 2y - 4z = 3$
$x - y - 2z = 0$

(g) $\begin{pmatrix} 1 & 2 & 1 \\ 3 & 1 & 0 \\ 5 & 5 & 2 \end{pmatrix} \begin{pmatrix} x \\ y \\ z \end{pmatrix} = \begin{pmatrix} 7 \\ -2 \\ 12 \end{pmatrix}$

(h) $\begin{pmatrix} 3 & -1 & 1 \\ -1 & 2 & 3 \\ 3 & -1 & 1 \end{pmatrix} \begin{pmatrix} x \\ y \\ z \end{pmatrix} = \begin{pmatrix} 2 \\ -2 \\ 12 \end{pmatrix}$

(i) $\begin{pmatrix} 2 & 1 & 1 \\ 4 & 2 & 2 \\ 5 & -2 & -4 \end{pmatrix} \begin{pmatrix} x \\ y \\ z \end{pmatrix} = \begin{pmatrix} 4 \\ 2 \\ 3 \end{pmatrix}$

(j) $\begin{pmatrix} 1 & 3 & -1 \\ 1 & -2 & 1 \\ -2 & 4 & -2 \end{pmatrix} \begin{pmatrix} x \\ y \\ z \end{pmatrix} = \begin{pmatrix} -3 \\ 4 \\ -8 \end{pmatrix}$

(k) $\begin{pmatrix} -1 & 2 & -1 \\ 1 & 4 & -3 \\ 1 & -10 & 7 \end{pmatrix} \begin{pmatrix} x \\ y \\ z \end{pmatrix} = \begin{pmatrix} -3 \\ -3 \\ 13 \end{pmatrix}$

(l) $\begin{pmatrix} 1 & -2 & 1 \\ 1 & 3 & -1 \\ -2 & 4 & -2 \end{pmatrix} \begin{pmatrix} x \\ y \\ z \end{pmatrix} = \begin{pmatrix} 4 \\ -3 \\ -2 \end{pmatrix}$

2. For each of the following pairs of equations find the values of $p$ and $q \in \mathbb{R}$ for which the system has (i) a unique solution (ii) an infinite number of solutions (iii) no solution

(a) $6x - py = 4$
$3x + 5y = q$

(b) $4x + 3y = q$
$px - 6y = 5$

(c) $x + y = p$
$x + qy = 8$

(d) $2x - 6y = p$
$px - 12y = q$

(e) $7x + 5y = 12$
$px + 2y = q$

(f) $py - 5x = 0$
$2y + x = q$

(g) $x + 11y = p$
$4x - qy = 2$

(h) $3y + px + 7 = 0$
$6y - 3x + q = 0$

3.  A pair of simultaneous equations is given by

$$\left.\begin{array}{l} 2x + py = 1 \\ 6x + 9y = p \end{array}\right\} \quad \text{where } p \in \mathbb{R}$$

Find the value of $p$, if any, for which the system of equations above has

(i)   an infinite number of solutions          (ii)   no solutions

4.  A pair of simultaneous equations is given by

$$\left.\begin{array}{l} 5x - 3y = p \\ 10x + py = q \end{array}\right\} \quad \text{where } p, q \in \mathbb{R}$$

(i)   Find the values of $p$ and $q$ for which the equations have an infinite number of solutions, and find those solutions.
(ii)  Interpret your answer geometrically.

## MATRICES AND TRANSFORMATIONS

Just as $2 \times 2$ matrices can represent transformations in the Cartesian plane, so $3 \times 3$ matrices can represent transformations in 3-dimensional space.

If $P'(x', y', z')$ is the image of $P(x, y, z)$ under the transformation $\mathbf{T}$, then $\mathbf{T P} = \mathbf{P'}$ or $\mathbf{T}\begin{pmatrix} x \\ y \\ z \end{pmatrix} = \begin{pmatrix} x' \\ y' \\ z' \end{pmatrix}$.

## Example 24.19

Given that the point $Q(1, 2, 7)$ is the image of $P(x, y, z)$ under the transformation $\begin{pmatrix} 0 & 1 & 1 \\ 1 & 1 & 0 \\ 1 & 0 & 1 \end{pmatrix}$, find the coordinates of P.

This can be done in one of 2 ways:

*either using inverse matrices:*

Since $\mathbf{TP} = \mathbf{Q}$, $\mathbf{T^{-1} T P} = \mathbf{T^{-1} Q}$, i.e. $\mathbf{P} = \mathbf{T^{-1} Q}$

➤  Find $\mathbf{T^{-1}}$: $\det \mathbf{T} = 0 - (1 - 0) + (0 - 1) = -2$, cofactor matrix of $\mathbf{T} = \begin{pmatrix} 1 & -1 & -1 \\ -1 & -1 & 1 \\ -1 & 1 & -1 \end{pmatrix}$

$$\mathbf{T^{-1}} = \frac{\text{adj} \mathbf{T}}{\Delta} = \frac{1}{-2}\begin{pmatrix} 1 & -1 & -1 \\ -1 & -1 & 1 \\ -1 & 1 & -1 \end{pmatrix} = \frac{1}{2}\begin{pmatrix} -1 & 1 & 1 \\ 1 & 1 & -1 \\ 1 & -1 & 1 \end{pmatrix}$$

> Find **P**: $\begin{pmatrix} x \\ y \\ z \end{pmatrix} = \frac{1}{2} \begin{pmatrix} -1 & 1 & 1 \\ 1 & 1 & -1 \\ 1 & -1 & 1 \end{pmatrix} \begin{pmatrix} 1 \\ 2 \\ 7 \end{pmatrix} = \begin{pmatrix} 4 \\ -2 \\ 3 \end{pmatrix}$

> Answer: P is point (4, 2, 3)

### *or using row reduction*

> Transformation of P into Q can be written as $\begin{pmatrix} 0 & 1 & 1 \\ 1 & 1 & 0 \\ 1 & 0 & 1 \end{pmatrix} \begin{pmatrix} x \\ y \\ z \end{pmatrix} = \begin{pmatrix} 1 \\ 2 \\ 7 \end{pmatrix}$

> The augmented matrix is $\begin{array}{c} r_1 \\ r_2 \\ r_3 \end{array} \left( \begin{array}{ccc|c} 0 & 1 & 1 & 1 \\ 1 & 1 & 0 & 2 \\ 1 & 0 & 1 & 7 \end{array} \right)$

> Rearrange rows: $\begin{array}{c} r_2 \\ r_1 \\ r_3 \end{array} \left( \begin{array}{ccc|c} 1 & 1 & 0 & 2 \\ 0 & 1 & 1 & 1 \\ 1 & 0 & 1 & 7 \end{array} \right)$

> Row reduce: $\begin{array}{c} r_2 \\ r_1 \\ r_2 - r_3 \end{array} \left( \begin{array}{ccc|c} 1 & 1 & 0 & 2 \\ 0 & 1 & 1 & 1 \\ 0 & 1 & -1 & -5 \end{array} \right)$

> $\begin{array}{c} s_1 \\ s_2 \\ s_3 \end{array} \left( \begin{array}{ccc|c} 1 & 1 & 0 & 2 \\ 0 & 1 & 1 & 1 \\ 0 & 0 & -2 & -6 \end{array} \right)$

> Find $x$, $y$ and $z$: from $s_3$, $-2z = -6$, $\therefore z = 3$; from $s_2$, $y + z = 1$, $\therefore y = -2$
>
> from $s_1$, $x + y = 2$, $\therefore x = 4$

> Answer: P is the point (4, –2, 3)

## More Problems involving Matrices – *Data storage and manipulation*

## Example 24.20

A catering firm supplies coffee, lunch and tea to a number of office canteens. Allen's receive 15 coffees, 20 lunches and 12 teas per day; Barclays receive 48 coffees, 40 lunches and 45 teas; Christie's receive 12 coffees and 12 teas; Davis' receive 18 coffees, 19 lunches and 5 teas.
Allen's and Christie's canteens open 5 days a week, Barclays 6 days and Davis' 4 days.

(a) Write down two matrices only such that the elements of their product under matrix multiplication give the total number of coffees, lunches and teas supplied each week. Find this product.

(b) If the cost of coffee, lunch and tea to the offices is $0.80, $2.50 and $1.00 respectively write this data as a matrix (**N**) which can be multiplied by **M**, your answer to (a). Find the product of **M** and **N** and interpret your answer.

(a) Let the first set of data be represented by the matrix

$$\begin{array}{cc} & \begin{array}{ccc} C & L & T \end{array} \\ \begin{array}{c} A \\ B \\ C \\ D \end{array} & \begin{pmatrix} 15 & 20 & 12 \\ 48 & 40 & 45 \\ 12 & 0 & 12 \\ 18 & 19 & 5 \end{pmatrix} \end{array}$$

Remembering that matrices can only be multiplied together if the number of columns in the first is the same as the number of rows in the second, and that the second set of data contains 4 elements (the number of days each of the canteens receives refreshments), we need to multiply a $1 \times 4$ matrix by the $4 \times 3$ matrix. This gives

$$(5 \quad 6 \quad 5 \quad 4) \begin{pmatrix} 15 & 20 & 12 \\ 48 & 40 & 45 \\ 12 & 0 & 12 \\ 18 & 19 & 5 \end{pmatrix} = (495 \quad 418 \quad 310) \begin{pmatrix} \text{giving the number of} \\ \text{each refreshment} \\ \text{supplied per week} \end{pmatrix}$$

(b) Taking $\mathbf{M}$ to be $(495 \quad 418 \quad 310)$ $\mathbf{N}$ must be a column matrix, i.e. $\begin{pmatrix} 0.80 \\ 2.50 \\ 1.00 \end{pmatrix}$ $\begin{pmatrix} \text{cost of each} \\ \text{refreshment} \end{pmatrix}$

$$\mathbf{MN} = (495 \quad 418 \quad 310) \begin{pmatrix} 0.80 \\ 2.50 \\ 1.00 \end{pmatrix} = (1719) \quad \text{showing that the weekly takings of the catering firm are}$$

$1719. **(NM also exists but is a $3 \times 3$ matrix, and has no practical significance.)**

# EXERCISE 24 I

1. Given that $\mathbf{T}$ is the transformation $\begin{pmatrix} 1 & 2 & 0 \\ 2 & 1 & 0 \\ 0 & 1 & 2 \end{pmatrix}$ and P is the point (5, 4, 8), Find

   (a) the coordinates of P′ such that P′ is the image of P under T,
   (b) the coordinates of Q, such that P is the image of Q under T.

2. Repeat question 1 with $\mathbf{T} = \begin{pmatrix} 0 & 3 & 2 \\ 1 & 0 & 1 \\ 2 & 1 & 0 \end{pmatrix}$ and P (12, 5, 6)

3. A factory manufactures 120 standard and 15 deluxe tennis rackets per week. This is represented by matrix $\mathbf{A} = (120 \quad 15)$. $\mathbf{B} = \begin{pmatrix} 4 & 0.8 \\ 6 & 1 \end{pmatrix}$ represents the fact that a standard racket requires 4 hours labour and 0.8 hours lamination etc. and $\mathbf{C} = \begin{pmatrix} 10.50 \\ 15 \end{pmatrix}$ that labour costs \$10.50 per hour and lamination \$15 per hour. Find the matrix $\mathbf{ABC}$ and interpret its meaning.

4.    Darren and Kesia went shopping and their purchases were as follows:

|        | crisps | biscuits | soda |
|--------|--------|----------|------|
| Darren | 6      | 4        | 3    |
| Kesia  | 8      | 5        | 2    |

(a)   Write this information in the form of a 2 × 3 matrix and call it **P**.

(b)   Crisps cost $ 0.20 per packet, biscuits cost $ 0.25 per packet and a bottle of soda  costs $ 0.35.  Write this information as a matrix **C** such that the matrix **PC** exists.

(c)   Work out **PC**.

(d)   Pre-multiply **PC** by the row matrix  (1   1)  and explain what the result means.


5.    Four companies each run a fleet of buses of three different sizes.  The table below shows how many buses of each size (measured by seating capacity) are run by each company, the laden weight of a bus of each size, and the fare (per passenger) charged by each company for a journey to X.

| number of seats→<br><br>name of company↓ | 12 | 40 | 52 | fare per passenger↓ |
|---|---|---|---|---|
| Archibald's | 10 | 5 | 1 | $ 12.00 |
| Barry's | 0 | 3 | 2 | $ 11.50 |
| Christopher's | 4 | 8 | 0 | $ 12.50 |
| Daniel's | 0 | 0 | 3 | $ 11.00 |
| Laden weight→ | 2 T | 5 T | 6 T | |

(a)   Write down two matrices such that the elements of their product under matrix multiplication give the total number of passengers carried by each company (assuming all seats on all buses are filled).  Find this product.

(b)   Use the product in part (a) and one other matrix to find the total amount of money received when all the filled buses travel to X.

One week the 4 companies plan a trip involving a ferry crossing.

(c)   Use matrix multiplication to find the total weight of the full buses for each company.

(d)   If the ferry can carry a maximum load of 150 T, can all the buses travel on the ferry?  Justify  your answer.


6.    In a soccer league 3 points are awarded for a win, 1 for a draw and 0 for a loss.  The results after 12 matches for four of the teams were: Athenians won 5 matches, drew 3 and lost 4;  Barbarians won 6 and lost 6;  Corinthians won 1, drew 3 and lost 8; and Delphians won 8, drew 2 and lost 2.

Represent this data in two matrices, such that their product will represent the number of points gained by each team.  Find this product, and hence which team was in the lead at this stage.


7.    Linen gift sets come in three varieties.  Set A contains 4 towels, 2 sheets, 8 pillowcases and 2 duvet covers; set B contains 2 sheets, 8 pillowcases and 4 duvet covers, and set C contains 6 towels, 6 sheets and 6 pillowcases.  Towels cost $5 each, sheets $10, pillowcases $2, and duvet covers $15.

(a)   Write down two matrices such that the elements of their product under matrix multiplication give the cost of each set, and find this product.

(b)   If a store buys 25 set A, 20 set B and 12 set C, use 2 matrices to find their total bill.

# EXERCISE 24 X

1. (a) A system of linear equations is given below

$$3x + 2y - z = 2$$
$$x - y + 2z = 14$$
$$-2x - 3y + z = 2$$

   (i) Write the system of equations in matrix form.

   (ii) Show that the determinant of the $3 \times 3$ matrix is 10. [5]

   (b) Given that the corresponding co-factor matrix is

$$\begin{pmatrix} 5 & -5 & -5 \\ 1 & 1 & 5 \\ 3 & -7 & -5 \end{pmatrix}$$

   find the values of $x, y$ and $z$. [5]

2. A pair of simultaneous equations is given by

$$\left.\begin{array}{r} px + 4y = 8 \\ 6x + 2y = q \end{array}\right\} \quad \text{where } p, q \in \mathbb{R}$$

   State the values of $p$ and $q$ for which the simultaneous equations have an infinite number of solutions. [2]

3. A pair of simultaneous equations is given by

$$\left.\begin{array}{r} p^2 x - 4y = 8 \\ 8x - 2y = p \end{array}\right\} \quad \text{where } p \in \mathbb{R}$$

   (a) Find the value of $p$ for which the equations have an infinite number of solutions. [4]

   (b) Find the solutions for this value of $p$. [2]

4. The transformation in three-dimensional space of a point, P, with coordinates $(x, y, z)$ is represented below.

$$\begin{pmatrix} 1 & 1 & 1 \\ 1 & -1 & 2 \\ 2 & 1 & -1 \end{pmatrix} \begin{pmatrix} x \\ y \\ z \end{pmatrix} = \begin{pmatrix} 7 \\ 9 \\ 1 \end{pmatrix}$$

   By row-reducing the augmented matrix, find the coordinates of P. [8]

5. The matrices **A** and **B** are given below.

$$\mathbf{A} = \begin{pmatrix} 1 & 2 & 3 \\ 4 & 5 & 6 \end{pmatrix}, \quad \mathbf{B} = \begin{pmatrix} 1 & -1 \\ 2 & 2 \\ 2 & -3 \end{pmatrix}$$

(a) Calculate

    (i) **AB** [2]

    (ii) **B$^T$A$^T$**. [3]

(b) Deduce that $(\mathbf{AB})^T = \mathbf{B}^T\mathbf{A}^T$. [2]

6. (a) Write the augmented matrix for the following system of equations:

$$x + 2y + 3z = 7$$

$$2x + 2y - z = 0$$

$$3x - 4y + 2z = 7$$ [2]

    (b) Reduce the augmented matrix to echelon form. [4]

    (c) Hence, solve the system of equations. [3]

7. Find the values of $x$ for which

$$\begin{vmatrix} x & 1 & 2 \\ 1 & x & 2 \\ 2 & 1 & x \end{vmatrix} = 0$$ [10]

8. A pair of simultaneous equations is given by

$$\left.\begin{aligned} 2x + py &= 13 \\ px + 32y &= 52 \end{aligned}\right\} \quad \text{where } p \in \mathbb{R}$$

Find the value(s) of $p$ for which the system of equations above has

(a) a unique solution [3]

(b) an infinite number of solutions. [3]

(c) no solution [3]

9. The determinant $\Delta$ is given by

$$\Delta = \begin{vmatrix} 1 & a & b+c \\ 1 & b & c+a \\ 1 & c & a+b \end{vmatrix}.$$

Show that $\Delta = 0$ for any $a$, $b$ and $c \in \mathbb{R}$. [6]

10. (a) Write the following system of equations in the form AX = D

$$x + y - z = 2$$
$$2x - y + z = 1$$
$$3x + 2z = 1$$ [2]

    (b) (i) Find the matrix B, the matrix of cofactors of the matrix A. [5]

       (ii) Calculate B$^T$A. [2]

       (iii) Deduce the value of $|A|$. [1]

11. If $A = \begin{pmatrix} 1 & 2 & -1 \\ -1 & 2 & 1 \\ 1 & -2 & 3 \end{pmatrix}$ and $B = \begin{pmatrix} 2 & -1 & 1 \\ 1 & 1 & 0 \\ 0 & 1 & 1 \end{pmatrix}$,

    (a) find AB                                                         [3]

    (b) deduce $A^{-1}$.                                                         [3]

12. A nursery sells three brands of grass-seed mix, P, Q and R. Each brand is made from three types of grass, C, Z and B. The number of kilograms of each type of grass in a bag of each brand is summarised in the table below.

| Grass Seed Mix | Type of Grass (Kilograms) | | |
|---|---|---|---|
| | C-grass | Z-grass | B-grass |
| Brand P | 2 | 2 | 6 |
| Brand Q | 4 | 2 | 4 |
| Brand R | 0 | 6 | 4 |
| Blend | $c$ | $z$ | $b$ |

A blend is produced by mixing $p$ bags of Brand P, $q$ bags of Brand Q and $r$ bags of Brand R.

    (a) Write down an expression in terms of $p$, $q$ and $r$, for the number of kilograms of Z-grass in the blend       [1]

    (b) Let $c$, $z$ and $b$ represent the number of kilograms of C-grass, Z-grass and B-grass respectively in the blend. Write down a set of THREE equations in $p$, $q$, $r$, to represent the number of kilograms of EACH type of grass in the blend.       [3]

    (c) Rewrite the set of THREE equations in (b) (ii) above in the matrix form MX = D where M is a 3 by 3 matrix, and X and D are column matrices.       [3]

    (d) Given that $M^{-1}$ exists, write X in terms of $M^{-1}$ and D.       [3]

    (e) Given that $M^{-1} = \begin{pmatrix} -0.2 & -0.2 & 0.3 \\ 0.35 & 0.1 & -0.15 \\ -0.05 & 0.2 & -0.05 \end{pmatrix}$,

         calculate how many bags of EACH brand, P, Q, and R, are required to produce a blend containing 30 kilograms of C-grass, 30 kilograms of Z-grass and 50 kilograms of B-grass.       [4]

13. (a)      Solve the equation using determinants

$$\begin{vmatrix} 5 & x & 3 \\ x+2 & 2 & 1 \\ -3 & 2 & x \end{vmatrix} = 0$$       [10]

    (b)      Solve the following set of equations

$$x_1 - 4x_2 + 2x_3 = 21$$
$$2x_1 + x_2 + 2x_3 = 3$$
$$3x_1 + 2x_2 - x_3 = -2$$       [10]

# What you have learnt in Chapter 24

## Some Definitions

A **matrix** (plural: matrices) is a rectangular array of numbers enclosed in round brackets

A matrix of **order** $m \times n$ has $m$ rows and $n$ columns.

If $m = n$ the matrix is a **square matrix**; if $m = 1$ it is a **row matrix**; if $n = 1$ it is a **column matrix**.

The **leading diagonal** in a square matrix is the diagonal from top left to bottom right.

A **zero matrix** is a matrix all of whose elements are zero.

A **unit matrix** is a square matrix with every element of its leading diagonal 1 and all other elements 0.

e.g.
$$\mathbf{O} = \begin{pmatrix} 0 & 0 & 0 \\ 0 & 0 & 0 \end{pmatrix} \qquad \mathbf{I} = \begin{pmatrix} 1 & 0 & 0 \\ 0 & 1 & 0 \\ 0 & 0 & 1 \end{pmatrix}$$

The **multiplicative inverse** of the square matrix $\mathbf{A}$ is $\mathbf{A}^{-1}$ where $\mathbf{AA}^{-1} = \mathbf{A}^{-1}\mathbf{A} = \mathbf{I}$.

The **determinant** of a square matrix $\mathbf{A}$ is a number associated with the matrix such that when $|\mathbf{A}|$ (or det $\mathbf{A}$) $\neq 0$ $\mathbf{A}^{-1}$ exists, but when $|\mathbf{A}| = 0$ $\mathbf{A}^{-1}$ does not exist and $\mathbf{A}$ is said to be **singular**.

## Laws of Association

Matrices of the same order can be **added or subtracted** to form another matrix of the same order, by adding/subtracting corresponding elements

(Matrix addition is associative, commutative and distributive)

A matrix can be multiplied by a scalar by multiplying every element of the matrix by that scalar

The **product** of an $m \times n$ matrix and an $n \times p$ matrix is an $m \times p$ matrix.

e.g.
$$\begin{pmatrix} a & b \\ c & d \\ e & f \end{pmatrix} \begin{pmatrix} p & q \\ r & s \end{pmatrix} = \begin{pmatrix} ap + br & aq + bs \\ cp + dr & cq + ds \\ ep + fr & eq + fs \end{pmatrix}$$

Matrices that can be added/multiplied are said to be **conformable** under matrix addition/multiplication.

Matrix multiplication is *not* commutative, though it *is* associative and distributive.

**Two matrices are equal if their corresponding elements are equal.**

For a $2 \times 2$ matrix $\mathbf{M} = \begin{pmatrix} a & b \\ c & d \end{pmatrix}$, $\mathbf{M}^{-1} = \dfrac{1}{|\mathbf{M}|}\begin{pmatrix} d & -b \\ -c & a \end{pmatrix}$ where $|\mathbf{M}| = ad - bc$

## $3 \times 3$ *Matrices*

If $\mathbf{M} = \begin{pmatrix} a_1 & b_1 & c_1 \\ a_2 & b_2 & c_2 \\ a_3 & b_3 & c_3 \end{pmatrix}$

the **minor** of an element $e$ is the determinant whose elements consist of the elements of the original matrix without the elements of the row and column which contain $e$.

a **cofactor** is a minor with the appropriate $+$ or $-$ sign attached to it:

The appropriate signs are $\begin{vmatrix} + & - & + \\ - & + & - \\ + & - & + \end{vmatrix}$

So the cofactors of the matrix $\mathbf{M}$ are

$$A_1 = \begin{vmatrix} b_2 & c_2 \\ b_3 & c_3 \end{vmatrix} \quad B_1 = -\begin{vmatrix} a_2 & c_2 \\ a_3 & c_3 \end{vmatrix} \quad C_1 = \begin{vmatrix} a_2 & b_2 \\ a_3 & b_3 \end{vmatrix}$$

$$A_2 = -\begin{vmatrix} b_1 & c_1 \\ b_3 & c_3 \end{vmatrix} \quad B_2 = \begin{vmatrix} a_1 & c_1 \\ a_3 & c_3 \end{vmatrix} \quad C_2 = -\begin{vmatrix} a_1 & b_1 \\ a_3 & b_3 \end{vmatrix}$$

$$A_3 = \begin{vmatrix} b_1 & c_1 \\ b_2 & c_2 \end{vmatrix} \quad B_3 = -\begin{vmatrix} a_1 & c_1 \\ a_2 & c_2 \end{vmatrix} \quad C_3 = \begin{vmatrix} a_1 & b_1 \\ a_2 & b_2 \end{vmatrix}.$$

Det $\mathbf{M}$ can therefore be written as $a_1 A_1 + a_2 A_2 + a_3 A_3$

$$\det \mathbf{M} = a_1 \begin{vmatrix} b_2 & c_2 \\ b_3 & c_3 \end{vmatrix} - b_1 \begin{vmatrix} a_2 & c_2 \\ a_3 & c_3 \end{vmatrix} + c_1 \begin{vmatrix} a_2 & b_2 \\ a_3 & b_3 \end{vmatrix}$$

or the sum of the products of the elements of any row or column of $\mathbf{M}$ with their cofactors

The cofactor matrix of $\mathbf{M}$ is $\begin{pmatrix} A_1 & B_1 & C_1 \\ A_2 & B_2 & C_2 \\ A_3 & B_3 & C_3 \end{pmatrix}$

and its transpose is the **adjoint** of $\mathbf{M}$: $\quad \text{adj } \mathbf{M} = \begin{pmatrix} A_1 & A_2 & A_3 \\ B_1 & B_2 & B_3 \\ C_1 & C_2 & C_3 \end{pmatrix}$

The inverse of a $3 \times 3$ matrix $\mathbf{M}$ is $\mathbf{M}^{-1} = \dfrac{\text{adj } \mathbf{M}}{\det \mathbf{M}}$

A **triangular** matrix is a matrix in which all the elements below (or above) the leading diagonal are 0

When a matrix has been manipulated so that it becomes a triangular matrix, it is said to be **reduced to echelon form**.

## Systems of Equations

The system $\begin{aligned} a_1x+b_1y+c_1z &= d_1 \\ a_2x+b_2y+c_2z &= d_2 \\ a_3x+b_3y+c_1z &= d_3 \end{aligned}\Bigg\}$ can be written in matrix form $\begin{pmatrix} a_1 & b_1 & c_1 \\ a_2 & b_2 & c_2 \\ a_3 & b_3 & c_3 \end{pmatrix}\begin{pmatrix} x \\ y \\ z \end{pmatrix} = \begin{pmatrix} d_1 \\ d_2 \\ d_3 \end{pmatrix}$.

The matrix $\left(\begin{array}{ccc|c} a_1 & b_1 & c_1 & d_1 \\ a_2 & b_2 & c_2 & d_2 \\ a_3 & b_3 & c_3 & d_3 \end{array}\right)$ is called the **augmented matrix** for the system.

When det $\mathbf{M} \neq 0$ the system is **consistent** and can be solved either by using the inverse matrix $\mathbf{M}^{-1}$ or by **row reduction** of the augmented matrix.

When det $\mathbf{M} = 0$ the consistency of the system can be determined by consideration of the augmented matrix:

If either row 2 or row 3 have all zeros in the triangular matrix but a non-zero number in the augmented section ($d_2$ or $d_3$) the system is inconsistent and has no solutions (A)

If (A) is not true and any row consists entirely of zeros, the system is consistent and has an infinite number of solutions.

## *Matrices and Transformations*

A $3 \times 3$ matrix can represent a transformation in 3-dimensional space.

If $P'\,(x',\ y',z')$ is the image of $P\,(x, y, z)$ under the transformation $\mathbf{T}$, then $\mathbf{T}\,P = P'$ or

$$\mathbf{T}\begin{pmatrix} x \\ y \\ z \end{pmatrix} = \begin{pmatrix} x' \\ y' \\ z' \end{pmatrix}$$

If the image, $P'$, is known, $P$ can be found either by row reduction or by using the inverse matrix

$$\mathbf{P} = \mathbf{T}^{-1}\mathbf{P}'$$

# CHAPTER 25
# 2.3c  Complex Numbers

## DEFINITIONS AND NOTATION

Earlier, when we considered the roots of a quadratic equation, we found that it was possible to have quadratic equations whose discriminants were negative, so that they did not have real roots.  Using the quadratic formula

for the equation $x^2 - 2x + 2 = 0$ produced roots of $1 \pm \sqrt{-1}$, and the graph of $y = x^2 - 2x + 2 = 0$ confirms that it has no real roots.

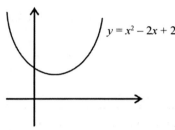

$y = x^2 - 2x + 2$

When we considered the real number system in chapter 1, we saw how the set of real numbers, $\mathbb{R}$, developed progressively from the sets, $\mathbb{N}$, $\mathbb{W}$, $\mathbb{Z}$, and $\mathbb{Q}$ as the need for zero, negative numbers, rational and irrational numbers became apparent.  At this point it is useful extend our number system further to accommodate square roots of negative numbers.

We define

$$i = +\sqrt{-1} \text{ , so that } i^2 = -1$$

Any square root of a negative number can then be written as a multiple of $i$ and is known as an **imaginary number**. e.g. $\sqrt{-4} = \sqrt{4 \times -1} = \sqrt{4} \times \sqrt{-1} = 2i$ and $\sqrt{-7} = \sqrt{7 \times -1} = \sqrt{7} \times \sqrt{-1} = i\sqrt{7}$.

When we look at the formula for the roots of a quadratic equation, we see that if $\Delta$ is negative then $\sqrt{b^2 - 4ac}$ is an imaginary number and the roots will be numbers of the form $p \pm iq$.

> A number of the form $z = p + iq$, $p, q \in \mathbb{R}$, is known as a **complex number**.
>
> The real part of $z$ is designated $\operatorname{Re} z$, so that $\operatorname{Re} z = p$
>
> The imaginary part of $z$ is designated $\operatorname{Im} z$, so that $\operatorname{Im} z = q$
>
> The **complex conjugate** of $z$ is designated $\bar{z}$ or $z^*$ and defined as $\bar{z} = p - iq$.

When $q = 0$, $z = p$, so $z$ is a real number,

when $p = 0$, $z = iq$, so $z$ is an imaginary number

so  the set of real numbers ($\mathbb{R}$)  and the set of imaginary numbers are both subsets of the set ($\mathbb{C}$) of complex numbers.

# THE ALGEBRA OF COMPLEX NUMBERS

## *Addition and subtraction of complex numbers*

Complex numbers can be added/subtracted by adding/subtracting their real and imaginary parts:

$$a + ib + (c + id) = a + c + i(b + d) \quad \text{and} \quad a + ib - (c + id) = a - c + i(b - d)$$

From this it follows that if $a + ib = c + id$, then $a + ib - (c + id) = 0$

so, $a - c + i(b - d) = 0$ giving $a = c$ and $b = d$.

> If $z = a + ib$ and $w = c + id$,
>
> then $\qquad z = w \Rightarrow a = c$ and $b = d$
>
> and, conversely $\qquad a = c$ and $b = d \Rightarrow z = w$
>
> i.e. two complex numbers are the same iff their real and
> imaginary parts are the same

## *Multiplication of complex numbers*

Complex numbers can be multiplied by expanding the brackets:

$$zw = (a + ib)(c + id) = ac + aid + ibc + i^2bd = (ac - bd) + i(ad + bd) \quad \text{(remembering that } i^2 = -1)$$

## Example 25.1

If $z = 3 + 2i$ and $w = 5 - 4i$,

(i) write down (a) Re $z$ (b) Im $z$ (c) Re $w$ (d) Im $w$

(ii) simplify (a) $z + w$ (b) $z^* + w^*$ (c) $z + z^*$ (d) $3z - 2w$ (e) $zw$ (f) $z^*w^*$ (g) $z^2$ (h) $zz^*$

   (i) (a) Re $z = 3$ (b) Im $z = 2$ (c) Re $w = 5$ (d) Im $w = -4$

   > ! Notice that the Im $z$ does not include $i$.

   (ii) (a) $z + w = 3 + 2i + (5 - 4i) = (3 + 5) + (2 - 4)i = 8 - 2i$
   (b) $z^* + w^* = 3 - 2i + (5 + 4i) = (3 + 5) + (-2 + 4)i = 8 + 2i$
   (c) $z + z^* = (3 + 2i) + (3 - 2i) = 6 + 0i = 6$
   (d) $3z - 2w = 3(3 + 2i) - 2(5 - 4i) = (9 - 25) + (6 + 8)i = -1 + 14i$
   (e) $zw = (3 + 2i)(5 - 4i) = 15 - 12i + 25i - 8i^2 = 15 + 8 - 2i = 23 - 2i$
   (f) $z^*w^* = (3 - 2i)(5 + 4i) = 15 + 12i - 25i - 8i^2 = 15 + 8 + 2i = 23 + 2i$
   (g) $z^2 = (3 + 2i)^2 = (3 + 2i)(3 + 2i) = 9 + 6i + 6i + 4i^2 = 9 - 4 + 12i = 5 + 12i$
   (h) $zz^* = (3 + 2i)(3 - 2i) = 9 - 6i + 6i - 4i^2 = 9 + 4 + 0i = 13$

# EXERCISE 25 A

1. If $v = 2 + 3i$, $w = 4 - 2i$, $z = -3 + 5i$, find

   (a) $v^*$ (b) $w^*$ (c) $z^*$ (d) $v + z$ (e) $w + z^*$
   (f) $2v - w$ (g) $3v + 2w^*$ (h) $5w - 2z$ (i) $vw$ (j) $wz$
   (k) $vz^*$ (l) $2zv$ (m) $v + v^*$ (n) $ww^*$ (o) $zz^*$

2. If $w = 5 + 2i$, $z = -7 + 4i$, find

   (a) $w^*$      (b) $z^*$      (c) $w + z$      (d) $w^* + z^*$   (e) $2w + 4z$

   (f) $5z^* - 3w$   (g) $zw$      (h) $z^*w^*$     (i) $w^*z$      (j) $wz^*$

   (k) $w + w^*$   (l) $w - w^*$   (m) $z + z^*$      (n) $ww^*$     (o) $zz^*$

3. If $p = 3 - 5i$ and $q = -4 + i$, find

   (a) $\text{Re}\,p$     (b) $\text{Im}\,q$   (c) $\text{Re}\,p\overline{q}$     (d) $\text{Im}\,pq$   (e) $\text{Im}\,q\overline{q}$

Before learning how to divide by a complex number we need to look in more detail at the

## Complex Conjugate

Compare the answers in Example 25.1 (c), (h), Exercise 25A 1 (n), 2 (k), (n), and (o) with all the other answers. You should notice that these answers alone produced real number solutions to problems of addition, subtraction and multiplication of complex numbers. In other words, the sum of a complex number and its complex conjugate is always a real number, and the product of a complex number with its complex conjugate is always a real number. Here is the proof.

$$\text{if } z = p + iq, \text{ then } z^* = p - iq$$
$$z + z^* = (p + iq) + (p - iq) = 2p$$
$$zz^* = (p + iq)(p - iq) = p^2 - ipq + ipq - i^2q^2 = p^2 + q^2$$

Since (as can be seen from the quadratic formula) the roots of a quadratic equation with real coefficients whose discriminant is negative are of the form $p \pm iq$, we see that they are complex conjugates, and therefore their sum and product are both real. The theory that we learnt in unit 1 about the sums and products of the roots of quadratic equations applies equally well to equations with complex roots as it does to equations with real roots.

However, if the roots of a quadratic equation are complex but are not complex conjugates, then the coefficients of the equation are not all real numbers.

These results can be summarised

> if $z = p + iq$, then $z^* = p - iq$   $p, q \in \mathbb{R}$
>
> $z + z^* = 2p$         (a real number)
>
> $zz^* = p^2 + q^2$     (a real number)
>
> if $\alpha$ and $\beta$ are the roots of the equation $ax^2 + bx + c = 0$
>
> then $\alpha + \beta = \dfrac{-b}{a}$ and $\alpha\beta = \dfrac{c}{a}$
>
> when $a, b, c \in \mathbb{R}$, $\alpha$ and $\beta$ are complex conjugates

## Division of complex numbers

## Example 25.2

Express $\dfrac{2 + 3i}{1 + 2i}$ in the form $a + ib$.

The 'problem' with this question is that we do not want any imaginary numbers in the denominator. In other words we need to 'realise' (make real) the denominator. This should remind you of the technique we used with surds to rationalise the denominator. Is there a number which, multiplied by the denominator, will give a real number?

Yes, its complex conjugate; in this case $1 - 2i$.

Realise the denominator: $\dfrac{2+3i}{1+2i} = \dfrac{2+3i}{1+2i} \times \dfrac{1-2i}{1-2i}$

$$= \dfrac{8-i}{5} = \dfrac{8}{5} - \dfrac{1}{5}i$$

## Example 25.3

Express $\dfrac{4+3i}{5-i}$ in the form $a+ib$.

➢ Multiply by the complex conjugate $5 + i$: $\quad \dfrac{4+3i}{5-i} = \dfrac{4+3i}{5-i} \times \dfrac{5+i}{5+i}$

➢ Simplify: $\qquad\qquad\qquad\qquad = \dfrac{20+4i+15i+3i^2}{5^2-i^2} = \dfrac{17+19i}{26}$

➢ Give the answer in the required form: $\qquad = \dfrac{17}{26} + \dfrac{19}{26}i$

## The square root of a complex number

Does a complex number have a square root, which is also a complex number? To find out we assume that it does and see whether we can find it.

## Example 25.4

Find the square root(s), if they exist, of the complex number $z = 3 + 4i$

➢ Assume that the square root is $a + ib$, $a, b \in \mathbb{R}$, then $(a+ib)^2 = 3 + 4i$

➢ Multiply out: $\qquad\qquad\qquad\qquad\qquad a^2 - b^2 + 2abi = 3 + 4i$

➢ Equate real and imaginary parts: $\qquad a^2 - b^2 = 3, \ 2ab = 4$

➢ Solve these simultaneous equations: $\quad a^2 - \left(\dfrac{2}{a}\right)^2 = 3 \ \rightarrow \ a^4 - 3a^2 - 4 = 0$

$\qquad\qquad\qquad\qquad\qquad\qquad\qquad (a^2 - 4)(a^2 + 1) = 0$

$\qquad\qquad\qquad\qquad\qquad\qquad\qquad$ since $a$ is real, $a = \pm 2$

$\qquad\qquad\qquad\qquad\qquad\qquad\qquad$ When $a = 2$, $b = 1$, when $a = -2$, $b = -1$

➢ Give the answer: $\qquad\qquad\qquad\qquad \sqrt{z} = \pm (2 + i)$

*Handwritten annotations:*
$a^4 - 4 = 3a^2$
$a^4 - 3a^2 - 4 = 0$
$a^2 - 4a + a^2 - 4 = 0$
$a^2(a^2 - 4) + (a^2 - 4) = 0$
$(a^2 + 1)(a^2 - 4)$
$a^2 = 4$
$a = \pm 2$

In simple problems, where $a$ and $b$ are both integers, solving the simultaneous equations for $a$ and $b$ can often be done by inspection:

## Example 25.5

Find the square roots of the complex number $z = 15 - 8i$

➢ Assume that the square root is $a + ib$, $a, b \in \mathbb{R}$, then $(a+ib)^2 = 15 - 8i$

➢ Multiply out: $\qquad\qquad\qquad\qquad\qquad a^2 - b^2 + 2abi = 15 - 8i$

➢ Equate real and imaginary parts: $\qquad a^2 - b^2 = 15, \ 2ab = -8$

➢ the product of $a$ and $b$ is $-4$, so guess: $\quad a = \pm 1, 2$ or $4$

➤ since, also, $a^2 - b^2 = 15$:        $a^2 = 16 \longrightarrow a = \pm 4$

➤ find corresponding values for $b$:     when $a = 4$, $b = -1$, when $a = -4$, $b = 1$

➤ give the answer:              $\sqrt{z} = \pm (4 - i)$

## Example 25.6

Solve the quadratic equation $x^2 + 4x + 13 = 0$

➤ Use the formula:   $x = \dfrac{-4 \pm \sqrt{16 - 52}}{2} = \dfrac{-4 \pm \sqrt{-36}}{2} = \dfrac{-4 \pm 6i}{2}$

➤ $x = -2 + 3i$ or $x = -2 - 3i$       [or: $x = -2 \pm 3i$]

## Example 25.7

Find the quadratic equations whose roots are   (a) $3 + i, 3 - i$    (b) $1 + i, 2 + i$

(a) ➤ Let roots be $\alpha$ and $\beta$. Then $\alpha + \beta = (3 + i) + (3 - i) = 6$, $\alpha\beta = (3 + i)(3 - i) = 9 + 1 = 10$

     ➤ Equation is: $x^2 - 6x + 25 = 0$

(b) ➤ Let roots be $\alpha$ and $\beta$. Then $\alpha + \beta = 3 + 2i$, $\alpha\beta = 1 + 3i$

     ➤ Since the coefficients will be complex, use $z$ rather than $x$ for the equation
equation is $z^2 - (3 + 2i)z + 1 + 3i = 0$

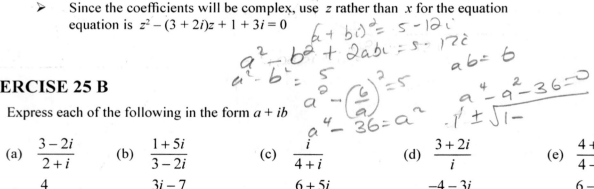

## EXERCISE 25 B

1.   Express each of the following in the form $a + ib$

   (a) $\dfrac{3 - 2i}{2 + i}$     (b) $\dfrac{1 + 5i}{3 - 2i}$     (c) $\dfrac{i}{4 + i}$     (d) $\dfrac{3 + 2i}{i}$     (e) $\dfrac{4 + i}{4 - i}$

   (f) $\dfrac{4}{8 - i}$     (g) $\dfrac{3i - 7}{2 + 5i}$     (h) $\dfrac{6 + 5i}{2i - 1}$     (i) $\dfrac{-4 - 3i}{8 + 3i}$     (j) $\dfrac{6 - 2i}{-3 - 5i}$

2.   Find the square roots of the following complex numbers giving your answers in the form $a + ib$
   (a) $5 - 12i$     (b) $35 + 12i$     (c) $16 - 30i$     (d) $21 - 20i$
   (e) $-3 + 4i$     (f) $7 - 24i$     (g) $-5 + 12i$     (h) $i$

3.   Use the quadratic formula to solve the following quadratic equations
   (a) $x^2 - 8x + 20 = 0$      (b) $4x^2 - 4x + 5 = 0$      (c) $x^2 + 2x + 5 = 0$

4.   Find the quadratic equations whose roots are:
   (a) $3 + i, 3 - i$     (b) $1 + 2i, 1 - 2i$     (c) $-1 + 3i, -1 - 3i$     (d) $\frac{1}{2} + \frac{1}{4}i, \frac{1}{2} - \frac{1}{4}i$

5.   A quadratic equation has real coefficients. One root is $3 - 2i$. Write down the other root, and hence write down the equation.

6.   Find the quadratic equations whose roots are:
   (a) $3 + i, 3 - 2i$      (b) $1 + 2i, 1 + i$      (c) $-5 + i, 2 - 3i$

**#7.** If $\alpha$ and $\beta$ are the roots of the quadratic equation $(1 + i)z^2 - 2iz + (3 + i) = 0$, express each of $\alpha + \beta$ and $\alpha\beta$ in the form $a + ib$, where $a$ and $b$ are real.

Find, in a form not involving $\alpha$ and $\beta$, the quadratic equation whose roots are $\alpha + 2\beta$ and $2\alpha + \beta$.

# THE ARGAND DIAGRAM

A complex number $z = a + ib$ is uniquely defined by its real and imaginary parts $a$ and $b$, so it could be represented by the ordered pair $(a, b)$. This is in fact what is done when a complex number is represented graphically. In an Argand diagram the horizontal ($x$-) axis is used to represent the real part of $z$ and the vertical ($y$-) axis to represent the imaginary part of $z$. Then $z$, like a position vector, is represented by the line joining the origin to $(a, b)$.

## Example 25.8

Represent on an Argand diagram the complex numbers

(a) $z_1 = 2 - 3i$, $z_2 = 3 + i$ and $z_3 = z_1 + z_2$

(b) $w_1 = 4 + 2i$, $w_2 = 3 - 4i$ and $w_3 = w_1 - w_2$

    (a)         $z_3 = z_1 + z_2 = 5 - 2i$

    (b)         $w_3 = w_1 - w_2 = 1 + 6i$

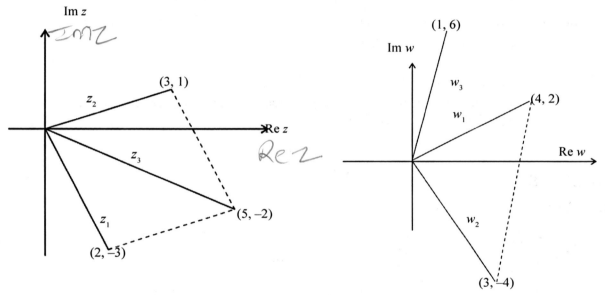

Notice the geometrical relationship, shown with dotted lines,
between two complex numbers and their sum, in (a),
and between two complex numbers and their difference, in (b).

# MODULUS AND ARGUMENT

You are probably aware that points in the plane can be located by means of coordinate systems other than the Cartesian ($x$-$y$) system. **Polar coordinates** are a means of locating a point with reference to a point and a line through that point.

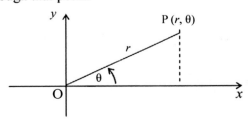

Given the line $Ox$, any point P is uniquely defined by its distance, $r = OP$, from O and the angle, $\theta$, that OP makes with $Ox$ (measured anti-clockwise). P is then the point $(r, \theta)$.

Comparing the Cartesian and polar coordinates of P we see that the $x$-coordinate of P is $r \cos \theta$, and the $y$-coordinate is $r \sin \theta$.

When polar coordinates are applied to a complex number $z = x + iy$ we get $z = r\cos \theta + ir\sin \theta$, or $z = r(\cos \theta + i\sin \theta)$.

When $z$ is represented in an Argand diagram, $r$ is the length of $z$, and $\theta$ is the angle that $z$ makes with the Real ($x$-) axis. We therefore define

> For the complex number $z = a + ib$,
> the **modulus of** $z$, $|z|$, is the length of $z$ in the Argand diagram, and
> the **argument of** $z$, Arg $z$, is the angle that $z$ makes with the positive $x$-axis in the Argand diagram.
> The argument is always measured in radians

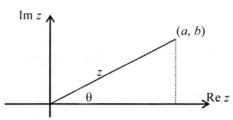

From the diagram, we can see that $|z| = \sqrt{(a^2 + b^2)}$

and $\tan \theta = \frac{b}{a}$ so Arg $z = \tan^{-1}(\frac{b}{a})$.

Now, there are an infinite number of angles, $\theta$, for which

$\tan \theta = \frac{b}{a}$, so Arg $z$ is not unique – there are an infinite number of arguments of $z$. However, it would be useful to have a unique argument, and so we define the **principal argument** of $z$, which we denote by arg $z$ (as opposed to Arg $z$), as that value of the argument for which $-\pi < \arg z \leq \pi$.

We need to be careful when finding the argument, however, since even within a range of $2\pi$ radians there are two angles with the same tan, and only one of them is correct. This is best seen if we look at an example:

Consider the modulus and argument of these complex numbers: $z_1 = 4 + 3i$, $z_2 = 4 - 3i$, $z_3 = -4 + 3i$ and $z_4 = -4 - 3i$. Clearly these are all different complex numbers, with the same modulus, 5, and they can be represented in Argand diagrams as follows:

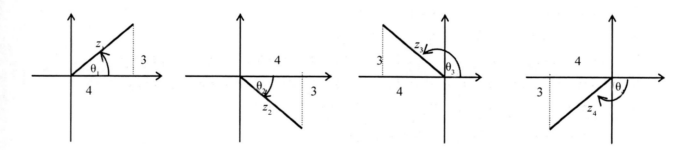

We can see from the diagrams that $\tan \theta_1 = \frac{3}{4}$, $\tan \theta_2 = \frac{-3}{4} = -\frac{3}{4}$, $\tan \theta_3 = \frac{3}{-4} = -\frac{3}{4}$ and $\tan \theta_4 = \frac{-3}{-4} = \frac{3}{4}$.

In other words, $\theta_1$ and $\theta_4$ are both $\tan^{-1}\frac{3}{4}$, but $\theta_1 = 0.6435$ while $\theta_4 = -2.2498$. Similarly $\theta_2$ and $\theta_3$ are both $\tan^{-1}\left(-\frac{3}{4}\right)$ but $\theta_2 = -0.6435$ and $\theta_3 = 2.2498$.

To find the argument of a complex number $z$ it is therefore necessary to sketch an Argand diagram to find which solution of $\theta = \tan^{-1}(\frac{b}{a})$ is the correct one.

As a general rule, if $z$ is in the first or fourth quadrants arg $z$ will be the value of $\tan^{-1}(\frac{b}{a})$ obtained from your calculator, if $z$ is in the second quadrant add $\pi$, and if $z$ is in the third quadrant subtract $\pi$ from the calculator solution.

## Example 25.9

Find the modulus and principal argument of the complex numbers  (a)  $w = 1 - \sqrt{3}i$,  (b) $z = -3 - 3i$.

(a)  ➢  for $w$, calculate modulus: $\qquad | w | = \sqrt{(1 + 3)} = 2,$

   ➢  find the argument: $\qquad \tan^{-1}(-\sqrt{3}) = -\tfrac{1}{3}\pi \text{ or } \tfrac{2}{3}\pi$

   ➢  check the Argand diagram: $w$ is in 4th quadrant, so arg $w = -\tfrac{1}{3}\pi$

   ➢  give the answer: $\qquad | w | = 2$, arg $w = -\tfrac{1}{3}\pi$

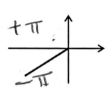

(b)  ➢  for $z$, calculate modulus: $\qquad | z | = \sqrt{(9 + 9)} = 3\sqrt{2},$

   ➢  find the argument: $\qquad \tan^{-1}(1) = \tfrac{1}{4}\pi \text{ or } -\tfrac{3}{4}\pi$

   ➢  check the Argand diagram: $w$ is in 3rd quadrant, so arg $w = -\tfrac{3}{4}\pi$

   ➢  give the answer: $\qquad | z | = 3\sqrt{2}$, arg $z = -\tfrac{3}{4}\pi$

## Modulus and Argument of Products and Quotients

## Example 25.10

Show that if $z_1$ and $z_2$ are two complex numbers,

then  (a)  $| z_1 z_2 | = | z_1 | | z_2 |$  and arg $z_1 z_2 = $ arg $z_1 + $ arg $z_2$

and  (b)  $\left| \dfrac{z_1}{z_2} \right| = \dfrac{|z_1|}{|z_2|}$  and  arg $\dfrac{z_1}{z_2} = $ arg $z_1 - $ arg $z_2$

🔑  Use polar (modulus and argument) form for $z_1$ and $z_2$:

(a)  ➢  Let  $z_1 = r_1(\cos \theta_1 + i\sin \theta_1)$ and  $z_2 = r_2(\cos \theta_2 + i\sin \theta_2)$

   ➢  then  $z_1 z_2 = r_1(\cos \theta_1 + i\sin \theta_1) \, r_2(\cos \theta_2 + i\sin \theta_2)$

   ➢  multiply out: $z_1 z_2 = r_1 r_2 (\cos \theta_1 \cos \theta_2 + i\sin \theta_1\cos \theta_2 + i\cos \theta_1\sin \theta_2 + i^2 \sin \theta_1 \sin \theta_2)$

   ➢  rearrange: $\quad z_1 z_2 = r_1 r_2\{\cos \theta_1 \cos \theta_2 - \sin \theta_1 \sin \theta_2 + i(\sin \theta_1\cos \theta_2 + \cos \theta_1\sin \theta_2)\}$

   ➢  simplify: $\quad z_1 z_2 = r_1 r_2\{\cos (\theta_1 + \theta_2) + i(\sin (\theta_1 + \theta_2)\}$

   ➢  find modulus: $\qquad | z_1 z_2 | = r_1 r_2 = | z_1 | | z_2 |$

   ➢  find argument:  arg $z_1 z_2 = \tan^{-1} \dfrac{\sin (\theta_1 + \theta_2)}{\cos (\theta_1 + \theta_2)} = \tan^{-1} \{\tan (\theta_1 + \theta_2)\} = \theta_1 + \theta_2$

$$= \text{arg } z_1 + \text{arg } z_2$$

similarly for (b)

$$\frac{z_1}{z_2} = \frac{r_1(\cos \theta_1 + i\sin \theta_1)}{r_2(\cos \theta_2 + i\sin \theta_2)}$$

$$= \frac{r_1(\cos \theta_1 + i\sin \theta_1)(\cos \theta_2 - i\sin \theta_2)}{r_2(\cos \theta_2 + i\sin \theta_2)(\cos \theta_2 - i\sin \theta_2)}$$

$$= \frac{r_1(\cos \theta_1 \cos \theta_2 + \sin \theta_1\sin \theta_2 + i(\sin \theta_1\cos \theta_2 + \cos \theta_1 \sin \theta_2)}{r_2(\cos^2 \theta_2 + \sin^2 \theta_2)}$$

$$= \frac{r_1}{r_2} \{\cos (\theta_1 - \theta_2) + i\sin (\theta_1 - \theta_2)\}$$

find modulus:

$$\left|\frac{z_1}{z_2}\right| = \frac{r_1}{r_2} = \frac{|z_1|}{|z_2|}$$

find argument:

$$\arg\left(\frac{z_1}{z_2}\right) = \tan^{-1}\frac{\sin(\theta_1 + \theta_2)}{\cos(\theta_1 + \theta_2)} = \tan(\theta_1 - \theta_2) = \arg z_1 - \arg z_2$$

for two complex numbers $z_1$ and $z_2$

$$|z_1 z_2| = |z_1||z_2| \text{ and } \arg z_1 z_2 = \arg z_1 + \arg z_2$$

$$\left|\frac{z_1}{z_2}\right| = \frac{|z_1|}{|z_2|} \text{ and } \arg\frac{z_1}{z_2} = \arg z_1 - \arg z_2$$

## Example 25.11

Given that $z = \sqrt{2} + i\sqrt{2}$ and $w = -3 + i3\sqrt{3}$, find the modulus and argument of each of $z$ and $w$ and hence find the modulus and argument of $zw$ and $\frac{z}{w}$.

$|z| = \sqrt{(2 + 2)} = 2$, ($z$ is in the first quadrant) $\arg z = \frac{1}{4}\pi$

$|w| = \sqrt{(9 + 27)} = 6$, ($w$ is in the second quadrant) $\arg w = \frac{2}{3}\pi$

$|zw| = |z||w| = 2 \times 6 = 12$, $\arg|zw| = \arg z + \arg w = \frac{11}{12}\pi$

$\left|\frac{z}{w}\right| = \frac{|z|}{|w|} = \frac{2}{6} = \frac{1}{3}$, $\arg\left|\frac{z}{w}\right| = \arg z - \arg w = -\frac{5}{12}\pi$

## EXERCISE 25 C

1. In the diagram, the lines OA, OB, OC and OD represent the complex numbers $z_1$, $z_2$, $z_3$ and $z_4$. Express $z_1$, $z_2$, $z_3$ and $z_4$ in the form $a + ib$.

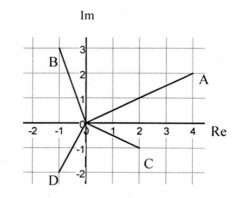

2. Represent each of the following complex numbers in an Argand diagram
   (a) $3 - 2i$    (b) $5 + 3i$    (c) $-2 + 4i$
   (d) $2i$    (e) $3i - 2$    (f) $-3 - i$

3. Find the modulus and principal argument of the following complex numbers, giving the argument in radians either as a multiple of $\pi$ or as a decimal correct to 2 d.p.
   (a) $1 + i$   (b) $3 - 4i$   (c) $-2 + 2i$   (d) $1 - \sqrt{3}i$   (e) $-5 - 12i$
   (f) $3i$   (g) $-4$   (h) $-2i$   (i) $5$   (j) $-2\sqrt{3} - 2i$

4. Find the modulus and principal argument of the following complex numbers, giving the argument in radians either as a multiple of $\pi$ or as a decimal correct to 2 d.p.

(a) $(1+i)(3-4i)$     (b) $\dfrac{1-i}{1+i}$    (c) $\dfrac{i\sqrt{3}-2}{-2-i\sqrt{3}}$    (d) $\dfrac{-1-7i}{4+3i}$    (e) $(1+i)(2-i)$

5. Given that $z=\sqrt{3}+i$, find the modulus and argument of (a) $z^2$ and (b) $\dfrac{1}{z}$

6. If $z_1=1-i$ and $z_2=7+i$, find the modulus of (a) $z_1-z_2$    (b) $z_1 z_2$    (c) $\dfrac{z_1-z_2}{z_1 z_2}$

7. If $a=3-i$ and $b=1+2i$, find the moduli of (a) $2a+3b$    (b) $\dfrac{a}{2b}$.

8. The complex number $z$ is given by $z=\dfrac{i}{1-i}$.

    (a) Express $z$ in the form $a+ib$, where $a$ and $b$ are real.

    (b) Hence, or otherwise, find the modulus and argument of $z$.

    (c) Draw an Argand diagram showing the points representing the complex numbers $i$, $1-i$ and $z$. Show that these three points form a right-angled triangle.

9. The complex numbers $w$ and $z$ are given by $w=4-2i$ and $z=1+3i$

    (a) Write down an expression for $\bar{z}$, the complex conjugate of $z$.

    (b) Express $\dfrac{\bar{z}}{w}$ in the form $x+iy$, where $x$ and $y$ are non-zero real numbers

    (c) Hence, find the modulus and principal argument of $\dfrac{\bar{z}}{w}$, giving the argument as a multiple of $\pi$.

10. The number $z$ is given by $z=-\sqrt{3}+i$. Find the value of

    (a) $|z|$     (b) $\arg z$,     (c) $\arg\left(\dfrac{i}{z}\right)$ giving your answer in radians.

11. If $z=5+5i$ and $w=7+i$, find

    (a) $z-w$     (b) $\dfrac{1}{z+w}$, expressing each in the form $x+iy$, where $x$ and $y$ are real.

    Verify that the real part of $\dfrac{z-w}{z+w}$ is zero.

12. The complex number $z$ satisfies the equation $\dfrac{z-2}{z+3i}=\lambda i$, $\lambda \in \mathbb{R}$.

    (a) Show that $z=\dfrac{(2-3\lambda)(1+\lambda i)}{1+\lambda^2}$.

    (b) In the case where $\lambda=1$ find $|z|$ and $\arg z$.

245

13. (a) If $z = 3 + 4i$, express $z + \dfrac{25}{z}$ in its simplest form.

(b) If $z = x + iy$, find the real part and the imaginary part of $z + \dfrac{1}{z}$.

## LOCUS ON THE ARGAND DIAGRAM

If the complex number $z = x + iy$ is represented in the Argand diagram by the point P $(x, y)$, and some constraint is placed on $z$, then the corresponding set of possible points P forms the locus of P. The locus of P can usually be written in Cartesian form (an equation connecting $x$ and $y$)

### Example 25.12

If P is the point in the complex plane represented by the complex number $z$, find the locus of P if $|z - 3| = 5$.
- ➤ Let $z = x + iy$
- ➤ Rewrite $|z - 3|$ in terms of $x$ and $y$: $\quad |x + iy - 3| = 5$
- ➤ Rearrange complex number: $\quad |x - 3 + iy| = 5$
- ➤ Find modulus and square: $\quad (x - 3)^2 + y^2 = 25$
- ➤ Give answer: $\quad$ locus of P is a circle, center $(3, 0)$ radius 5.

### Example 25.13

If P is the point in the complex plane represented by the complex number $z$, find the locus of P if $|z - 3| = |z|$.
- ➤ Let $z = x + iy$

Rewrite in terms of $x$ and $y$: $\quad |x + iy - 3| = |x + iy|$
- ➤ Find modulus and square: $\quad (x - 3)^2 + y^2 = x^2 + y^2$
- ➤ Rearrange: $\quad 9 - 6x = 0 \;\rightarrow\; x = \frac{3}{2}$
- ➤ Give answer: $\quad$ locus of P is the straight line $x = \frac{3}{2}$

### Example 25.14

If P is the point in the complex plane represented by the complex number $z$, find the locus of P if $\arg(z - 1) = \frac{3}{4}\pi$.
- ➤ Let $z = x + iy$
- ➤ Rewrite $z - 1$ in terms of $x$ and $y$: $\quad z - 1 = x + iy - 1 = x - 1 + iy$

- ➤ Find $\arg(z - 1)$: $\qquad \arg(z - 1) = \tan^{-1}\left(\dfrac{y}{x - 1}\right) = \frac{3}{4}\pi$

- ➤ tan both sides: $\qquad \dfrac{y}{x - 1} = -1 \;\rightarrow\; y = -x + 1$

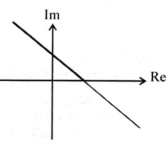

- ➤ draw an Argand diagram:
- ➤ $\frac{3}{4}\pi$ is an obtuse angle, so only the section of the line
  $y = 1 - x$ that is above the real axis has an argument of $\frac{3}{4}\pi$,
  (the lower part has an argument of $-\frac{1}{4}\pi$), so
- ➤ give answer: locus of P is line $y = 1 - x$, $y \geq 0$

# EXERCISE 25 D

1.  If P is the point in the complex plane represented by the complex number $z$, find the locus of P if

    (a) $|z-2|=3$     (b) $|z-2|=|z-3|$     (c) $|z-(1+i)|=1$

    (d) $|z|=|z-4i|$     (e) $\arg z = \frac{1}{4}\pi$     (f) $\arg(z-2)=\frac{1}{6}\pi$

2.  If $z = x + iy$, find an equation connecting $x$ and $y$, and state the curve it represents when

    (a) $\left|\dfrac{z+1}{2z-i}\right|=1$     (b) $\text{Im}\left(\dfrac{z}{z+4i}\right)=2$     (c) $\text{Re}\left(\dfrac{z+1}{z-1}\right)=0$

    (d) $\arg(z+3)=-\frac{1}{3}\pi$     (e) $|z-1|=|z-i|$     (f) $2|z|=|2z-i|$

3.  What is the locus given by $zz^* + 2(z + z^*) = 0$, where $z = x + iy$ and $z^*$ is its complement?

# DE MOIVRE'S THEOREM

We saw in example 25.10 that $(\cos\theta_1 + i\sin\theta_1)(\cos\theta_2 + i\sin\theta_2) = \cos(\theta_1 + \theta_2) + i(\sin(\theta_1 + \theta_2)$

Putting $\theta_1 = \theta_2 = \theta$ gives $\qquad (\cos\theta + i\sin\theta)^2 = \cos 2\theta + i\sin 2\theta \qquad$ ①

Multiplying ① by $\cos\theta + i\sin\theta$ gives $(\cos\theta + i\sin\theta)^3 = (\cos 2\theta + i\sin 2\theta)(\cos\theta + i\sin\theta)$
$$= \cos 3\theta + i\sin 3\theta$$

this suggests that $\qquad (\cos\theta + i\sin\theta)^n \equiv \cos n\theta + i\sin n\theta$

This is in fact true, when $n$ is an integer, and can be proved by induction. It is known as de Moivre's theorem. (when $n$ is a real number that is not an integer the theorem is subject to some limitation)

> de Moivre's theorem states that
> $(\cos\theta + i\sin\theta)^n \equiv \cos n\theta + i\sin n\theta$ where $n$ is an integer

De Moivre's theorem has a number of uses including finding formulae for multiple angles, finding powers of complex numbers and finding all the complex roots any number. It is used extensively in physics, electrical engineering and optics.

## Example 25.15

Use de Moivre's theorem to find identities for $\cos 4\theta$, $\sin 4\theta$ and $\tan 4\theta$ in terms of $\cos\theta$, $\sin\theta$ and $\tan\theta$

> Put $n = 4$ in de Moivre's theorem: $\cos 4\theta + i\sin 4\theta \equiv (\cos\theta + i\sin\theta)^4$

> then, using the binomial theorem:
$$\cos 4\theta + i\sin 4\theta \equiv \cos^4\theta + 4\cos^3\theta.i\sin\theta + 6\cos^2\theta.i^2\sin^2\theta + 4\cos\theta.i^3\sin^3\theta + i^4\sin^4\theta$$
$$\equiv \cos^4\theta - 6\cos^2\theta\sin^2\theta + \sin^4\theta + i(4\cos^3\theta\sin\theta - 4\cos\theta\sin^3\theta)$$

> equating real and imaginary parts:
$\cos 4\theta \equiv \cos^4\theta - 6\cos^2\theta\sin^2\theta + \sin^4\theta$ ①     $\sin 4\theta \equiv 4\cos^3\theta\sin\theta - 4\cos\theta\sin^3\theta$ ②

> use the identity $\cos^2\theta + \sin^2\theta = 1$ to put ① in terms of $\cos\theta$ only:
$$\cos 4\theta \equiv \cos^4\theta - 6\cos^2\theta(1 - \cos^2\theta) + (1 - \cos^2\theta)^2$$
$$\equiv 8\cos^4\theta - 8\cos^2\theta + 1$$
$$\sin 4\theta \equiv 4\sin\theta\cos\theta(\cos^2\theta - \sin^2\theta)$$

➤ find tan 4θ:  $\qquad \tan 4\theta \equiv \dfrac{\sin 4\theta}{\cos 4\theta}$

$$\equiv \dfrac{4\sin\theta\cos\theta\,(\cos^2\theta - \sin^2\theta)}{8\cos^4\theta - 8\cos^2\theta + 1}$$

➤ divide by $\cos^4\theta$:  $\qquad \tan 4\theta = \dfrac{4\tan\theta\,(1 - \tan^2\theta)}{8 - 8\sec^2\theta + \sec^4\theta}$

➤ simplify denominator:  $\qquad \tan 4\theta_1 = \dfrac{4\tan\theta\,(1 - \tan^2\theta)}{1 - 6\tan^2\theta + \tan^4\theta}$

## Example 25.16

Use de Moivre's theorem to find $(1 + i\sqrt{3})^6$

➤ to use de Moivre's theorem we need to express the question in the form $(\cos\theta + i\sin\theta)^6$

➤ now, $\cos\frac{1}{3}\pi = \frac{1}{2}$ and $\sin\frac{1}{3}\pi = \frac{\sqrt{3}}{2}$, so $(1 + i\sqrt{3}) = 2(\cos\frac{1}{3}\pi + i\sin\frac{1}{3}\pi)$

➤ $(1 + i\sqrt{3})^6 = 2^6(\cos\frac{1}{3}\pi + i\sin\frac{1}{3}\pi)^6$

➤ $\qquad\qquad = 2^6(\cos 2\pi + i\sin 2\pi)$  using de Moivre's theorem ($n = 6$)

$\qquad\qquad = 2^6 \times 1 = 64$

## Example 25.17

If $f(x)$ is a polynomial of degree $n$, then the equation $f(x) = 0$ has $n$ roots. Not all of these roots will necessarily be real. Consider the equation $x^3 - 1 = 0$. The obvious root of this equation is $x = 1$. However, if we factorise the equation we get $(x - 1)(x^2 + x + 1) = 0$

In other words, the roots are $x = 1$, $x = \dfrac{-1}{2} \pm \dfrac{\sqrt{3}}{2}i$

There are therefore three cube roots of 1; one real and two complex.

We could find all the roots in this case because we could factorise and then use the quadratic formula. This is not possible when finding higher roots of 1, e.g. the fifth root. This can be found using de Moivre's theorem, but for the sake of simplicity and to show that the theorem works we shall use de Moivre's theorem to find $\sqrt[3]{1}$

➤ since $\cos 2n\pi = 1$ and $\sin 2n\pi = 0$, $\cos 2n\pi + i\sin 2n\pi = 1$

➤ $\therefore \sqrt[3]{1} = (\cos 2n\pi + i\sin 2n\pi)^{\frac{1}{3}}$

$\qquad = \cos\frac{2}{3}n\pi + i\sin\frac{2}{3}n\pi$ (by de Moivre's theorem)

➤ $\qquad = \cos\frac{2}{3}\pi + i\sin\frac{2}{3}\pi,\ \cos\frac{4}{3}\pi + i\sin\frac{4}{3}\pi,$ (higher values for $n$ do not give different values)

$\qquad = -\frac{1}{2} + i\frac{\sqrt{3}}{2},\ -\frac{1}{2} - i\frac{\sqrt{3}}{2}$

## Example 25.18

If $z = \cos\theta + i\sin\theta$, find an expression for $z + \dfrac{1}{z}$

Hence find an expression for $z^n + \dfrac{1}{z^n}$

since $z = \cos\theta + i\sin\theta$, $\dfrac{1}{z} = \dfrac{\cos\theta - i\sin\theta}{(\cos\theta + i\sin\theta)(\cos\theta - i\sin\theta)} = \cos\theta - i\sin\theta$

so $\quad z + \dfrac{1}{z} = \cos\theta + i\sin\theta + \cos\theta - i\sin\theta = 2\cos\theta$

by deMoivre's theorem $\quad z^n = \cos n\theta + i\sin n\theta$

so $\quad z^n + \dfrac{1}{z^n} = 2\cos n\theta$

# THE EXPONENTIAL FORM OF A COMPLEX NUMBER
### (Relationships between the Exponential, Sine and Cosine Functions)

In chapter 19 (example 19.13 and exercise 19C question 1) we found that, using Maclaurin's theorem, the exponential, sine and cosine functions could be written as series of ascending powers of $x$ (or $\theta$).

These series are:
$$e^\theta = 1 + \frac{\theta}{1!} + \frac{\theta^2}{2!} + \frac{\theta^3}{3!} + \frac{\theta^4}{4!} + \frac{\theta^5}{5!} + \frac{\theta^6}{6!} + \frac{\theta^7}{7!} + \cdots \qquad \text{①}$$

$$\cos\theta = 1 - \frac{\theta^2}{2!} + \frac{\theta^4}{4!} - \frac{\theta^6}{6!} + \cdots \qquad \text{②}$$

$$\sin\theta = \frac{\theta}{1!} - \frac{\theta^3}{3!} + \frac{\theta^5}{5!} - \frac{\theta^7}{7!} + \cdots \qquad \text{③}$$

If ① is applied to the function $e^{i\theta}$ we get

$$e^{i\theta} = 1 + \frac{i\theta}{1!} + \frac{i^2\theta^2}{2!} + \frac{i^3\theta^3}{3!} + \frac{i^4\theta^4}{4!} + \frac{i^5\theta^5}{5!} + \frac{i^6\theta^6}{6!} + \frac{i^7\theta^7}{7!} + \cdots$$

$$= 1 + i\frac{\theta}{1!} - \frac{\theta^2}{2!} - i\frac{\theta^3}{3!} + \frac{\theta^4}{4!} + i\frac{\theta^5}{5!} - \frac{\theta^6}{6!} - i\frac{\theta^7}{7!} + \cdots$$

$$= 1 - \frac{\theta^2}{2!} + \frac{\theta^4}{4!} - \frac{\theta^6}{6!} + \cdots + i\left(\frac{\theta}{1!} - \frac{\theta^3}{3!} + \frac{\theta^5}{5!} - \frac{\theta^7}{7!} + \cdots\right)$$

$$= \cos\theta + i\sin\theta \qquad \text{④}$$

In the same way $\quad e^{-i\theta} = \cos\theta - i\sin\theta \qquad \text{⑤}$

and eliminating first $\sin\theta$ and then $\cos\theta$ from ④ and ⑤ we get
$$\cos\theta = \tfrac{1}{2}(e^{i\theta} + e^{-i\theta}) \quad \text{and} \quad \sin\theta = \tfrac{1}{2}(e^{i\theta} - e^{-i\theta})$$
From ④ we see that the complex number $z = x + iy$ can also be written in modulus and argument form as $z = r(\cos\theta + i\sin\theta) = re^{i\theta}$. $z = re^{i\theta}$ is known as the **exponential form** for the complex number $z$.
Summarising:

$$z = x + iy = r(\cos\theta + i\sin\theta) = re^{i\theta}$$

$$e^{i\theta} = \cos\theta + i\sin\theta \quad \text{and} \quad e^{-i\theta} = \cos\theta - i\sin\theta$$

$$\cos\theta = \tfrac{1}{2}(e^{i\theta} + e^{-i\theta}) \quad \text{and} \quad \sin\theta = \frac{1}{2i}(e^{i\theta} - e^{-i\theta})$$

It was asserted in chapter 17 that the general solution of a second order linear differential equation, when the roots of the auxiliary equation were complex, was of the form $y = e^{px}(A\cos qx + B\sin qx)$ rather than $y = Ae^{\alpha x} + Be^{\beta x}$. You can see now where this comes from.

Many of proofs/problems for which we used de Moivre's theorem, can be equally proved/solved using the exponential form – which is often quicker and simpler

## Example 25.19

Express in exponential form     (a)    $\cos \frac{1}{3}\pi + i \sin \frac{1}{3}\pi$          (b)    $1 + i$

    (a)   $r = 1$, $\theta = \frac{1}{3}\pi$, so   $\cos \frac{1}{3}\pi + i \sin \frac{1}{3}\pi = e^{\frac{1}{3}i\pi}$

    (b)   find the modulus and argument:   $r = \sqrt{1^2 + 1^2} = \sqrt{2}$ ;   $\theta = \tan^{-1}(\frac{1}{1}) = \frac{1}{4}\pi$

      so $1 + i = \sqrt{2}\, e^{\frac{1}{4}i\pi}$

## Example 25.20

Use the exponential form of a complex number to prove that   $(\cos \theta + i \sin \theta)^4 = \cos 4\theta + i \sin 4\theta$

$$(\cos \theta + i \sin \theta)^4 = (e^{i\theta})^4$$
$$= e^{i(4\theta)}$$
$$= \cos 4\theta + i \sin 4\theta$$

## EXERCISE 25E

de Moivre's theorem should be used in questions $1 - 7$

1.   Show that $\sin 3\theta \equiv 3 \sin \theta - 4 \sin^3 \theta$ and $\cos 3\theta \equiv 4 \cos^3 \theta - 3 \cos \theta$.
    Obtain an expression for $\tan 3\theta$ in terms of $\tan \theta$.

2.   Show that $\sin 5\theta \equiv 16 \sin^5 \theta - 20 \sin^3 \theta + 5 \sin \theta$

3.   Simplify the following expressions:
    (a)   $(\cos \theta + i \sin \theta)^5$     (b)   $(\cos \theta + i \sin \theta)^{-3}$   (c)   $(\cos \theta - i \sin \theta)^4$     (d)   $(\sin \theta + i \cos \theta)^3$

4.   Simplify:
    (a)   $(1 + i)^5$          (b)   $(1 - i)^7$          (c)   $(1 - i\sqrt{3})^4$        (d)   $(\sqrt{3} + i)^{-2}$

5.   Find $(\sqrt{2} + i\sqrt{2})^6$

6.   Solve the equation $x^3 + 1 = 0$

7.   Express in polar $(r, \theta)$ form
    (a)   $e^{2i\theta}$           (b)   $2e^{4i\theta}$           (c)   $5e^{-i\theta}$

8.   By first expressing in polar form, express the following in the form $a + ib$
    (a)   $e^{i\pi}$           (b)     $e^{-\frac{1}{4}i\pi}$         (c)      $e^{1 + \frac{1}{2}i\pi}$

9.   Express in exponential form $(re^{i\theta})$
    (a)   $\cos 3\theta + i \sin 3\theta$     (b)   $1 - i$           (c)   $i$             (d)     $2 - 2i\sqrt{3}$

10. Use the exponential form of the complex numbers $z_1$ and $z_2$ to show that $|z_1 z_2| = |z_1||z_2|$ and $\arg(z_1 z_2) = \arg z_1 + \arg z_2$

11. Use the exponential form of a complex number to prove that $(\cos\theta + i\sin\theta)^5 = \cos 5\theta + i\sin 5\theta$

# EXERCISE 25 X

1. $z_1$ and $z_2$ are the complex numbers $4 + 2i$ and $3 + i$ respectively.
   (i) Express the product $w = z_1 z_2$ in the form $a + bi$. [2]
   (ii) Find the modulus and principal argument of $w$. [3]
   (iii) Write down the complex numbers $iw$ and $\overline{w}$ (the complex conjugate of $w$). [2]
   (iv) Find the distance between the points on the Argand diagram represented by $iw$ and $\overline{w}$. [2]

2. The complex number $z$ is expressed in the form $x + iy$, where $x, y \in \mathbb{R}$.

   Express the complex number $\dfrac{z-1}{z+1}$ in a similar form. [7]

   The argument of the complex number $\dfrac{z-1}{z+1}$ is $\frac{1}{4}\pi$.
   (i) Find the equation connecting $x$ and $y$. [3]
   (ii) Show that the equation represents a circle, C. [3]
   (iii) Determine the centre and radius of C. [3]

3. (a) Find the two square roots of the complex number $5 - 12i$ in the form $x + yi$, where $x, y \in \mathbb{R}$. [8]

   (b) (i) If $z = x + yi$, where $x, y \in \mathbb{R}$, $y \neq 0$, find the real and imaginary parts of $z + \dfrac{1}{z}$. [5]

   (ii) Find and identify the locus of the points for which the imaginary part of $z + \dfrac{1}{z}$ is zero [5]

4. (a) A complex number, $z$ is expressed in the form $x + iy$, where $x, y \in \mathbb{R}$.

   Express the complex number, $w = \dfrac{z-1}{z+2}$, in similar form. [8]
   (b) The argument of $w$ is $\frac{1}{4}\pi$.
   (i) Find the equation connecting $x$ and $y$ in the form
   $ax^2 + by^2 + cx + dy + f = 0$ where $a, b, c, d, f$ are integers. [4]
   (ii) Show that the equation in (i) represents a circle, C. [3]
   (iii) Determine the centre and radius of the circle C. [4]

5. Find ALL the complex numbers $x + yi$ such that
   $(x + yi)^2 = -3 + 4i$ [8]

6. (a) The roots of the quadratic equation $x^2 + 6x + k$ are $-3 + 2i$ and $-3 - 2i$
   Find the value of the constant $k$. [2]

   (b) Find the real numbers $u$ and $v$ such that $\dfrac{u + 2i}{3 - 4i} \equiv 1 + vi$. [6]

7.  Let $x$ and $y$ be non-zero real numbers, $z$ be the complex number given by $z = x + iy$ and $\overline{z}$ be the conjugate of $z$. Find the values of $x$ and $y$ if

$$\frac{1}{z} + \frac{3}{\overline{z}} = i + 1.$$  [6]

8.  Find the range of values of $k \in \mathbb{R}$ for which the equation $x^2 + kx + 1 = 0$ has complex roots.  [2]

9.  Find the range of values of $k$ for which the equation $x^2 + 2kx + 9 = 0$ has complex roots.  [4]

10. Express the complex number $\dfrac{2 + 3i}{3 + 4i}$ in the form $x + iy$, where $x$ and $y$ are real numbers.  [4]

11. Find the modulus of the complex number $z = \dfrac{25(2 + 3i)}{4 + 3i}$.  [4]

12. Let $z_1 = 2 - 3i$ and $z_2 = 3 + 4i$.
    (a)  Find in the form $z = a + bi$, $a, b \in \mathbb{R}$,
        (i)   $z_1 + z_2$  [1]
        (ii)  $z_1 z_2$  [3]
        (iii) $\dfrac{z_1}{z_2}$  [5]

    (b)  Find the quadratic equation whose roots are $z_1$ and $z_2$  [2]

# What you have learnt in Chapter 25

## Definitions:

the imaginary number $i$ is defined by $i^2 = -1$, so that $i = \sqrt{-1}$

a complex number is a number of the form $z = p + iq$, $p, q \in \mathbb{R}$

The real part of $z$ is $\text{Re}\, z = p$

The imaginary part of $z$ is $\text{Im}\, z = q$

The complex conjugate of $z$ is $\bar{z}$ or $z^* = p - iq$.

the modulus of $z$ is the length of $z$ in the Argand diagram: $|z| = \sqrt{p^2 + q^2}$

and the argument of $z$, is the angle that $z$ makes with the positive $x$-axis in the Argand diagram:

$$\text{Arg}\, z = \tan^{-1}\left(\frac{q}{p}\right)$$

the principal argument, $\arg z$, is the argument for which $-\pi < \arg z \leq \pi$

(The argument is always measured in radians)

## Polar and Exponential Forms:

If $r$ and $\theta$ are the modulus and argument of $z$, then $z = r\cos\theta + ir\sin\theta = re^{i\theta}$

if $z_1$ and $z_2$ are two complex numbers, then

$$|z_1 z_2| = |z_1||z_2| \quad \text{and} \quad \arg z_1 z_2 = \arg z_1 + \arg z_2$$

$$\left|\frac{z_1}{z_2}\right| = \frac{|z_1|}{|z_2|} \quad \text{and} \quad \arg\frac{z_1}{z_2} = \arg z_1 - \arg z_2$$

$\cos 2n\pi + i \sin 2n\pi = 1$

## Quadratic Equations:

A quadratic equation with **real** coefficients which does not have real roots has complex roots which are complex conjugates.

sum of roots: $z + z^* = 2p$ and product of roots: $zz^* = p^2 + q^2$ are both real numbers

## de Moivre's theorem:

$(\cos\theta + i\sin\theta)^n \equiv \cos n\theta + i\sin n\theta$ where $n$ is an integer

# APPENDIX 3
# Multiple Choice Questions

You are already familiar with multiple choice questions.

Here are a few points to remember when sitting a CAPE multiple choice paper:

❖ There are 45 questions to be answered in 90 minutes – i.e. an average of 2 minutes per question. In a ninety minute exam it is easy to run out of time if you spend too long on one or two questions that you are struggling with. So try not to spend more than 2 minutes on any particular question at a first attempt. Come back to it later if you have time.

❖ The examiners will usually have chosen common mistakes for the wrong alternative answers. So be careful, it's not only guesswork that will lead to the wrong answer, so can a silly mistake.

❖ Usually you will need to work a question as if it were a normal question (though no one is going to see your working, so don't waste time on presentation – there can be no method marks). However, occasionally it's easier to work a question backwards – for instance if you had an integration you weren't sure about you could differentiate the 4 answers to see which one led to the question. Likewise, to find the square root of a complex number you could try squaring all the possible answers.

❖ Some questions include 'none of these' as the 4th alternative answer. This may be a way for the examiners to avoid thinking up a 4th alternative! but sometimes they have deliberately not included the correct answer – in other words it is a genuine possible answer.

The exercise that follows includes questions from the whole syllabus, but has been divided into 3 sections corresponding to the 3 modules of the syllabus.

# MODULE A
## (CHAPTERS 15 – 18)

1. For the function $f(x) = 2^{-x}$, which of the following statements are true:
   I     $f(x)$ is an increasing function
   II    $f(x)$ is a decreasing function
   III   the graph of $y = f(x)$ passes through the point $(0, 1)$
   IV   the graph of $y = f(x)$ passes through the point $(0, 2)$

   (A) I and III only

   (B) I and IV only

   (C) II and III only

   (D) II and IV only

2. $\log_a b$ is

(A) $\dfrac{\log_b c}{\log_a c}$

(B) $\log_a c . \log_c b$

(C) $\dfrac{\log_c b}{\log_c a}$

(D) none of the above

3. If $5^x = 7$ then $x$ is equal to

(A) 0.827
(B) 1.4
(C) 1.54
(D) 1.21

4. The solution of the equation $e^{2x} - 2e^x + 1 = 0$ is

(A) 1
(B) 0
(C) ln 2
(D) none of these

5. If $3(27)^x = 9^{2x+1}$, then $x$ is equal to

(A) −1
(B) 1
(C) 3
(D) $1\frac{1}{2}$

6. If $7^x = 3^{2x-1}$ then $x$ is equal to

(A) 3
(B) 0.229
(C) 4.37
(D) None of these

7. If $f(x) = e^{-x}$, then $f^{-1}(x)$ is

(A) $\ln(-x)$
(B) $e^x$
(C) $-\ln x$
(D) $\ln \frac{1}{x}$

8. $\lg(100ab)$ can be written as

(A) $2 + \lg a + \lg b$
(B) $2(\lg a + \lg b)$
(C) $10 \lg ab$
(D) $20(\lg a + \lg b)$

9. If $7^x = 11$, then $x$ is equal to

(A) $\dfrac{\log 7}{\log 11}$

(B) $\dfrac{\log 11}{\log 7}$

(C) $\log(\frac{7}{11})$

(D) $\log(\frac{11}{7})$

10. $\log_c a\sqrt{b}$ can be written as

(A) $\frac{1}{2}\log_c ab$
(B) $\frac{1}{2}\log_c (a + b)$
(C) $\log_c a + \frac{1}{2}\log_c b$
(D) $\log_a c + \frac{1}{2}\log_b c$

11. If $\log_3 5 = x$, then

(A) $x^3 = 5$
(B) $3^x = 5$
(C) $3^5 = x$
(D) $5^x = 3$

12. $\dfrac{d}{dx}(x\cos x)$ is

(A) $-\sin x$
(B) $-x\sin x$
(C) $x\sin x$
(D) none of these

13. $\dfrac{d}{dx}(\sin^{-1} 2x)$ is

(A) $\dfrac{2}{4 + x^2}$

(B) $\dfrac{2}{\sqrt{4 - x^2}}$

(C) $\dfrac{1}{\sqrt{4 - x^2}}$

(D) $\dfrac{2}{\sqrt{1 - 4x^2}}$

14. $\dfrac{d}{dx}(e^{\cos x})$ is

(A) $e^{-\sin x}$
(B) $-\sin x\, e^{\cos x}$
(C) $\cos x\, e^{-\sin x}$
(D) $\sin x\, e^{\sin x}$

15. $\dfrac{d}{dx}(x^2\,e^{2x})$ is

(A) $2x(x+1)e^{2x}$
(B) $4xe^{2x}$
(C) $2x + 2e^{2x}$
(D) $x(x+4)e^{2x}$

16. $\dfrac{d}{dx}\left(\dfrac{e^x}{5+e^x}\right)$ is

(A) $\dfrac{5e^x + 2e^{2x}}{(5+e^x)^2}$

(B) $\dfrac{5+2e^x}{25+e^{2x}}$

(C) $\dfrac{5e^x}{(5+e^x)^2}$

(D) none of these

17. A curve is defined parametrically by the equations $x = \cos t,\ y = 1 - \sin t$. $\dfrac{dy}{dx}$ is

(A) $\operatorname{cosec} t - \tan t$
(B) $\cot t$
(C) $\tan t$
(D) $-\cot t$

18. A curve is defined parametrically by the equations $x = t^3,\ y = 2t^2 - 3$. $\dfrac{dy}{dx}$ is

(A) $\dfrac{4}{3t}$

(B) $\dfrac{3t}{4}$

(C) $12t^3$
(D) $t(4-3t)$

19. A curve is defined parametrically by the equations $y = 2t,\ x = \sin^{-1}t$. $\dfrac{dy}{dx}$ is

(A) $\dfrac{1}{2\sqrt{1-t^2}}$

(B) $2\sqrt{1-t^2}$

(C) $\dfrac{1}{2(1+t^2)}$

(D) $2(1+t^2)$

20. For the curve $x^2 + y^2 = 7$, $\dfrac{dy}{dx}$ is

(A) $2x$
(B) $-2x$
(C) $-\dfrac{x}{y}$

(D) $\dfrac{7-2x}{2y}$

21. The gradient of the curve $2xy + x^2 = 5y + 8$ at the point $(2, -4)$ is

(A) $3$
(B) $7$
(C) $-4$
(D) $-2$

22. $\dfrac{d}{dx}[\ln(x^2 - 4)]$ is

(A) $\dfrac{2x}{x^2 - 4}$

(B) $\dfrac{x^2 - 4}{2x}$

(C) $\dfrac{2}{x^2}$

(D) $\dfrac{2}{x}$

23. The gradient of the curve $xy + y^2 = \ln x$ at the point $(1, -1)$ is

(A) Undefined
(B) $-2$
(C) $0$
(D) $1$

24. If $f(x) = x^2 - \cos x$, then $f''(x) =$

(A) $\sin x + 2$
(B) $\cos x$
(C) $2x + \sin x$
(D) $2 + \cos x$

25. $\dfrac{2}{(x-5)(x-3)}$ decomposes into the partial fractions

(A) $\dfrac{1}{x-3} - \dfrac{1}{x-5}$

(B) $\dfrac{1}{x-5} - \dfrac{1}{x-3}$

(C) $-\dfrac{1}{x-5} - \dfrac{1}{x-3}$

(D) $\dfrac{1}{x-5} + \dfrac{1}{x-3}$

26. $\displaystyle\int \dfrac{dx}{1-x^2}$ , $|x| < 1$, is

(A) $\sin^{-1}x + c$

(B) $\tan^{-1}x + c$

(C) $-\ln 2x + c$

(D) $\dfrac{1}{2}\ln\dfrac{1+x}{1-x} + c$

27. $\displaystyle\int \ln x \, dx$ is

(A) $1 - x \ln x + c$

(B) $\dfrac{1}{x} + c$

(C) $x \ln x - x + c$

(D) $\dfrac{1}{2}\ln x^2 + c$

28. $\displaystyle\int \sin^2 x \, dx$ is

(A) $2\sin x \cos x + c$

(B) $-\dfrac{1}{3}\sin^3 x + c$

(C) $\dfrac{1}{4}(2x - \sin 2x) + c$

(D) $-\dfrac{1}{2}\cos^2 x + c$

29. $\displaystyle\int_{-3}^{5} \dfrac{1}{x}\,dx$ is

(A) $\ln\left|\dfrac{5}{-3}\right|$

(B) $0$

(C) $\dfrac{1}{25} - \dfrac{1}{9}$

(D) none of these

30. $\displaystyle\int x \sin x \, dx$ is

(A) $\sin x + x \cos x$

(B) $\sin x - x \cos x$

(C) $-\sin x - x \cos x$

(D) $-\sin x + x \cos x$

31. $\displaystyle\int \dfrac{2x}{x^2+1}\,dx$ is

(A) $2x \tan^{-1}x$

(B) $\ln(x^2 + 1)$

(C) $x \tan^{-1}x$

(D) $-(x^2 + 1)^{-1}$

32. The general solution of the differential equation $y'' - 3y' + 2y = 0$ is

(A) $y = (Ax + B)e^{2x}$

(B) $y = A\cos x + B\sin 2x$

(C) $y = e^{-3x}(A\cos 2x + B\sin 2x)$

(D) $y = Ae^x + Be^{2x}$

33. The general solution of the differential equation $y'' - 6y' + 9y = 0$ is

(A) $y = e^{3x}(Ax + B)$

(B) $y = Ae^{3x} + Be^x$

(C) $y = Ae^{3x} + B$

(D) $y = A\sin 3x + B\cos 3x$

34. The general solution of the differential equation $y'' - 2y' + 5y = 0$ is

(A) $Ae^x + Be^{5x}$

(B) $Ae^{2x} + Be^{5x}$

(C) $e^x(A\sin 2x + B\cos 2x)$

(D) $e^{2x}(A\sin x + B\cos x)$

35. The rate of decrease in the population of tigers in India is proportional to the number of tigers. This situation can be modelled by the equation

(A) $\dfrac{dP}{dt} = \dfrac{k}{P}$

(B) $\dfrac{dP}{dt} = kP$

(C) $\dfrac{dP}{dt} = \dfrac{-k}{P}$

(D) $\dfrac{dP}{dt} = -kP$

# MODULE B
## (CHAPTERS 18 – 22)

36. The sequence $1 + (\frac{1}{2})^0, 1 + (\frac{1}{2})^1, 1 + (\frac{1}{2})^2,$ $1 + (\frac{1}{2})^3, 1 + (\frac{1}{2})^4 \ldots$

   (A)   is divergent
   (B)   converges to 0
   (C)   converges to 1
   (D)   converges to 2

37. For the sequence
   $A = \{12, 12\frac{1}{2}, 12\frac{1}{4}, 12\frac{1}{8} \ldots \}$ which of
   the following statements are true?
   I  $\{a_n\}$ is arithmetic      II $\{a_n\}$ is geometric
   III $\{a_n\}$ is convergent     IV $\{a_n\}$ is divergent

   (A)   II and III only
   (B)   III only
   (C)   I and IV only
   (D)   II and IV only

**Questions 38 and 39** refer to the sequence
$B = \{1, -\frac{1}{3}, \frac{1}{5}, -\frac{1}{7} \ldots \}$

38. For the sequence B which of the following
   statements are true?
   I  $\{a_n\}$ is oscillating      II $\{a_n\}$ is finite
   III $\{a_n\}$ is convergent
   (A)   I and II
   (B)   I only
   (C)   I and III
   (D)   II and III

39. The $r$ th term of B is

   (A)   $\dfrac{(-1)^r}{r}$

   (B)   $\dfrac{(-1)^{r+1}}{2r - 1}$

   (C)   $\dfrac{(-1)^r}{2r - 1}$

   (D)   $\dfrac{-1}{2r + 1}$

**Questions 40 to 42** refer to the series

$$S = 5 + \frac{5}{2} + \frac{5}{4} + \frac{5}{8} + \cdots$$

40. The $r$th term of S is

   (A)   $5 \times (\frac{1}{2})^r$
   (B)   $10 \times (\frac{1}{2})^r$
   (C)   $5 - \frac{5}{2}r$
   (D)   none of the above

41. The sum of the first 20 terms of S is

   (A)   $10(1 - (\frac{1}{2})^{20})$
   (B)   $5(1 - (\frac{1}{2})^{19})$
   (C)   $5(1 - (\frac{1}{2})^{20})$
   (D)   $10(1 - (\frac{1}{2})^{19})$

42. The sum to infinity of S is

   (A)   $5\frac{1}{2}$
   (B)   10
   (C)   15
   (D)   20

**Questions 43 to 45** refer to
$S_n = 5 + 17 + 29 + 41 + \ldots + u_n \quad n \le 50$

43. $S_n$ is

   (A)   a finite sequence
   (B)   a finite series
   (C)   an infinite sequence
   (D)   an infinite series

44. If $u_n = 185$, $n$ is equal to

   (A)   14
   (B)   15
   (C)   16
   (D)   49

45. If $u_n = 185$, $S_n$ is equal to
    (A) 3040
    (B) 1616
    (C) 1520
    (D) 1480

46. The sum of the <u>second</u> twenty integers is
    (A) 610
    (B) 410
    (C) 820
    (D) 590

47. The 7th term of an AP is 17 and the 12th term is 27. The first term and common difference are
    (A) $a = -1, d = 3$
    (B) $a = -6, d = 3$
    (C) $a = 5, d = 2$
    (D) $a = 3, d = 2$

48. A GP has first term 48 and 4th term $-6$. The sum to infinity is
    (A) 96
    (B) 72
    (C) 32
    (D) 16

49. The first 3 non-zero terms in the Maclaurin expansion of $\ln(1+x)$ are
    (A) $x + \frac{1}{2}x^2 - \frac{1}{3}x^3$
    (B) $1 + x + \frac{1}{2}x^2$
    (C) $1 + x - \frac{1}{2}x^2$
    (D) $x - \frac{1}{2}x^2 + \frac{1}{3}x^3$

50. $\sum_{r=5}^{20} (3 - 4r)$ is
    (A) $-755$
    (B) $-752$
    (C) $-735$
    (D) $-732$

Questions 51 and 52 refer to an AP whose first term is 51 and common difference is $-4$.

51. What is the minimum value of $n$ for which the $n$th term of the series is negative?
    (A) 12
    (B) 13
    (C) 14
    (D) 15

52. What is the minimum value of $n$ for which the sum of the series is negative?
    (A) 25
    (B) 26
    (C) 27
    (D) 28

53. $^nC_r =$
    (A) $^nC_{r-1} + {}^nC_{r+1}$
    (B) $^{n-1}C_{r-1} + {}^{n-1}C_r$
    (C) $^{n+1}C_{r-1} + {}^{n+1}C_r$
    (D) $^{n+1}C_r + {}^{n+1}C_{r+1}$

54. $\dfrac{1}{\sqrt{0.95}} = (1+x)^r$ where
    (A) $x = 0.05, r = -\frac{1}{2}$
    (B) $x = -0.05, r = 2$
    (C) $x = 0.05, r = \frac{1}{2}$
    (D) $x = -0.05, r = -\frac{1}{2}$

55. The coefficient of $x^4$ in the expansion of $(2-x)^7$ is
    (A) $-280$
    (B) 280
    (C) $-560$
    (D) 35

56. The coefficient of $x^3$ in the expansion of $(x+4)(1-3x)^5$ is
    (A) $-990$
    (B) 1170
    (C) $-270$
    (D) 90

57. The coefficient of $x^2$ in the expansion of $(1-2x)^{\frac{1}{2}}$ is

    (A) $-\frac{1}{2}$
    (B) $\frac{1}{8}$
    (C) $-\frac{1}{8}$
    (D) $\frac{1}{2}$

58. When $r$ is not a positive integer, the binomial expansion of $(a+bx)^r$ is subject to what restrictions?

    (A) none
    (B) $a=1, |x|<1$
    (C) $a=1, |bx|<1$
    (D) $\left|\frac{b}{a}x\right|<1$

59. If $f(x) \equiv x^3 + x^2 - 3x + 2$, the equation $f(x) = 0$ has roots in the intervals
    I $[-3,-2]$   II $[-2,-1]$   III $[-1,0]$
    IV $[0,1]$   V $[1,2]$

    (A) I only
    (B) I, IV and V only
    (C) III, IV and V only
    (D) II and IV only

60. If $f(x) \equiv 2x^2 - 5x - 6$, the equation $f(x) = 0$ has roots in the intervals
    I $[-2,-1]$   II$[-1,0]$   III $[2,3]$
    IV $[3,5]$

    (A) I and III only
    (B) III and IV only
    (C) III only
    (D) II and IV only

61. It is known that the equation $3x^3 - 4x^2 - 3 = 0$ has a root, $\alpha$, between 1.5 and 2.

    Use the intermediate value theorem to find $\alpha$ correct to 1 d.p.

    (A) 1.5
    (B) 1.6
    (C) 1.7
    (D) 1.8

62. If an application of the Newton-Raphson method fails the reason could be

    I    there is a discontinuity between the approximation and the root

    II   there is a critical point between the approximation and the root
    III  the root is too close to 0

    (A) I and II only
    (B) I and III only
    (C) II and III only
    (D) I, II and III

**Questions 63 to 64** refer to the situation described below:
    The front cover of a book is actually 20.7 cm wide and 27.8 cm high
    A librarian estimated the cover to be 21 cm by 28 cm

63. The error in the estimated area of the cover is

    (A) 0.5 cm²
    (B) 0.6 cm²
    (C) 1 cm²
    (D) 12.54 cm²

64. To 2 significant figures, the % error in the estimated area of the cover is

    (A) 2.1%
    (B) 2.2%
    (C) 0.02
    (D) 2.13%

**Questions 65 to 67** refer to the situation described below:
    In order to provide lace edging, a square handkerchief is measured and found to have sides of length 14cm to the nearest cm.

65. The maximum perimeter of the handkerchief is

    (A) 54 cm
    (B) 56 cm
    (C) 58 cm
    (D) 57.96 cm

66. If 20 handkerchiefs are to be edged, the maximum difference in the amount of lace needed is

    (A) 80 cm
    (B) 39.2 cm
    (C) 78.4 cm
    (D) 40 cm

67. The minimum area of a single handkerchief is
   (A)   182.25 cm²
   (B)   196 cm²
   (C)   189 cm²
   (D)   none of the above

# MODULE C
## (Chapters 23 – 25)

68. 5 children stand in line for a photograph. If Jacqui and Duane must stand next to each other, how many possible arrangements are there?

   (A)   24
   (B)   120
   (C)   60
   (D)   48

69. How many different arrangements are there of the letters of the word PANAMA?

   (A)   720
   (B)   120
   (C)   24
   (D)   36

70. How many different arrangements are there of the letters of the word COFFEE in which the Fs remain together and the Es remain together?

   (A)   96
   (B)   720
   (C)   24
   (D)   6

71. In how many ways can 4 children be chosen from a family of 7 to go to the pantomime.

   (A)   24
   (B)   840
   (C)   35
   (D)   none of the above

**Questions 72 to 74** refer to a class of 23 students of whom 15 are female.

72. What is the probability that the student with the highest grades is not female

   (A)   15/38
   (B)   8/15
   (C)   15/23
   (D)   8/23

73. 4 students are chosen to sit on the student representative committee. In how many ways can this be done if there must be equal numbers of male and female students
   (A)   $^{23}C_4$
   (B)   $^{23}C_2 \times {}^{21}C_2$
   (C)   $^{8}C_2 \times {}^{15}C_2$
   (D)   $\dfrac{23!}{2!2!}$

74. 3 students are chosen at random to represent the class in a quiz team. What is the probability that they are all female?

   (A)   0.257
   (B)   0.277
   (C)   0.224
   (D)   0.13

75. If $P(A) = \frac{1}{2}$, $P(B) = \frac{1}{3}$ and $P(A \cap B) = \frac{1}{4}$, which of the following statements are true?
   I   A and B are independent events
   II  $P(A \cup B) = \frac{7}{12}$
   III $P(A \cup B) = \frac{5}{6}$

   (A)   I and II
   (B)   I and III
   (C)   II only
   (D)   III only

76. There are 18 boys in a class and 11 must be chosen for the class cricket team to play on Saturday. Darren is the only boy who can keep wicket so must be included in the team; and Trevor is not available on Saturday. In how many ways can the team be chosen?

    (A) $^{16}C_{10}$
    (B) $^{18}C_{11}$
    (C) $^{17}C_{11}$
    (D) $^{17}C_{10}$

77. How many different arrangements are there of the letters of the word CARIBBEAN?

    (A) 1260
    (B) 90720
    (C) 362880
    (D) 181440

78. How many different arrangements are there of the letters of the word COFFEE?

    (A) 720
    (B) 180
    (C) 6
    (D) 96

79. If $P(X) = \frac{1}{4}$, $P(Y) = \frac{1}{2}$ and $P(X \cap Y) = \frac{1}{8}$, events X and Y are

    (A) complementary
    (B) mutually exclusive
    (C) independent
    (D) none of the above

80. If $P(X) = \frac{2}{5}$, $P(Y) = \frac{1}{3}$ and $P(X \cup Y) = \frac{2}{3}$, then $P(X \cap Y) =$

    (A) $\frac{1}{15}$
    (B) $\frac{2}{15}$
    (C) $\frac{1}{30}$
    (D) impossible

**Questions 81 and 82** refer to the situation and table below:

100 people, whose gender and age are represented in the table, respond to a questionnaire

|  | 20 and under | over 20 |
|---|---|---|
| male | 15 | 32 |
| female | 13 | 40 |

81. One person is picked at random to be interviewed. What is the probability that the person chosen is a male over 20?

    (A) 87%
    (B) 72%
    (C) 47%
    (D) 32%

82. Given that the person chosen is female what is the probability that she is under 20?

    (A) $\frac{28}{100}$
    (B) $\frac{13}{100}$
    (C) $\frac{13}{28}$
    (D) $\frac{13}{53}$

83. F and G are independent events. Which of the following statements are true?
    I $P(F \cap G) = 0$
    II $P(F | G) = P(F)$
    III $P(F \cap G) = P(F) \cdot P(G)$

    (A) I and II only
    (B) II and III only
    (C) I only
    (D) II only

84. 5 sisters stand in a line for a photograph. If they are positioned at random what is the probability that Khalia and Serena stand next to each other?

    (A) $\frac{2}{5}$
    (B) $\frac{1}{2}$
    (C) $\frac{1}{3}$
    (D) $\frac{3}{5}$

85. 7 books are placed at random on a shelf. Three of these books are volumes 1, 2 and 3 of a dictionary.

    What is the probability, to 3 d.p. that the 3 volumes of the dictionary are adjacent to each other and in the right order?

    (A)  0.143
    (B)  0.024
    (C)  0.005
    (D)  0.004

86. Given the equations
    I   $2x - 3y = 7$,
    II  $3x + 2y = 5$,
    III $6y = 4x - 3$,
    which pairs of equations do NOT form a consistent system?

    (A)  I and III
    (B)  I and II
    (C)  II and III
    (D)  all pairs are consistent

87. **M** is a matrix, $\mathbf{M^T}$ is its transpose, $\mathbf{M^{-1}}$ is its inverse and **I** is the unit matrix
    Which of the following statements are true?
    I   $(\mathbf{M^T})^T = \mathbf{M}$,
    II  $(\mathbf{M^{-1}})^{-1} = \mathbf{M}$,
    III $\mathbf{MM^{-1}} = \mathbf{I}$

    (A)  I and II only
    (B)  I and III only
    (C)  II and III only
    (D)  I, II and III

88. Which of the following matrices are singular?

    $I = \begin{pmatrix} 1 & -1 \\ 3 & 5 \end{pmatrix}$  $II = \begin{pmatrix} -2 & 4 \\ 3 & -6 \end{pmatrix}$  $III = \begin{pmatrix} 2 & 3 \\ 3 & 2 \end{pmatrix}$

    (A)  none of them
    (B)  II only
    (C)  II and III
    (D)  III only

**Questions 89 and 90** refer the the following matrices

$$\mathbf{M} = \begin{pmatrix} 1 \\ 2 \\ 3 \end{pmatrix} \quad \mathbf{N} = (5 \quad 3 \quad 2) \quad \mathbf{P} = \begin{pmatrix} 2 & 1 & 7 \\ -3 & 4 & 0 \\ 0 & 2 & -1 \end{pmatrix}$$

$$\mathbf{Q} = \begin{pmatrix} 1 & 0 \\ 2 & -4 \\ -1 & 2 \end{pmatrix}$$

89. Which of these products are possible under matrix multiplication?
    I   **MN**   II   **NM**   III   **PQ**

    (A)  I and II
    (B)  II and III
    (C)  I, II and III
    (D)  I and III

90. Which of these sums are possible under matrix addition?
    I   $\mathbf{M + N}$;   II   $\mathbf{N + P}$;   III   $\mathbf{P + Q}$;   IV   $\mathbf{M + Q}$

    (A)  II and III
    (B)  I and III
    (C)  none
    (D)  III and IV

91. The determinant of the matrix $\begin{pmatrix} 0 & 1 & 2 \\ 1 & 0 & 5 \\ 2 & 1 & 0 \end{pmatrix}$ is

    (A)  12
    (B)  8
    (C)  −8
    (D)  −12

92. The solution of the determinant equation
    $$\begin{vmatrix} x & x^2 & x^3 \\ 1 & 0 & 1 \\ 0 & 1 & 0 \end{vmatrix} = 0 \text{ is}$$

    (A)  $x = 0, 1, -1$
    (B)  $x = 0$
    (C)  $x = 0, i, -i$
    (D)  it can't be solved

93. $\sqrt{3 - 4i} =$

    (A)  $\sqrt{3} - 2i$
    (B)  $\pm(2 - i)$
    (C)  $\sqrt{3} + 2i$
    (D)  $\pm(2 + i)$

94. $\dfrac{1 - 2i}{i - 3} =$

    (A)  $\frac{1}{4}(5i - 1)$
    (B)  $\frac{1}{15}(5i - 1)$
    (C)  $\frac{1}{4}(i - 5)$
    (D)  $\frac{1}{2}(i - 1)$

95. The roots of the equation $x^2 - 2x + 5 = 0$ are

    (A)  $1 \pm 2i$
    (B)  $-1$ and $5$
    (C)  $2$ and $3$
    (D)  $1 \pm i\sqrt{5}$

96. One root of a quadratic equation with real coefficients is $2 - i$.
The equation can be written as

    (A)  $x^2 + 4x + 5 = 0$
    (B)  $x^2 - 5x + 4 = 0$
    (C)  $x^2 - 4x + 5 = 0$
    (D)  $x^2 + 5x + 4 = 0$

97. $\cos 5\theta + i \sin 5\theta$ can be expressed as
    I    $(\cos \theta + i \sin \theta)^5$
    II   $(\cos \theta - i \sin \theta)^5$
    III  $e^{5i\theta}$
    IV  $5e^{i\theta}$

    (A)  I and III
    (B)  I and IV
    (C)  II and III
    (D)  II and IV

98. Given that $z = 1 + i\sqrt{3}$

    (A)  $|z| = 2$, arg $z = \frac{1}{6}\pi$
    (B)  $|z| = \sqrt{10}$, arg $z = \frac{1}{6}\pi$
    (C)  $|z| = 2$, arg $z = \frac{1}{3}\pi$
    (D)  $|z| = \sqrt{10}$, arg $z = \frac{1}{3}\pi$

99. If $z_1 = \sqrt{3} + i$ and $z_2 = i - 1$, arg $z_1 z_2 =$

    (A)  $\frac{1}{8}\pi$
    (B)  $\frac{11}{12}\pi$
    (C)  $\frac{1}{12}\pi$
    (D)  $\frac{7}{12}\pi$

100. If $|z - 5| = 3$, where $z = x + iy$ and P is the point $(x, y)$, the locus of P in the complex plane is

    (A)  circle, centre $(0, 5)$ radius 9
    (B)  circle, centre $(0, 5)$ radius 3
    (C)  circle, centre $(5, 5)$ radius 3
    (D)  circle, centre $(5, 0)$ radius 3

# APPENDIX 4
# A Unit 2 Cape Exmination Paper

## (Specimen)
### PAPER 1 - $1\frac{1}{2}$ HOURS
### Answer ALL questions
### Section A (Module 1)

**Item 1** refers to the following graph

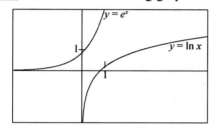

1.  The graphs of $y = e^x$ and $y = \ln x$ in the diagram above are reflections of each other in the line
    (A) $y = 0$      (B) $x = 0$      (C) $y = -x$      (D) $y = x$

2.  Which of the graphs in the diagrams below represents $y = 3^x$ and $y = e^x$

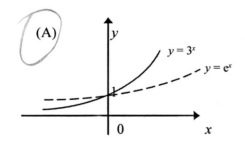

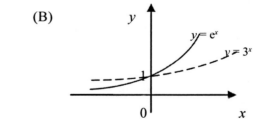

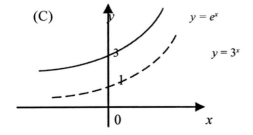

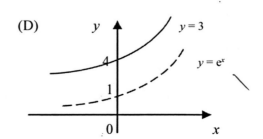

3. $\frac{1}{2} \ln 4 - \frac{1}{2} \ln 3$ can also be expressed as
   I  $\frac{1}{2} \ln 1$
   II  $\frac{1}{2} \ln \frac{4}{3}$
   III  $\ln 2 - \ln \sqrt{3}$
   IV  $\ln \frac{2}{\sqrt{3}}$

   (A)  I and II only
   (B)  I, II and III only
   (C)  II III and IV only
   (D)  III and IV only

4. If $5^x = 20$, then $x =$

   (A)  $\log(\frac{5}{20})$
   (B)  $\log(\frac{20}{5})$
   (C)  $\frac{\log 5}{\log 20}$
   (D)  $\frac{\log 20}{\log 5}$

5. $\log_c\left(\frac{a^2}{b}\right)$ can be written as

   (A)  $2\log_c a - \log_c b$
   (B)  $2 \log_c\left(\frac{a}{b}\right)$
   (C)  $2 \log_c\left(\frac{b}{a}\right)$
   (D)  $2 \log_c b^a$

6. $\frac{d}{dx}(e^{3x^2+2x+1})$ is

   (A)  $(6x + 2)e^{6x+2}$
   (B)  $(6x + 2)e^{3x^2+2x+1}$
   (C)  $(3x^2 + 2x + 1) e^{6x+2}$
   (D)  $(3x^2 + 2x + 1)e^{3x^2+2x+1}$

7. A curve is defined parametrically by the equations $x = t^2, y = t(1 - t^2)$. The gradient of the curve, in terms of $t$, is

   (A)  $\frac{2t}{1 - 3t^2}$
   (B)  $\frac{1 - 3t^2}{2t}$

(C)  $2t(1 - 2t)$
(D)  $2t(1 + 2t)$

8. For $x^2y - 3 = -6x$, $\frac{dy}{dx}$ at the point where $x = 1$ and $y = -3$ is equal to

   (A)  $-15$
   (B)  $\frac{2}{3}$
   (C)  $3$
   (D)  $11$

9. Given $y = \ln (2x + 3)^3$, then $\frac{dy}{dx}$ is

   (A)  $\frac{2x}{2x + 3}$
   (B)  $\frac{2}{2x + 3}$
   (C)  $\frac{6x}{2x + 3}$
   (D)  $\frac{6}{2x + 3}$

10. If the function f$(x)$ is defined by f$(x) = \cos x$ then f$''(x)$ is
    (A)  $-\cos x$
    (B)  $-\sin x$
    (C)  $\cos x$
    (D)  $\sin x$

11. The partial fractions expression for $\frac{5}{(x + 2)(x - 3)}$ may be written as

    (A)  $\frac{1}{x + 2} + \frac{1}{x - 3}$
    (B)  $\frac{-1}{x + 2} + \frac{1}{x - 3}$
    (C)  $\frac{1}{x + 2} + \frac{-1}{x - 3}$
    (D)  $\frac{-1}{x + 2} + \frac{-1}{x - 3}$

12. Which of the following functions, when integrated wrt $x$, gives the result $x - \ln x^2 + k$

    (A)  $\frac{1}{1 - x^2}$

(B) $\dfrac{1-2x}{x^2}$

(C) $\dfrac{x-2}{x}$

(D) $1-\dfrac{2}{x^2}$

13. $\int \cos^2 x \, dx$ is equal to
    (A) $\frac{1}{2}\sin^2 x + c$
    (B) $\frac{1}{3}\cos^3 x + c$
    (C) $\frac{1}{4}\sin 2x + c$
    (D) $\frac{1}{4}\sin 2x + \frac{1}{2}x + c$

14. $\int \dfrac{x}{x^2+3}\, dx$ is equal to

    (A) $\frac{1}{2}\ln(x^2+3) + c$
    (B) $2\ln(x^2+3) + c$
    (C) $2x\ln(x^2+3) + c$
    (D) $(x^2+3)\ln x + c$

15. $\int xe^{2x}dx$ may be expressed as

    (A) $2xe^{2x} + e^{2x} + c$
    (B) $2xe^{2x} - 4e^{2x} + c$
    (C) $\frac{1}{2}xe^{2x} - \frac{1}{4}e^{2x} + c$
    (D) $\frac{1}{2}xe^{2x} + \frac{1}{2}x^2e^{2x} + c$

**Item 16** refers to the diagram below

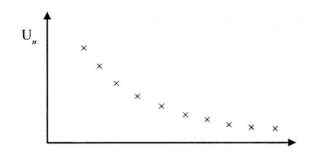

16. The term that best describes the behaviour of the sequence $\{U_n\}$ shown above is
    (A) periodic
    (B) finite
    (C) divergent
    (D) convergent

**Items 17 – 19** refer to S as defined below

$$S = 2^0 - 2^1 + 2^2 - 2^3 + 2^4$$

17. S is best described as a
    (A) finite series
    (B) infinite series
    (C) finite sequence
    (D) infinite sequence

18. The general term in S is best defined by
    (A) $(-1)^r 2^r$
    (B) $(-1)^r (-2)^r$
    (C) $(-1)^{r-1}(2)^r$
    (D) $(-1)^{r-1}(-2)^{r-1}$

19. S may be written as

    (A) $\displaystyle\sum_{r=0}^{r}(-1)^r(2^r)$

    (B) $\displaystyle\sum_{r=1}^{r}(-1)^r(2)^{r-1}$

    (C) $\displaystyle\sum_{r=0}^{4}(-1)^r(2)^r$

    (D) $\displaystyle\sum_{r=1}^{4}(-1)^{r-1}(-2)^{r-1}$

**Items 20 – 22** refer to E as defined below

$$E = \sum_{n=19}^{30} \frac{3^n}{n}$$

20. The number of terms in the expansion of E is

    (A) 10
    (B) 11
    (C) 12
    (D) 13

21. The $r$th term of E is $3^{24}$. The value of $r$ is
    (A) 6
    (B) 8
    (C) 9
    (D) 10

22. Which of the series below has its $n$th term equal to $\dfrac{a}{2^n}$?

   (A) $\displaystyle\sum_{r=1}^{n} \dfrac{a}{2^{-r}}$

   (B) $\displaystyle\sum_{r=1}^{n} \dfrac{a}{2^{r}}$

   (C) $\displaystyle\sum_{r=1}^{n} \dfrac{a}{2^{r-1}}$

   (D) $\displaystyle\sum_{r=0}^{n} \dfrac{a}{2^{r-1}}$

23. $\displaystyle\sum_{r=1}^{30} (10 - 2r) =$

   (A) $-1410$
   (B) $-220$
   (C) $-10$
   (D) $220$

24. The first term of an AP is $a$ and its common difference is $-1$. The sum of the first 10 terms is equal to

   (A) $5(2a - 9)$
   (B) $5(2a + 9)$
   (C) $10(2a + 11)$
   (D) $10(2a - 11)$

25. The second and fifth terms of a convergent geometric series with first term $\frac{1}{2}$ are 27 and 8 respectively. The sum to infinity of this series is

   (A) $\frac{2}{3}$
   (B) $\frac{3}{2}$
   (C) $\frac{27}{2}$
   (D) $\frac{81}{2}$

26. The sum to infinity of the geometric series $a + a^2 + a^3 + \dots$ is $4a$ $(a \neq 0)$. The common ratio is

   (A) $-\frac{3}{4}$
   (B) $\frac{3}{4}$
   (C) $\frac{4}{3}$
   (D) $\frac{5}{4}$

27. ${}^{n}C_{r-1} =$

   (A) $\dfrac{n!}{(n-r-1)!}$

   (B) $\dfrac{n!}{(n-r+1)!}$

   (C) $\dfrac{n!}{(n-r+1)!(r-1)!}$

   (D) $\dfrac{n!}{[n-(r-1)]!\,r!}$

28. The 3rd term in the expansion of $(2 - \frac{1}{2}x)^6$ in ascending powers of $x$ is

   (A) $-60x^2$
   (B) $-20x^3$
   (C) $60x^2$
   (D) $20x^3$

29. The coefficient of $x^2$ in the expansion of $(2 - 3x)^5$ is

   (A) $-720$
   (B) $-240$
   (C) $240$
   (D) $720$

30. The function $f(x) = x^3 - 3x - 3$ has a root in the closed interval

   (A) $[-10, -8]$
   (B) $[-2, 0]$
   (C) $[2, 3]$
   (D) $[5, 6]$

31. The number of ways in which a committee of four men and six women can be seated in a row if they can sit in any position is

   (A) $2!$
   (B) $4!$
   (C) $6!$
   (D) $10!$

32. The number of ways in which 3 boys and 2 girls can sit so that no two persons of the same sex sit next to each other is

   (A) $2 \times 3$
   (B) $3! + 2!$
   (C) $3! \times 2!$
   (D) $5!$

33. A team of 11 players is to be chosen from a squad of 16 players. Given that 2 players must be chosen, the number of ways in which the team can be chosen is
(A) $^{14}C_{11}$
(B) $^{14}C_{9}$
(C) $^{16}C_{11}$
(D) $^{16}C_{9}$

34. In how many ways can 2 persons be selected from a group of 10 persons?
(A) 20
(B) 45
(C) 90
(D) 100

35. What is the probability that an integer chosen at random from 1, 2, 3, 4, 5, 6, 7, 9, 11 and 15 is prime?
(A) $\frac{3}{10}$
(B) $\frac{4}{10}$
(C) $\frac{5}{10}$
(D) $\frac{6}{10}$

36. The letters P, Q, R, S, T and U are arranged randomly on a line. What is the probability that P and Q are next to each other?
(A) $\frac{2 \times 5!}{6!}$
(B) $\frac{2 \times 6!}{5!}$
(C) $\frac{5 \times 2!}{6!}$
(D) $\frac{5 \times 6!}{2!}$

37. What is the probability that an integer chosen at random from 1, 2, 3, 4, 5, 6, 7, 9, 11 and 15 is divisible by 3?
(A) $\frac{1}{10}$
(B) $\frac{2}{10}$
(C) $\frac{3}{10}$
(D) $\frac{4}{10}$

38. A and B are two independent events. Given $P(A) = 0.3$ and $P(B)$ 0.4, which of the following are true?
I. $P(A \cup B) = 0.7$
II. $P(A \cap B) = 0.12$
III. $P(A \mid B) = 0.3$
IV. $P(A \mid \bar{B}) = 0.4$

(A) I and II only
(B) II and III only
(C) II and IV only
(D) III and IV only

39. A fair die is tossed twice. What is the probability that at least one toss results in a 5?
(A) $\frac{2}{36}$
(B) $\frac{10}{36}$
(C) $\frac{11}{36}$
(D) $\frac{25}{36}$

**Item 40** refers to the matrices P and Q below
$$P = [\, a \ b \ c\,] \quad Q = \begin{bmatrix} x \\ y \\ z \end{bmatrix}$$

40. The product PQ is
(A) $[\, ax + by + cz\,]$
(B) $\begin{bmatrix} ax \\ by \\ cz \end{bmatrix}$
(C) $[\, ax \ by \ cz\,]$
(D) not possible

**Items 41 – 42** refer to the matrix
$$\begin{bmatrix} -2 & 0 & 1 \\ 0 & 1 & 0 \\ 1 & 0 & \boxed{-2} \end{bmatrix}$$

41. The cofactor of the boxed element, $\boxed{-2}$, is
(A) $-2$
(B) $-1$
(C) 0
(D) 2

269

42. The determinant of the given matrix is
    (A) −5
    (B) −3
    (C) 3
    (D) 5

43. If $z$ is the complex number $2 - i$, then $z^2$ equals
    (A) 3
    (B) 4
    (C) $3 - 4i$
    (D) $4 - 3i$

44. Which Argand diagram best represents the complex number $z = 1 - i\sqrt{8}$

    (A)

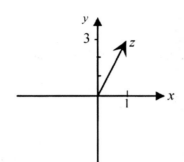

    (B)

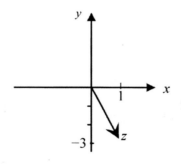

    (C)

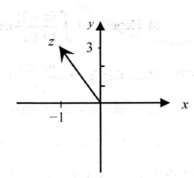

    (D)

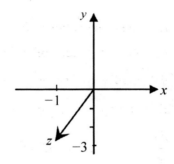

45. Determine $\mathrm{Im}\left(\dfrac{1}{z}\right)$, where $z = \dfrac{3 - i}{1 + i}$
    (A) $-\frac{2}{5}$
    (B) $-\frac{1}{5}$
    (C) $\frac{1}{5}$
    (D) $\frac{2}{5}$

**PAPER 2**

$2\frac{1}{2}$ **HOURS**

**Answer ALL questions**

**Section A (Module 1)**

1. (a) (i) Using the fact that $e^{-x} = \dfrac{1}{e^x}$ or otherwise, show that $\dfrac{d}{dx}(e^{-x}) = -e^{-x}$. [2 marks]

   (ii) Hence, evaluate $\int x^2 e^{-x} dx$. [4 marks]

   (b) (i) a) Find $\dfrac{dy}{dx}$ when $y = \tan^{-1}(3x)$ [4 marks]

b) Hence find $\displaystyle\int \frac{(x+2)}{1+9x^2}\,dx$ [4 marks]

(ii) Show that if $y=\dfrac{\ln(5x)}{x^2}$ then $\dfrac{dy}{dx}=\dfrac{1-\ln(25x^2)}{x^3}$

(c) Solve the first order differential equation

$$y\tan x\frac{dy}{dx}=(4+y^2)\sec^2 x.$$ [5 marks]

2. (a) In 1950, the world population was 2.5 billion and it grew to 5 billion in 1987. The world's population grows exponentially so that at time $t$ years the population is N $= 2.5e^{kt}$ where $t=0$ corresponds to the year 1950 and N is measured in billions of people.

Find

(i) the <u>exact</u> value of $k$ [3 marks]

(ii) the <u>exact</u> value of N in 2003 [2 marks]

(iii) the year in which N $=10$. [5 marks]

(b) Given that $y=u\cos 3x+v\sin 3x$ is a particular integral of the differential equation

$$\frac{d^2 y}{dx^2}+4\frac{dy}{dx}+3y=-30\sin 3x,$$

Find

(i) the values of the constants $u$ and $v$ [10 marks]

(ii) the general solution of the differential equation. [5 marks]

### Section B (Module 2)

3. (a) (i) Find constants $A$ and $B$ such that $\dfrac{1}{(2r-1)(2r+1)}=\dfrac{A}{2r-1}+\dfrac{B}{2r+1}.$ [5 marks]

(ii) Hence, find the value of $S$ where $S=\displaystyle\sum_{r=1}^{n}\frac{1}{(2r-1)(2r+1)}.$ [5 marks]

(iii) Deduce the sum to infinity of $S$. [3 marks]

(b) (i) Find the $n$th term of the series $1\,(2)+2\,(5)+3\,(8)+\ldots$ [2 marks]
 (ii) Prove, by Mathematical Induction, that the sum to $n$ terms of the series in (b) (i) above is $n^2(n+1)$. [10 marks]

4. (a) Given the series $\dfrac{1}{2}+\dfrac{1}{2^4}+\dfrac{1}{2^7}+\dfrac{1}{2^{10}}+\ldots$

(i) show that the series is geometric [4 marks]

(ii) find the sum of the series to $n$ terms [4 marks]

(b) Use Maclaurin's Theorem to find the <u>**first**</u> three non-zero consecutive terms

in the power series expansion of $\cos 2x$.

271

(c) (i) Expand up to and including the term in $x^3$

$$\sqrt{\left(\frac{1+x}{1-x}\right)}\ ,$$ stating the values of $x$ for which the expansion is valid. [6 marks]

(ii) By taking $x = 0.02$ find an approximation for $\sqrt{51}$ correct to 5 decimal places. [4 marks]

### Section C (Module 3)

5. (a) Two cards are drawn without replacement from ten cards which are numbered 1 to 10. Find the probability that
   (i) the numbers on **BOTH** cards are even [4 marks]
   (ii) the number on one card is odd and the number on the other card is even. [4 marks]

   (b) A journalist reporting on criminal cases classified 150 criminal cases crime by age (in years) of the criminal and by the type of crime committed, violent or non-violent. The information is presented in the table below.

| Type of Crime | Age (in years) | | |
|---|---|---|---|
| | Less than 20 | 20 to 39 | 40 or older |
| Violent | 27 | 41 | 14 |
| Non-violent | 12 | 34 | 22 |

What is the probability that a case randomly selected by the journalist

   (i) a) is a violent crime? [2 marks]
   b) was committed by someone LESS than 40 years old? [4 marks]
   c) is a violent crime OR was committed by a person LESS than 20 years old? [5 marks]
   d) is a violent crime that was committed by a person LESS than 20 years old? [2 marks]

   (ii) Two criminal cases are randomly selected for review by a judge. What is the probability that BOTH cases are violent crimes? [4 marks]

6. (a) Solve the following equation using determinants

$$\begin{vmatrix} 5 & x & 3 \\ x+2 & 2 & 1 \\ -3 & 2 & x \end{vmatrix} = 0$$ [10 marks]

   (b) Solve the following set of equations

$$x_1 - 4x_2 - 2x_3 = 21$$
$$2x_1 + x_2 + 2x_3 = 3$$
$$3x_1 + 2x_2 - x_3 = -2$$ [10 marks]

   (c) (i) Express the complex number $\dfrac{4-2i}{1-3i}$ in the form $a + bi$

   where $a$ and $b$ are real numbers. [4 marks]

   (ii) Show that the argument of the complex number in (c)(i) above is $\frac{1}{4}\pi$. [1 mark]

# ANSWERS TO EXERCISES

## CHAPTER 14

### Exercise 14B (page 5)

1.

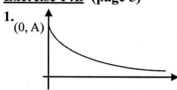

(0, A)

2. 1250

3. 82, $R^n : R^n + 1$

4. (a) $1000\pi \left( \dfrac{1}{1+t} - \dfrac{1}{4} \right)$

   (b)

   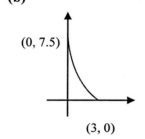

   (0, 7.5)

   (3, 0)

   (c) $t = 3$

   (d) $\displaystyle\int_0^3 1000\pi \left( \dfrac{1}{1+t} - \dfrac{1}{4} \right) dt$

   (e) $2000 \text{ m}^3$

5. $x = a + (A - a)e^{r/at}$, $A < a$

6. (a) $745
   (b) $298 / hr

9. B = 4,
   (b) 0.773 hrs

10. (a) $ 70 620
    (b) 1062

11. 8.7 billion

12. (a) (i) $ 3370
        (ii) $ 5680

    (b) $2000 \displaystyle\sum_{r=1}^{12} (1.11)^r$

    (c) $ 45426

13. 62 hours

14. (a) $\dfrac{dh}{dt} = p - kh$

    (b) $kh = p - (p - 2k)e^{-kt}$

    (c) if $k < \frac{1}{2}p$ water level rises
        if $k > \frac{1}{2}p$ water level falls

15. (a) 2
    (b) $4.81 \times 10^{-4}$

16. $t = \dfrac{1}{\beta V} \ln 3$

17. (a) $x = kt$, 4 p.m.

    (b) $\dfrac{dx}{dt} = \dfrac{k}{x}$ ; $\frac{1}{2}x^2 = kt$;
        $x = 2\sqrt{t}$; 6 p.m.

    (c) the temperature remains
        constant

18. (a) $\dfrac{dr}{dt} = \dfrac{k}{r^2}$

    (b) $k = 5000$; 141 m

    (c) $\dfrac{dr}{dt} = \dfrac{k}{r^2(t+2)}$ ; $k = 10^4$

    (d) 104 m

19. $r = 19.96$, A = 7513, less

20. (b)
    $t = \ln (x^2 + 25) + 0.8\tan^{-1}(\frac{1}{5}x) - \ln 25$

## CHAPTER 15

### Exercise 15A (page 13)

1. (a) $a^d = c$
   (b) $b^t = s$
   (c) $10^q = p$
   (d) $e^y = x$

2. (a) 4
   (b) 3
   (c) 6

   (d) 256
   (e) 0.01
   (f) 128
   (g) $1\frac{2}{3}$
   (h) $\frac{3}{2}$
   (i) $\frac{1}{8}$

3. (a) 5      (g) 1
   (b) 2      (h) $\frac{1}{2}$
   (c) 3      (i) $-\frac{1}{3}$
   (d) $-1$   (j) $\frac{2}{3}$
   (e) $-2$   (k) $\frac{2}{3}$
   (f) 0      (l) $-\frac{1}{3}$

4. (a) $\log a + \log b + \log c$
   (b) $2 \log a + \log b + \log c$
   (c) $\log a + 3\log b + \log c$
   (d) $\log b + \frac{1}{2} \log c$
   (e) $\log a + \log b - \log c$
   (f) $\log b - \log a - 3\log c$
   (g) $- \log a - \log b$
   (h) $\log a + \frac{1}{2}\log b - \frac{3}{2}\log c$
   (i) $1 + \lg a$
   (j) $\frac{1}{2} + \frac{1}{2} \lg a - 2\lg b$

5. (a) $\log_a 10$
   (b) $\log_a 45$
   (c) $\log_a 2$
   (d) $\log_a 20$

6. (a) $\log_a x^3$
   (b) $\log_a [x(x - 1)]$
   (c) $\log_a [\frac{1}{2}(x + 2)]$
   (d) $\log_a (x + 1)$

   (e) $\log_a \dfrac{x^2}{x - 1}$

   (f) $\log_a [x^2(x + 1)]$

   (g) $\log_a \dfrac{x}{2 + x}$

   (h) $\log_a (x + 1)$

### Exercise 15B (page 15)

1. (a) 2.63
   (b) 0.431
   (c) 1.46

(d) −1.16
(e) 1.46
(f) 1.64
(g) 1.23

2. (a) 0.792
   (b) 1.23
   (c) 0.671
   (d) 0.845
   (e) 1.18

3. (a) 3.17
   (b) 1.46
   (c) −2.71
   (d) 4.42
   (e) 0.326

4. (a) 1, 1.58
   (b) 0, 1.89
   (c) 3, 2.32
   (d) 1, 0.431
   (e) 0.431, 0.683
   (f) −1, 1
   (g) 1, 0.631
   (h) 0, 1.58

## Exercise 15C (page 15)

1. for all the curves: curve is continuous and increasing, passes through point (0, 1); $x$-axis is an asymptote and range is $y > 0$.

## Exercise 15D (page 20)

1. $\frac{1}{27}$

2. 2

3. 0.86

4. (a) 3
   (b) $\frac{1}{2}$

5. 2.32

6. 0.631

7. −0.631

8. $\frac{1}{9}$

9. $3^{(2a+b)}$, $3^{(2a-b)}$, $a = \frac{3}{2}$, $b = 2$

10. 5

11. $\sqrt{27}$, $1/\sqrt{27}$

12. $\log_b \frac{1}{5}$; −1

13. $\frac{2}{9}$

14. 3.82

15. $\frac{1}{4}$, 16

16. 0, 2.81

17. (a) $y$
    (b) $1/y$

18. 0, ln 2

19.
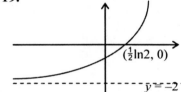

20. (a)
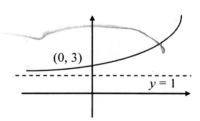

(b)
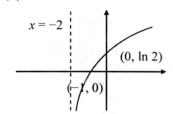

21. (a)
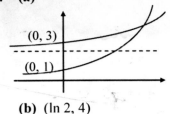

    (b) (ln 2, 4)

22. (a)
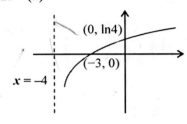

    (b) $f^{-1}(x) = e^x - 4$
    (c) $f^{-1}(x) > -4$
    (d) 4

23. $x = 4$, $y = 16$

24. $x = 2\frac{1}{2}$, $y = 20$

25. $x = 3$, $y = 2$

26. $x = \frac{3}{22}$, $y = \frac{24}{11}$

27. $p = 6\frac{1}{2}$, $q = 1\frac{1}{2}$

## Exercise 15X (page 21)

1. proof

2. (a) $y$
   (b) $1/y$; 16 or $\frac{1}{4}$

3. (0, 2), (ln 2, 4), (0, 1)

4. 0, ln 2

5. 2, 4

6. 1.71

7. (i) for f, (0, 1); for g, (1, 0)
   (ii) f = g$^{-1}$ or g = f$^{-1}$ or f and g are inverse functions

8. 51

9. $\ln 3$ (1.10)

10. $x = 2, y = 2$

11. $y = x^3$ or $y = x^{-1}$

12. $a = \frac{1}{2}\ln 3$ (0.549), $b = 3$

13. $\ln 4$ (1.38)

14. (a) $\frac{2}{3}$
 (b) 3 (–2 is not a valid solution)
 (e) (i) 0
 (ii) 2

15. (a) 3.86
 (b) 2, 4

16. (a) $f'(x) = a_1 + 2a_2x + 3a_3x^2 + \ldots$
 (b) $a_1 = \lambda a_0$, $a_2 = \frac{1}{2}\lambda^2 a_0$, $a_3 = \frac{1}{6}\lambda^3 a_0$

17. (a) $m - 1$
 (b) $2m/n$

18. 1.77

19. (b) inverse functions
 (c) (i) $r = 33.1$
 (ii) $p = -1$

20. proof

21. 0, $\ln 3$

22. (a) 2, 4
 (b) 3.15

# CHAPTER 16

## Exercise 16A (page 28)

1. (a) $10x$
 (b) $-3x^{-4}$
 (c) $\frac{1}{3}x^{-\frac{2}{3}}$
 (d) $\frac{1}{2}x^{-\frac{1}{2}}$
 (e) $-10x^{-3}$
 (f) $-\sin x$
 (g) $-9x^{-4}$
 (h) $\frac{2}{3}x^{-\frac{2}{3}}$
 (i) $-2x^{-\frac{3}{2}}$
 (j) $25x^{\frac{3}{2}}$
 (k) $-\frac{1}{5}x^{-1.2}$
 (l) $\sec^2 x$

2. (a) $3e^{3x}$
 (b) $-5e^{-5x}$
 (c) $8e^{8x+5}$
 (d) $2xe^{x^2}$
 (e) $ae^{ax+b}$
 (f) $-e^{\cos x}\sin x$
 (g) $(6x^2 - 1)e^{2x^3 - x}$
 (h) $-e^{-\tan x}\sec^2 x$

3. (a) $12(3x - 1)^3$
 (b) $21x^2(x^3 - 2)^6$
 (c) $20(x - 1)(2x^2 - 4x)^4$
 (d) $-40(3 - 5x)^7$
 (e) $x(x^2 - 1)^{-\frac{1}{2}}$
 (f) $-2x(x^2 + 3)^{-2}$
 (g) $(3 - 2x)^{-\frac{3}{2}}$
 (h) $(n + 1)p(px + q)^n$
 (i) $4 + 10(2x - 7)^4$
 (j) $4x(1 + x^2)^{-3}$
 (k) $5\cos 5x$
 (l) $-7\sin 7x$
 (m) $2\sec^2 2x$
 (n) $4\cos 4x$
 (o) $2x\cos x^2$
 (p) $2\sin x \cos x$
 (q) $-\sin(x + \frac{1}{4}\pi)$
 (r) $-\sec^2(\frac{1}{3}\pi - x)$
 (s) $p\cos(px + q)$
 (t) $-p\sin(px + q)$
 (u) $p\sec^2(px + q)$

4. (a) $10x - 2.5x^{\frac{3}{2}}$
 (b) $x\cos x + \sin x$
 (c) $2\cos x + (3 - 2x)\sin x$
 (d) $\sin x(1 + \sec^2 x)$
 (e) $\dfrac{4x}{(x^2 + 1)^2}$
 (f) $\dfrac{-7}{2\sqrt{x}\left(\sqrt{x} + 5\right)^2}$
 (g) $\dfrac{(2 + x)\cos x - \sin x}{(2 + x)^2}$
 (h) $\dfrac{\cos x + x\sin x}{\cos^2 x}$ or $\dfrac{1 + x\tan x}{\cos x}$
 (i) $e^x(x + 1)$
 (j) $e^{2x}(2\sin x + \cos x)$

 (k) $\dfrac{e^{x^2}(2x\sin x - \cos x)}{\sin^2 x}$
 (l) $\dfrac{2x(e^{2x} + 1) - 2e^{2x}(x^2 + 1)}{(e^{2x} + 1)^2}$

5. $-e^{-x}$

6. $f'(x)\,e^{f(x)}$

## Exercise 16B (page 29)

1. (a) $5(x - 2)^4$
 (b) $-21(2 - 3x)^6$
 (c) $3(2x - 5)(x^2 - 5x)^2$
 (d) $x(x^2 + 1)^{-\frac{1}{2}}$
 (e) $an(ax + b)^{n-1}$

2. (a) $3\cos 3x$
 (b) $-4\sin 4x$
 (c) $2\sec^2(2x + \frac{1}{4}\pi)$
 (d) $-2x\sin x^2$
 (e) $a\cos(ax + b)$

3. (a) $7e^{7x}$
 (b) $2e^{2x+3}$
 (c) $2xe^{x^2-1}$
 (d) $e^{\sin x}\cos x$
 (e) $pe^{px+q}$

4. (a) $3(10x - 3)(5x^2 - 3x)^2$
 (b) $-2e^{-2x}$
 (c) $-7\sin 7x$
 (d) $-4(2x + 4)^{-3}$
 (e) $e^{\tan x}\sec^2 x$
 (f) $15x^2(x^3 - 1)^4$
 (g) $2x\cos(x^2)$
 (h) $-2\sin x \cos x$ or $-\sin 2x$
 (i) $3e^{3x-1}$
 (j) $-8xe^{-x^2}$

## Exercise 16C (page 32)

1. (a) $\dfrac{1}{x}$
 (b) $\dfrac{3}{x}$
 (c) $e^x + \dfrac{1}{x}$

**(d)** $\dfrac{5}{x}$

**2.** **(a)** $\dfrac{2x}{x^2 - 5}$

**(b)** $\dfrac{6}{2x + 7}$

**(c)** $\cot x$

**(d)** $-\tan x$

**(e)** $\dfrac{2 - 3x^2}{2x - x^3}$

**(f)** $\dfrac{56x}{4x^2 - 1}$

**(g)** $\dfrac{4}{8x - 5}$

**(h)** $\dfrac{1}{1 - x}$

**3.** proofs

**4.** **(a)** $\dfrac{1}{\sqrt{4 - x^2}}, \ |x| < 2$

**(b)** $\dfrac{3}{9 + x^2}, \ x \in \mathbb{R}$

**(c)** $\dfrac{-1}{\sqrt{16 - x^2}}, \ |x| < 4$

**(d)** $\dfrac{1}{\sqrt{\frac{1}{16} - x^2}}$ or

$\dfrac{4}{\sqrt{1 - 16x^2}}, \ |x| < \frac{1}{4}$

**(e)** $\dfrac{2}{1 + 4x^2}, \ x \in \mathbb{R}$

**(f)** $\dfrac{-3}{\sqrt{1 - 9x^2}}, \ |x| < \frac{1}{3}$

## Exercise 16D   (page 34)

**1.** $e^x(\tan x + \sec^2 x)$

**2.** $2x \cos 3x - 3x^2 \sin 3x$

**3.** $e^x(\ln x + 1/x)$

---

**4.** $3x^2 \sin 2x + 2x^3 \cos 2x$

**5.** $(2x + 1)^2(6\sin x + (2x + 1)\cos x)$

**6.** $\dfrac{3e^x}{e^x + 3}$

**7.** $\dfrac{-2\{(1 + \tan 2x)\sin 2x + \sec 2x\}}{(1 + \tan 2x)^2}$

**8.** $\dfrac{2x(2 - x^2)e^{x^2}}{\left(1 - x^2\right)^2}$

**9.** $e^{5x}\left(5\ln 3x + \frac{1}{x}\right)$

**10.** $-\sin x \cos(\cos x)$

**11.** $-(1 + 1/x)\sin(x + \ln x)$

**12.** $(\sin x + x\cos x)e^{x\sin x}$

**13.** $\sin^{-1} x + \dfrac{x}{\sqrt{1 - x^2}}$

**14.** $2x \tan^{-1} x + \dfrac{x^2}{1 + x^2}$

**15.** $e^x\left(\tan^{-1} x + \dfrac{1}{1 + x^2}\right)$

## Exercise 16E   (page 35)

**1.** **(a)** $-6\cos(3x - 1)\sin(3x - 1)$
   **(b)** $e^{2x}(2\ln 3x + 1/x)$
   **(c)** $10x \sec^2(5x^2 - 3)$
   **(d)** $\ln x$
   **(e)** $2x(\cos x^2 - x^2 \sin x^2)$
   **(f)** $\dfrac{2(x - 1)}{x(x - 2)}$
   **(g)** $4\tan 2x \sec^2 2x$
   **(h)** $e^{5x}(5\sin x + \cos x)$
   **(i)** $2x^3(x^2 + 3)^4(7x^2 + 6)$

   **(j)** $e^{x^3}\left(3x^2 \ln x + \dfrac{1}{x}\right)$

   **(k)** $e^{\sin x}(\sin x + \sec^2 x)$

---

**(l)** $7\cos x \sin^6 x$
**(m)** $e^{-x}(\cos x - \sin x)$
**(n)** $-\tan x$

**(o)** $-\dfrac{1 + \cos x + \sin x}{(1 + \cos x)(1 + \sin x)}$

or $-\left(\dfrac{\sin x}{1 + \cos x} + \dfrac{\cos x}{1 + \sin x}\right)$

**2.** **(a)** $1, \ y = x$
   **(b)** $-7, \ 7x + y + 11 = 0$
   **(c)** $0.2, \ 10y = x + 7$

**3.** **(a)** $\frac{3}{2}, \ 2x + 3y = 16$
   **(b)** $2\frac{1}{4}, \ 4x + 9y = 22$
   **(c)** $1, \ x + y = 0$

**4.** $-1; \ y = \frac{1}{2}\pi - x$

**5.** proof

**6.** $\tan x \sec x$

**7.** $(1 + \cos t)^{-1}$

**8.** $4(e^x + e^{-x})^{-2}$

**9.** $2(1 - x^2)^{-1}$

**10.** $60(2x + 1)(2x + 3)^3$

**11.** $4(1 - x)^{-3}$

**12.** $4$

**13.** proof

**14.** proof

**15.** proof

**16.** **(i)** $1/x$
   **(ii)** $e^x(\ln x + 1/x)$

**17.** proof

## Exercise 16F   (page 40)

**1.** **(a)** $t^{-1}, \frac{1}{2}$
   **(b)** $\frac{1}{4}t^{-\frac{3}{2}}, \frac{1}{4}$
   **(c)** $-\frac{1}{2}t^{-2}, -\frac{1}{8}$
   **(d)** $4t, -8$

(e) $-2\frac{1}{2}\cot t, -2\frac{1}{2}$

(f) $-\sin t, -\frac{1}{2}$

(g) $\dfrac{-2\cos 2t}{\sin t}, -2$

(h) $-2t^2, -\frac{1}{2}$

(i) $\dfrac{\cos t - \sin t}{\cos t + \sin t}, 1$

(j) $-\tan t \cdot e^{\cos t - \sin t}, -1$

(k) $\frac{1}{4}t^{\frac{3}{2}}(1+2\sqrt{t}), \frac{3}{4}$

(l) $\dfrac{1}{t\cos t}, 6/\pi$

2. (a) $\dfrac{4x}{3y}$

(b) $\dfrac{2x}{2y-1}$

(c) $\dfrac{2}{2y+3}$

(d) $\dfrac{y-1}{2y-x-1}$

(e) $\dfrac{2e^x}{y}$

(f) $10xe^{-y}$

(g) $\dfrac{\cos x}{1+2y}$

(h) $\dfrac{2x}{\cos y}$

(i) $\dfrac{1-e^y}{xe^y}$

(j) $\dfrac{3x^2+6xy}{4y-3x^2}$

(k) $\dfrac{\cos y + y\cos x}{x\sin y - \sin x}$

(l) $\dfrac{-y\ln y}{2y^2+x}$

3. $16y=5x+4$

4. $-4/37$

5. $\dfrac{x+8}{2-y}$

(a) $5y=12x-63$
(b) $(-21, 2), (5, 2)$

6. $(\pi, -1)$ or $(-\pi, -1)$ **not** $(0, 1)$ where gradient is undefined

7. proof

8. $2x+y=4,\ x+y+4=0$

9. $3y+x=28$

10. (a) $\dfrac{2e^{2t}}{1-2e^{2t}}$
(b) $\frac{1}{2}\ln\frac{3}{8}$

11. (a) $\dfrac{t^2+1}{t^2-1}$

12. (a) $-2\sin t + 2\cos 2t,$ $-\sin t - 4\cos 2t$
(b) $\frac{1}{2}$
(c) $4x+2y=5\sqrt{2}$

13. (a) $-\frac{2}{3}$
(b) $2x+3y=21$

14. proof

15. (a) $2$
(b) $\frac{3}{16}t^{-5}$ or $\frac{3}{16}y^5$
(c) $-4$

16. $\dfrac{dy}{dx}=e^{-t}\cos t,$,
$\dfrac{d^2y}{dx^2}=-e^{-2t}(\cos t+\sin t)$
proof

17. $\dfrac{dy}{dx}=-\tan t+1,$
$\dfrac{d^2y}{dx^2}=-\sec^3 t$,
$(\frac{1}{2}\sqrt{2}, \sqrt{2})$ max,
$(-\frac{1}{2}\sqrt{2}, -\sqrt{2})$ min.

18, 19. proofs

20. $\dfrac{x}{1-y}$; proof

21. proof

22. $-\frac{4}{11}$

23. (a) $(0, 2)$ max
(b) $(1, 2)$ min

**Exercise 16X** (page 42)

1. (a) $4\sin(2x+3)\cos(2x+3)$ or $2\sin(4x+6)$
(b) $e^{3x}(3\ln 2x + 1/x)$

2. $1\frac{1}{3}$

3. (a) $e^{x^2}\left\{\dfrac{1}{x}+2x\ln\left(\dfrac{x}{2}\right)\right\}$
(b) $\dfrac{5x^2+4x-10}{(2+5x)^2}$

4. (a) $e^{\sin x}(\cos^2 x - \sin x)$
(b) $6x\sec^2(3x^2+5)$

5. (a) $2x\tan 3x + 3x^2\sec^2 3x$
(b) $\ln x$

6. (a) $2x^3\cos x^2 + 2x\sin x^2$
(b) $-3\frac{1}{3}$

7. (iii) $\dfrac{1}{\sqrt{1-x^2}}$

8. (ii) $e^x(\ln x + 1/x)$

9. (ii) $0, -3$
(iii) $y=-e^{-1}$

11. (ii) $(\frac{1}{12}\pi, 1)$
(iii) $2y(1+y^2)$

16. (a) $e^{2x}(2\sin 3x + 3\cos 3x)$
(b) $2x^{-1}\ln x$

17. (i) $2x\tan 3x + 3x^2\sec^2 3x$
(ii) $\dfrac{1-t^2}{6t(1+t^2)^2}$

277

**18.** (a) $2/x$

(b) $2\sin x \cos^2 x - \sin^3 x$

**19.** $-\frac{1}{2}$

**20.** (a) $\dfrac{xe^x}{(x+1)^2}$

(b) $6x^2\sec^2(x^3)\tan(x^3)$

**21.** (a) $\frac{1}{2}(2t-1)$

(b) $4x + 4y + 3 = 0$

**22.** (a) $16\sin(4x+1)\cos(4x+1)$

(b) $e^{2x+3}(3\sin 3x - 2\cos 3x)$

**24.** (a) $-\sin x\, e^{\cos x}$

(b) $2x(1+x)^{\frac{1}{2}} + \frac{1}{2}x^2(1+x)^{-\frac{1}{2}}$

**27.** (a) $\dfrac{3t^2 - 2}{4(2t-3)}$

(b) $-0.4$

**28.** (a) $(2 + \cos x)e^{2x + \sin x}$

(b) $3\sec^2 3x + \dfrac{2x}{x^2 + 4}$

**29.** (a) $-\frac{1}{2}$

(b) $y = 2x + 5$

**30.** (a) $2\cos 2x - 2\sin 2x$

# CHAPTER 17

### Exercise 17A (page 49)

**1.** (a) $\frac{1}{5}e^{5x} + c$

(b) $-4e^{-3x} + c$

(c) $\frac{1}{4}e^{4x-3} + c$

(d) $c - 2e^{5-x}$

(e) $3x + e^{-x} + c$

(f) $e^x - e^{-x} + c$

(g) $\frac{1}{4}e^{4x-1} + \frac{1}{3}x^3 + c$

(h) $\frac{1}{4}e^{2x+1} + c$

(i) $-\frac{1}{3}\cos 3x + c$

(j) $\frac{1}{5}\sin 5x + c$

(k) $2\sin \frac{1}{2}x + c$

(l) $-(\pi)^{-1}\cos \pi x + c$

(m) $\frac{1}{4}\tan 4x + c$

(n) $3\sin \frac{1}{3}x - \frac{1}{2}\cos 2x + c$

(o) $\frac{1}{3}\tan(3x + \frac{1}{2}\pi) + c$

(p) $\frac{1}{4}\sin(4x - \frac{1}{3}\pi) + c$

(q) $\frac{1}{25}(5x-3)^5 + c$

(r) $\frac{2}{9}(3x+4)^{\frac{3}{2}} + c$

(s) $-4(3x+5)^{-1} + c$

(t) $2/5(5x+2)^{\frac{1}{2}} + c$

(u) $5\ln|x| + c$

(v) $\ln A|x^2 - 1|$

(w) $\frac{1}{2}\ln(2x^2 + 6x - 1) + c$

(x) $\frac{1}{2}\ln|5 + e^{2x}| + c$

**2.** (a) $-\frac{1}{2}\cos(2x + \frac{1}{4}\pi) + c$

(b) $e^x + \ln|x| + c$

(c) $4\ln|x^2 - 1| + c$

(d) $-2(x-1)^{-1} + c$

(e) $\ln\dfrac{Ax^7}{(2x+3)^3}$ or

$7\ln|x| - 3\ln|2x+3| + c$

(f) $4(2x-5)^{2.5} + c$

(g) $4x - e^{-(x+5)} + c$

(h) $\frac{1}{2}x^4 - x^3 + 2\ln|x| + c$

(i) $\ln A|1 + \sin x|$

(j) $-\frac{1}{18}(4 - 3x)^6 + c$

(k) $\frac{1}{3}(2x-7)^{\frac{3}{2}} + c$

(l) $-\frac{1}{3}\ln A|2 - x^3|$

(m) $\sin^{-1} x + c$

(n) $4\tan^{-1} x + c$

(o) $3\sin^{-1}\frac{1}{3}x + c$

(p) $15\tan^{-1}\frac{1}{5}x + c$

**3.** (a) $1\frac{1}{2}$

(b) $3.19$

(c) $10$

(d) $0.5$

(e) $1.10$ (ln 3)

(f) $4$

(g) $8$

(h) $\ln 3$

(i) $\pi$

(j) $\frac{1}{2}\pi$

### Exercise 17B (page 53)

**1.** (a) $\dfrac{1}{x} - \dfrac{1}{x+1}$

(b) $\dfrac{1}{x+2} - \dfrac{1}{x+3}$

(c) $\dfrac{3}{x} - \dfrac{5}{1-x}$

(d) $\dfrac{1}{x-2} + \dfrac{3}{x-3}$

(e) $\dfrac{4}{x-4} - \dfrac{3}{x+3}$

(f) $\dfrac{2}{2x-3} - \dfrac{1}{x+2}$

(g) $\dfrac{6}{x+3} - \dfrac{4}{x+2}$

(h) $\dfrac{2}{x-3} + \dfrac{3}{x+4}$

**2.** (a) $\dfrac{4}{x} - \dfrac{4x}{x^2 + 2}$

(b) $\dfrac{1}{x+1} + \dfrac{1-x}{x^2 + 4}$

(c) $\dfrac{3}{x-2} + \dfrac{2-3x}{x^2 + 1}$

(d) $\dfrac{3}{2x} - \dfrac{x}{2(x^2 + 2)}$

(e) $\dfrac{1}{x-1} - \dfrac{x+1}{x^2 + 1}$

(f) $\dfrac{2}{x-2} - \dfrac{2x+3}{x^2 + 2}$

(g) $\dfrac{1}{x+1} + \dfrac{2x-2}{x^2 - 2}$

(h) $\dfrac{1}{x-1} + \dfrac{9-x}{x^2 + 9}$

**3.** (a) $\dfrac{1}{x+1} - \dfrac{1}{x-1} + \dfrac{2}{(x-1)^2}$

(b) $\dfrac{2}{x-1} - \dfrac{2}{x} - \dfrac{1}{x^2}$

(c) $\dfrac{1}{3(x+1)} - \dfrac{1}{3(x-2)} + \dfrac{2}{(x-2)^2}$

(d) $\dfrac{1}{x-1} - \dfrac{1}{x+2} - \dfrac{2}{(x+2)^2}$

**(e)** $\dfrac{2}{1-2x}+\dfrac{1}{x+1}+\dfrac{2}{(x+1)^2}$

**(f)** $\dfrac{1}{x-1}+\dfrac{1}{x-2}-\dfrac{2}{(x-2)^2}$

**(g)** $\dfrac{3}{x}-\dfrac{9}{3x-1}+\dfrac{9}{(3x-1)^2}$

**(h)** $\dfrac{2}{x}-\dfrac{1}{x^2}-\dfrac{3}{2x+1}$

## Exercise 17C   (page 55)

**1.**  **(a)** $1+\dfrac{1}{x-1}-\dfrac{1}{x+1}$

**(b)** $1+\dfrac{5}{2(x+2)}-\dfrac{17}{2(x+4)}$

**(c)** $1-\dfrac{3}{x-1}+\dfrac{5}{x+2}$

**(d)** $3-\dfrac{1}{x}+\dfrac{5}{x-3}$

**(e)** $2x+1+\dfrac{2}{x-1}+\dfrac{1}{x+1}$

**(f)** $1+\dfrac{2}{x^2+1}-\dfrac{1}{x+1}$

**(g)** $2+\dfrac{1}{x+1}+\dfrac{1}{(x+1)^2}$

**(h)** $x-2+\dfrac{3}{x}+\dfrac{3}{x^2}-\dfrac{5}{x-1}$

**2.**  **(a)** $\dfrac{3}{2x-1}-\dfrac{3}{2x+1}$

**(b)** $\dfrac{1}{x-1}+\dfrac{9-x}{x^2+9}$

**(c)**
$\dfrac{1}{5(x-2)}-\dfrac{1}{5(x+3)}-\dfrac{1}{(x+3)^2}$

**(d)** $\dfrac{1}{5(2-x)}+\dfrac{2}{5(1+2x)}$

**(e)** $2+\dfrac{7}{2(x-1)}-\dfrac{3}{2(x+1)}$

**(f)** $\dfrac{4-2x}{x^2+3}+\dfrac{2}{x+2}$

---

**(g)** $\dfrac{1}{4(x+3)}-\dfrac{1}{4(x-1)}+\dfrac{1}{(x-1)^2}$

**(h)** $\dfrac{1}{2x-1}+\dfrac{1}{x+2}$

**(i)** $x-2+\dfrac{4}{x+1}+\dfrac{2}{x+4}$

**(j)** $\dfrac{5}{2(x+1)}-\dfrac{1}{2(x-1)}$

**(k)** $1+\dfrac{1}{x-3}-\dfrac{1}{2x+5}$

**(l)** $2+\dfrac{x-5}{x^2+4}+\dfrac{5}{x-3}$

## Exercise 17D   (page 57)

**1.**  **(a)** $\ln A|(x-1)(x+2)^3|$

**(b)** $\ln\left|\dfrac{x-3}{x+1}\right|+\dfrac{4}{x+1}+c$

**(c)** $\dfrac{1}{8}\ln A\left|\dfrac{x-5}{x+3}\right|$

**(d)** $\ln A(x^2+1)(x+1)^3$

**(e)** $\tan^{-1}x-\ln(x+1)+c$

**(f)** $\ln|x+1|-\dfrac{6}{x-3}+c$

**(g)** $\ln\dfrac{A(2x-1)}{x+1}$

**(h)** $3\tan^{-1}x+\ln(x-1)$

**(i)** $\dfrac{1}{2}\ln\dfrac{A(x^2-3)}{(x-1)^2}$

**(j)** $\tan^{-1}\tfrac{1}{2}x+\ln(x+1)$

**(k)** $\ln\dfrac{A(x-1)(x+2)^2}{(x+1)^3}$

**2.**  **(a)** $\tfrac{1}{3}\ln 2$

**(b)** $\tfrac{4}{3}+\ln\tfrac{3}{5}$

**(c)** $\ln\tfrac{32}{3}-\tfrac{5}{24}$

**(d)** $\ln\tfrac{1}{2}$

**(e)** $0.661$

**(f)** $\ln 135$

**3.**  because integrand is undefined when $x=1$, which is between limits of integration

---

## Exercise 17E   (page 60)

**1.**  **(a)** $\tfrac{2}{9}(x^3+5)^{\frac{3}{2}}+c$

**(b)** $\tfrac{1}{12}(x^2-4)^6+c$

**(c)** $\tfrac{1}{10}(x-2)^{10}+\tfrac{2}{9}(x-2)^9+c$

**(d)** $-2\sin(3-x^2)+c$

**(e)** $-2\cos(x^3-\tfrac{1}{3}\pi)+c$

**(f)** $\tfrac{2}{3}(1+e^x)^{\frac{3}{2}}+c$

**(g)** $\tfrac{1}{32}(2x-3)^8+\tfrac{1}{4}(2x-3)^7+c$

**(h)** $\tfrac{1}{6}\sin^6x+c$

**(i)** $\tfrac{1}{5}\tan^5x+c$

**(j)** $-e^{\cos x}+c$

**(k)** $e^{x^3}+c$

**(l)** $\tfrac{2}{3}(x+2)^{\frac{3}{2}}-4(x+2)^{\frac{1}{2}}+c$

**2.**  **(a)** $\tfrac{1}{3}(2x^2+4)^6+c$

**(b)** $\tfrac{1}{6}(x-5)^6+(x-5)^5+c$

**(c)** $\tfrac{2}{3}(x^2+3x)^{\frac{3}{2}}+c$

**(d)** $-2(1-x^2)^{\frac{1}{2}}+c$

**(e)** $e^{x^4}+c$

**(f)** $-\tfrac{1}{3}(x+2)^{-3}+\tfrac{1}{2}(x+2)^{-4}+c$

**3.**  **(a)** $\tfrac{1}{9}(16-6\sqrt{3})$

**(b)** $\tfrac{1}{5}$

**(c)** $97/168=0.577$

**(d)** $\tfrac{2}{3}$

**(e)** $25.9$

**(f)** $\tfrac{1}{5}$

**4.**  $\tfrac{2}{3}(x+1)^{\frac{3}{2}}-2(x+1)^{\frac{1}{2}};\ 2\tfrac{2}{3}$

**5.**  $\ln\left(\dfrac{Ax}{x+1}\right),\ x-\ln(1+e^x)+c$

## Exercise 17F   (page 61)

**1.**  $\tfrac{1}{2}\sin^4x+c$

**2.**  $1/24(2x^2+4)^6+c$

**3.**  $\tfrac{1}{10}(1+2e^x)^5+c$

**4.**  $\tfrac{1}{2}e^{x^2+3}+c$

**5.**  $e^{x^3-x}+c$

**6.**  $-\tfrac{1}{2}(1+\cos x)^6$

**7.**  $2\ln|x^2-x|+c$

**8.**  $\tfrac{1}{4}(1+\ln x)^4+c$

**9.**  $\ln|1-\cos x|+c$

## Exercise 17G (page 63)

1. $e^x(x-1)+c$

2. $\frac{1}{3}x(x+4)^6 - 1/21(x+4)^7 + c$ or $2/21(3x-2)(x+4)^6$

3. $\frac{1}{3}x\sin 3x + \frac{1}{9}\cos 3x + c$

4. $-x^2\cos x + 2x\sin x + 2\cos x + c$

5. $-e^{-x}(x+1)+c$

6. $2/5(x-3)^{\frac{3}{2}}(x+2)+c$ or $2/5(x-3)^{5/2} + 2(x-3)^{3/2} + c$

7.
$x^3\sin x + 3x^2\cos x - 6x\sin x - 6\cos x + c$

8. $e^x(2x-1)+c$

## Exercise 17H (page 64)

1. (a) $\frac{1}{9}x^3(3\ln x - 1)+c$
   (b) $\frac{1}{5}e^x(\sin 2x - 2\cos 2x)+c$
   (c) $\frac{1}{2}(\ln x)^2 + c$
   (d) $x(\ln 2x - 1)+c$
   (e) $\frac{1}{5}e^{2x}(\sin x + 2\cos x)+c$
   (f) $\frac{1}{2}x + \frac{1}{2}\sin x \cos x + c$
       or $\frac{1}{4}(2x + \sin 2x)+c$

2. (a) $\frac{1}{8}\pi - \frac{1}{4}$ or 0.143
   (b) $\frac{1}{4}(3 - \ln 4)$ or 0.403
   (c) 1
   (d) $\pi^2 - 4$ or 5.87
   (e) 32/15 or 2.13

## Exercise 17I (page 67)

1. (a) $\frac{1}{4}(2x - \sin 2x)+c$
   (b) $\sin x - \frac{1}{3}\sin^3 x + c$
   (c) $1/24\sin 12x + \frac{1}{4}\sin 2x + c$
   (d) $\frac{1}{2}\sin x - 1/14\sin 7x + c$
   (e) $\frac{3}{8}x + \frac{1}{4}\sin 2x + \frac{1}{32}\sin 4x + c$
   (f) $\tan x - x + c$
   (g) $-\cos x + \frac{2}{3}\cos^3 x - \frac{1}{5}\cos^5 x + c$
   (h) $-2(1 + \tan \frac{1}{2}x)^{-1} + c$
       or $\tan x - \sec x + c$

2. (a) $-\cos^3 x + c$
   (b) $\frac{1}{5}\tan 5x + c$
   (c) $\frac{1}{8}\cos 4x - \frac{1}{12}\cos 6x + c$
   (d) $\frac{1}{4}\tan^4 x + c$
   (e) $-\frac{1}{2}\cos 2x + \frac{1}{6}\cos^3 2x + c$
   (f) $\frac{1}{4}\sin 2x - \frac{1}{8}\sin 4x + c$
   (g) $\frac{1}{3}\ln|\sec 3x| + c$
   (h) $\frac{1}{6}\sin^6 x + c$
   (i) $x(\sin x - \cos x) + \cos x + \sin x + c$
   (j) $-\frac{1}{7}\cos 7x + c$
   (k) $\frac{1}{2}e^x(\sin x - \cos x) + c$
   (l) $-\frac{1}{6}\cos 3x - \frac{1}{2}\cos x + c$

## Exercise 17J (page 70)

1. $I_n = \dfrac{n-1}{n}I_{n-2}$; $\dfrac{3}{16}\pi$

2. $\dfrac{1}{n-1} - I_{n-2}$; $\frac{1}{4}\pi - \frac{2}{3}$

## Exercise 17K (page 72)

1. (a) $\frac{1}{2}e^{2x} + c$
   (b) $\sin x - x\cos x + c$
   (c) $e^{\sin x} + c$
   (d) $5/2\cos 2x - 5/4\cos 4x + c$
   (e) $\frac{1}{3}x^3 - 2\frac{1}{2}x^2 + 6x + c$
   (f) $\frac{1}{2}\ln|x^2 + 10x - 3| + c$
   (g) $7\ln|x| + c$
   (h) $\frac{1}{2}e^{x^2} + c$
   (i) $\ln\left(\dfrac{x}{x+1}\right)^2 + c$
   (j) $\sqrt{4x+5} + c$
   (k) $2\tan x - 2x + c$
   (l) $2\sin \frac{1}{2}x + c$
   (m) $\frac{4}{5}\ln|5x - 2| + c$
   (n) $-e^{-x}(x+1) + c$
   (o) $\frac{1}{9}(2x^3 - 7)^{\frac{3}{2}} + c$
   (p) $3/20\sin 10x + \frac{3}{8}\sin 4x + c$
   (q) $c - 2(1-x)^{-1}$
   (r) $\ln A\left|\dfrac{x-1}{x+1}\right|$
   (s) $\ln A|x^2 - 1|$
   (t) $\sin^{-1}x + c$

   (u) $\frac{1}{4}x^4\ln|x| - \frac{1}{16}x^4 + c$
   (v) $\sin\theta - \frac{1}{3}\sin^3\theta + c$
   (w) $\frac{1}{6}x^6 + c$
   (x) $\frac{1}{10}(1 + e^{2x})^5 + c$
   (y) $\frac{1}{5}e^x(\sin 2x - 2\cos 2x) + c$
   (z) $e^x(x+2) + c$
   (aa) $4\tan^{-1}\frac{1}{2}x + c$
   (bb) $2\sqrt{x}(\ln x - 2)$

2. (a) $\frac{3}{2}\ln 3$
   (b) 0.48
   (c) $\frac{1}{2}\pi$
   (d) 1
   (e) $(\ln 4) - 1$
   (f) $\frac{4}{3}\pi$ (4.12)
   (g) $\sqrt{3}$
   (h) 4
   (i) $\frac{1}{12}\pi + \frac{1}{8}\sqrt{3}$ (0.478)
   (j) $-4$

## Exercise 17L (page 73)

1. $\dfrac{dm}{dt} = -km$

2. $\dfrac{dv}{dt} = kv$

3. $\dfrac{dh}{dt} = \dfrac{k}{H-h}$

4. $\dfrac{dy}{dx} = k(x+y)$

5. $\dfrac{dc}{dt} = -kc^2$

## Exercise 17M (page 76)

1. $y = e^{x^2} + c$

2. $\sin y = 3x + c$

3. $y = Ax(y+1)$

4. $y = A(x+1)$

5. $e^x + e^{-y} = c$

6. $\tan y = c - \cos x$

7. $e^{-y} + \sin x = c$

**8.** $y = \sin(2x^2 + c)$

**9.** $e^{2x} + 2e^{-y} = c$

**10. (a)** $x^2 + 2y^1 = c$

**11. (a)** $y = A(1 + x)^2$
**(b)** $y = 2(1 + x^2)$

**12. (a)** $y = c - e^{-x}(1 + x)$
**(b)** $2e^{-x}(1 + x) = \cos 2y + 2$

**13. (a)** $x \sin x + \cos x + c$
**(b)** $\frac{1}{2}y + \frac{1}{4}\sin 2y + c$
**(c)** $x \sin x + \cos x = \frac{1}{2}y + \frac{1}{4}\sin 2y + c$

## Exercise 17N    (page 79)

**1. (a)** $y = x - 4 + ce^{-x}$
**(b)** $y = e^{-2x}(\sin x + c)$
**(c)** $y = e^{-x}\ln\{A(1 + e^x)\}$
**(d)** $y = \frac{1}{4}e^{2x} + ce^{-2x}$
**(e)** $y = e^{-x}\{\sin(e^x) + c\}$
**(f)** $y = Ae^{2x} - \frac{1}{4}$

**2. (a) (i)** $5(1 - e^{-3t})$
**(ii)** $4.75$ A
**(b) (i)** $4(1 - e^{-5t})$
**(ii)** $1.57$ A
**(c)** $\frac{1}{5}e^{20t}(\sin 60t - 3\cos 60t) + Ae^{-20t}$

**3.** $109,000$

**4.** $y^3 = 3x^2 + 27$

**5.** $\dfrac{dx}{dt} = k(m - x); \quad \dfrac{m}{m - x} = e^{kt}, \, k = 1$

**(i)** $\frac{7}{8}m$
**(ii)** $\ln 4$

**6.** $\dfrac{dp}{dt} = kp(100 - p)$, $4.87$ weeks

**7.** proof

**8.** $\dfrac{dx}{dt} = -2x$, $x = ae^{-2t}$

**9.** $y = 20e^{\frac{1}{(1+t^2)} - 1}$, $y \to \dfrac{20}{e}$

**10. (a) (i)** separate the variables, then use partial fractions – assuming $k$ is known.
**(ii)** $50$ ms$^{-1}$
**(b) (i)** $6\frac{2}{3}$ms$^{-1}$
**(ii)** $v = \frac{2}{3}(10 + 65e^{-1.5t})$ms$^{-1}$

## Exercise 17O    (page 85)

**1. (a)** $y = 7 + 2x^2 - x^3$
**(b)** $y = x^3 - 2x^2 + 5$
**(c)** $y = 5x^3 - 10x + 4$

**2. (a)** $y = Ae^{2x} + Be^{-3x}$
**(b)** $y = Ae^x + Bxe^x$
**(c)** $y = e^{3x}(A\cos 2x + B\sin 2x)$
**(d)** $y = e^{-4x}(A\cos 5x + B\sin 5x)$
**(e)** $y = Ae^{4x} + Be^{2x}$
**(f)** $y = (A + Bx)e^{-x}$
**(g)** $y = 2e^{-3x}(2\cos 2x + 3\sin 2x)$
**(h)** $y = (2 + 3x)e^{4x}$
**(i)** $y = 2e^{-3x} + 3e^{2x}$

**3.**
**(a)** $y = e^{-x}(A\cos x + B\sin x) + 4$
**(b)** $y = Ae^x + Be^{-2x} + \frac{2}{5}\sin x - \frac{6}{5}\cos x$
**(c)** $y = e^{2x}(A + Bx) + 3$
**(d)** $y = e^x(A\cos 2x + B\sin 2x) + 2x + 1$
**(e)** $y = Ae^{-2x} + Be^{-x} + 2x^2 - 6x + 7$
**(f)** $y = (A + Bx)e^{-3x} + x + 2$
**(g)** $y = A + Be^{2x} + 4\cos 4x - 2\sin 4x$
**(h)** $y = A\cos 2x + B\sin 2x - \frac{3}{2} + \frac{1}{4}x + \frac{1}{2}x^2$

**4. (a)** $y = A\cos 3x + B\sin 3x + 2$
**(b)** $y = 2\cos x + 3\sin x$
**(c)** $y = y = (A + Bx)e^{-4x} + 2x^2 - 2x + 7$
**(d)** $y = e^{3x} - e^x$
**(e)** $y = Ae^x + Be^{-4x} + x + 2$
**(f)** $y = e^{-2x}(Ax + B)$
**(g)** $y = \cos 2x + 2\sin 2x + \frac{1}{2}x^2 - 1$
**(h)** $y = A + Be^{2x}$
**(i)** $y = A\cos x + B\sin x - \frac{1}{3}\cos 2x$
**(j)** $y = e^{5x}(2 - x)$
**(k)** $Ae^{5x} + Be^{-2x}$
**(l)** $y = e^{3x} - e^{-3x} - 2$
**(m)** $Ae^{2x} + Be^{-2x} - \frac{1}{3}\sin 2x$

**Exercise 17P   (page 86)**

1.   $x = 2\cos 10t$ metres

2.   $x = 4(3e^{-2t} - 2e^{-4t})$ cm

3.   $\theta = 2(te^{-2t} + 1)$

4.   $x = [s + (u + 3s)t]e^{-3t}$

5.   $x = \dfrac{e^{-t}}{25}(11\sin 2t + 2\cos 2t) - \dfrac{2}{25}(2\sin 5t + \cos 5t)$

**Exercise 17Q   (page 89)**

1.   0.696
2.   **(a)**   3.28
     **(b)**   3.46, which both give 3 to the nearest
              integer
     **(c)**   use more ordinates
3.   2.85
4.   1.12
5.   0.352
6.   0.820
7.   **(a)**   9
     **(b)**   8.44
     **(c)**   6.4, Tanya

**Exercise 17X   (page 89)**

1.   $x - 1 + \dfrac{1}{2(x+1)} + \dfrac{1-x}{2(x^2+1)}$

3.   $\dfrac{1}{3(x-1)} + \dfrac{2(x+1)}{3(x^2+2)}$

5.   $\dfrac{1}{2x} - \dfrac{1}{2(x+2)}$

6.   $\dfrac{3}{2} + \dfrac{14}{5(x-2)} - \dfrac{11}{10(2x+1)}$

7.   **(i)**   $e^x(x^3 - 3x^2 + 6x - 6) + c$
     **(ii)**  $\frac{1}{16}x^4(4\ln 4x - 1) + c$

8.   **(i)**   $2e^x(x - 1) + c$
     **(ii)**  $c - \frac{1}{5}(2\ln|x+4| + 3\ln|x-6|)$

9.   $c - e^{-x}(x^2 + 2x + 2)$

10.  $\dfrac{1}{x} - \dfrac{2x}{x^2+1}$

11.  $\frac{1}{3}\cos^3 x - \cos x + c$

12.  $\sin x - \frac{1}{3}\sin^3 + c$

13.  $\dfrac{1}{x+1} + \dfrac{1}{x^2} - \dfrac{1}{x}$, $\ln\left|\dfrac{x+1}{x}\right| - \dfrac{1}{x} + c$

14.  **(a)**   $\frac{2}{5}$
     **(b)**   $\frac{26}{15} = 1.73$

15.  $\dfrac{1}{x+2} + \dfrac{2x}{x^2+1}$, ( A= 1, B = 2, C = 0)
     $\ln|K(x+2)(x^2+1)|$

16.  **(a)**   $\dfrac{1}{2x} + \dfrac{1}{2(x+2)}$
     **(b)**   $\frac{1}{2}\ln A\,|x(x+2)|$

17.  $\frac{1}{9}x^3(3\ln x - 1) + c$

18.  $e^{\sin x} + c$

19.  $\dfrac{x+2}{3(x^2+5)} - \dfrac{1}{3(x+1)}$

20.  **(a)**   $\ln|\tan\theta| + c$
     **(b)**   $\frac{1}{27}e^{3x}(9x^2 - 6x + 2)$

21.  **(a)**   **(ii)**   $c - e^{-x}(x^2 + 2x + 2)$
     **(b)**   **(ii)**   $\frac{3}{4}$

22.  **(a)**   $\frac{1}{9}\sin^9 x + c$
     **(b)**   $\frac{1}{10}(2x + 1)^{2.5} - \frac{1}{6}(2x + 1)^{1.5} + c$
            or   $\frac{1}{30}(6x - 2)(2x + 1)^{\frac{3}{2}} + c$

23.  $\frac{1}{4}\tan^4\theta$

24.  **(a)**   $3\ln A(x^2 + 1)$
     **(b)**   $-\frac{1}{2}e^{-x^2}$

25.  **(a)**   $\dfrac{1}{x} + \dfrac{1}{x^2} - \dfrac{1}{x+1}$
     **(b)**   $\frac{1}{2} + \ln\frac{4}{3}$  (or 0.788)

**26.** $\dfrac{dP}{dt} = 0.03\,P, \quad P = P_0\,e^{0.03t}$

**27.** $\dfrac{dr}{dt} = 4kr$

**28.** (a) 502
(b) 153 days

**29.** (a) (i) $\dfrac{dx}{dt} = kx$

     (iii) $200 \ln 3 = 220$ s

   (b) (ii) $x = \dfrac{h + (kp - h)e^{kt}}{k}$

     (iii) if $kp = h$ the quantity remains static, if $kp < h$ the quantity decreases

**30.** (ii) (a) 1.26
      (b) 1.59

**31.** $\dfrac{dx}{dt} = -k\sqrt{x}$

**32.** (a) the rate of decrease in the volume is proportional to the volume (exponential decay)
   (d) 20

**33.** (i) $\dfrac{dP}{dt} = 0.035\,P,$

     where P is the population at time $t$ (years)

   (ii) $\dfrac{dn}{dt} = kn - 5,$ where $k = \frac{1}{40}\ln 2;$

     $n$ is the number of rabbits after $t$ days

**34.** (i) $n = me^{-kt}$

   (ii) $k = \frac{1}{76}\ln 2$

**35.** (a) (i) $\frac{2}{5}Q$

     (ii) $\frac{14}{25}Q$

   (b) 100 units

**36.** (i) 10
   (ii) 4.98 per day
   (iii) 500

**37.** (i) $\dfrac{dp}{dt} = kp(1 - p)$

   (ii) $\dfrac{p}{1 - p} = Ae^{kt}$

   (iii) 9/25 (0.36)

**38.** (a) $\dfrac{dr}{dt} = \frac{1}{3}kr$

**39.** (i) 800
   (ii) $k = 15, r = \ln 5$
   (iii) 714

**40.** (a) $\dfrac{1}{x - 1} - \dfrac{x}{x^2 + 1}$

   (b) $\ln(x - 1) - \frac{1}{2}\ln(x^2 + 1) + c \ \text{ or } \ \ln\dfrac{A(x - 1)}{\sqrt{x^2 + 1}}$

**41.** (a) 1
   (c) $6 - 2e$

# CHAPTER 18

## Exercise 18A   (page 102)

**1.** (a) 0, 3, 8, 15
   (b) $-1, 1, -1, 1$
   (c) $4, 2\frac{1}{2}, 2, 1\frac{3}{4}$
   (d) $-4, -1, 2, 5$
   (e) $\frac{1}{3}, \frac{1}{9}, \frac{1}{27}, \frac{1}{81}$
   (f) $1, -3, 9, -27$
   (g) 1, 8, 27, 64
   (h) $5, 2, -1, -4$
   (i) $-1\frac{1}{2}, 4, 1\frac{1}{4}, 6/7$
   (j) $\frac{1}{2}, -\frac{1}{2}, -1, -\frac{1}{2},$
   (k) $1, \frac{1}{8}, \frac{1}{27}, 1/64$
   (l) $\sqrt{3}, 2\sqrt{3}, 3\sqrt{3}, 4\sqrt{3}$

**2.** (a) $a_n = 4n - 1$
   (b) $a_n = 2 \times 3^{n-1}$

   (c) $a_n = \dfrac{1}{n(n+1)}$

   (d) $a_n = 18 - 3n$
   (e) $a_n = 192(\frac{1}{2})^{n-1}$ or $384(\frac{1}{2})^n$
   (f) $a_n = n^2 + 1$
   (g) $a_n = 2(-1)^{n+1}$

   (h) $a_n = \dfrac{n}{(n+2)}$

   (i) $a_n = \frac{1}{16}(-2)^n$

**3.** (a) (3) 5, 9, 17
   (b) (0) $1, \frac{1}{2}, \frac{2}{3}$
   (c) (2) $-6, 18, -54$
   (d) (2) 9, 100, 10404
   (e) (1) $1, 2, \frac{3}{2}$
   (f) (0) 1, 3, 6

4.  (a) $u_{r+1} = u_r + 2, u_1 = 6$
    (b) $u_{r+1} = 3u_r, u_1 = 2$
    (c) $u_{r+2} = u_r + u_{r+1}, u_1 = 2, u_2 = 5$
    (d) $u_{r+1} = 2u_r + 1, u_1 = 1$
    (e) $u_{r+1} = u_r - 4, u_1 = 25$
    (f) $u_{r+1} = 3u_r - 1, u_1 = 1$

5.  1, 1, −1, −1, 1, 1.; period 4; $a_9 = a_1 = 1$;
    $a_{30} = a_2 = 1$; $a_{52} = a_4 = -1$

6.  (a) 1, 3, 2, −1, −3, −2, 1, 3 …
    (b) period 6; $a_{13} = a_1 = 1$;
        $a_{27} = a_3 = 2$; $a_{48} = a_6 = -2$

7.  $u_1 = 1, u_3 = 1$; $u_1 = -\frac{1}{2}, u_3 = -\frac{1}{2}$

8.  4, −1

9.  proof

10. (a) $u_2 = 2, u_3 = 1\frac{1}{2}$
    (b) $\frac{1}{2}(1 + \sqrt{5})$

11. 1, 1, 2, 3, 5, 8… Fibonacci sequence,
    $u_{n+2} = u_n + u_{n+1}$, $u_1 = u_2 = 1$

## Exercise 18B (page 104)

1.  (a) 1
    (b) $\frac{1}{2}$
    (c) 0
    (d) 0
    (e) 3

2.  (a) convergent, $\frac{2}{3}$
    (b) convergent, 0
    (c) divergent
    (d) convergent, 0
    (e) divergent
    (f) convergent, 0
    (g) divergent
    (h) convergent, $\frac{2}{5}$
    (i) convergent, 0
    (j) convergent, 0

3.  (i) a, d, f, g, h, l
    (ii) c,1; e, 0; i, $\frac{1}{3}$; k, 0
    (iii) b, 2; j, 6
    (iv) d, 3; h, −3; l, $\sqrt{3}$
    (v) e, $\frac{1}{3}$; f, −3; b , −1

4.  (i) a, b, d, f, i
    (ii) c, 0; e, 0; h, 1
    (iii) g, 2
    (iv) a, 4; d, −3
    (v) b, 3; e, $\frac{1}{2}$; g, −1; i, −2

5.  (a) $\{t_n\}$ converges to 1 $(1/n \to 0)$
    (b) periodic $\{2, \frac{1}{2}, -1, 2, \frac{1}{2}, \ldots\}$

6.  $u_2 = \dfrac{1}{1-a}$, $u_3 = \dfrac{a-1}{a}$, $u_4 = a$; periodic

7.  5, $\frac{2}{5}$, 5, $\frac{2}{5}$. periodic.

8.  $u_n \to 1, v_n \to \infty, w_n \to 2$

9.  proof

10. proof; $\frac{1}{2}(1 \pm \sqrt{5})$; $\frac{1}{2}(1 + \sqrt{5})$ because $l$ must be positive

## Exercise 18X (page 107)

1.  $\frac{3}{2}$

2.  −1, 3

3.  proof

4.  $p = 2, q = -3$

5.  (i) divergent, arithmetic
    (ii) convergent
    (iii) periodic

6.  (a) proof
    (b) $\{u_n\} = \{1, 2, 1, 2, \ldots\}$;
        $\{a_n\} = \{3, 3, 3, 3, \ldots\}$;
        $\{b_n\} = \{1, -1, 1, -1, \ldots\}$
    (c) $\{u_n\}$ and $\{b_n\}$ are periodic $\{a_n\}$ is convergent.

7.  proof.

8.  $u_1 = 4, u_2 = 8, u_3 = 1.6$, or $u_1 = -1$,
    $u_2 = -2, u_3 = -1.6$

9.  (b) 1, −3, −1, 3

10. proofs

# CHAPTER 19

## Exercise 19A    (page 114)

1.  (a)  $0 + 2 + 6 + 12 + 20$
    (b)  $5 + 7 + 9 + 11 + 13$
    (c)  $\sqrt{2} + \sqrt{5} + \sqrt{10} + \sqrt{17}$
    (d)  $1 + \frac{1}{2} + \frac{1}{3} + \frac{1}{4} + \ldots$

2.  (a)  $\sum\limits_{r=1}^{16}(2r + 1)$

    (b)  $\sum\limits_{r=1}^{\infty} r^3$

    (c)  $\sum\limits_{r=1}^{6} 2r(2r + 1)$

3.  (a)  $S_1 = 0, S_2 = 2, S_3 = 8, S_4 = 20$
    (b)  $0, \frac{1}{3}, \frac{1}{2}, 3/5$
    (c)  $1$

4.  (a)  no
    (b)  (i)    $a = 7, d = 8$
         (ii)   $55$
         (iii)  $612$
    (c)  (i)    $a = 25, d = -8$
         (ii)   $-23$
         (iii)  $-228$
    (d)  no
    (e)  no
    (f)  (i)    $a = -12, d = 3$
         (ii)   $6$
         (iii)  $54$

5.  (a)  $19; 646$
    (b)  $12; 1578$
    (c)  $11, -308$
    (d)  $15, 315$

6.  (a)  $78$
    (b)  $\frac{1}{2}n(n + 1)$

7.  $2500$

8.  $2550$

9.  $25050; 50200$

10. $a = 6, d = 4$

11. $245$

12. $a = 12, d = -2\frac{1}{2}$

13. $a = 25, d = -3$

14. $31$

15. $a = 8, d = 6; 128; 1428$

16. $a = 7, d = -3; -65; -725$

17. $a = 4, d = 3; 125$

18. $5, 10$

19. $26, 22, 18, 14; a = 26, d = -4; -30$

20. $u_{n+1} - u_n = 3; 730$

21. $(18\text{th}) -3$

22. $9$

23. $18$

24. $d = \log_a 5, u_n = \log_a(5)^n$ or $n\log_a 5; \frac{1}{2}n(n + 1)\log_a 5$

25. $n \log_a b + \frac{1}{2}n(n + 1) \log_a c$

26. (a)  (i)    $3n^2 - n - 2$
         (ii)   $6n + 2$
    (b)  $8, 6$

## Exercise 19B    (page 119)

1.  (a)  $2$
    (b)  no
    (c)  $\frac{1}{3}$
    (d)  $-2$
    (e)  no
    (f)  $-\frac{1}{2}$
    (g)  $3$
    (h)  no

2.  (a)  $\frac{1}{2}; (\frac{1}{2})^{11}; 2.00$
    (b)  $3; 45927; 620011$
    (c)  $-2; 5120; -1747625$
    (d)  $\frac{1}{3}; 0.0229; 675$
    (e)  $-\frac{1}{5}; -0.064; -833$
    (f)  $2; 768; 1533$
    (g)  $\frac{1}{5}; 5^{-7}; 1.25$
    (h)  $-1\frac{1}{2}; 324; -3296$

(a) 2;
(d) 675;
(e) $-833\frac{1}{3}$;
(g) $1\frac{1}{4}$

4.   (a) 8
    (b) 9
    (c) 9

5.   3; 4/9; 118098

6.   ±3

7.   (a) $|x| < 1$
    (b) $-4 < x < -2$
    (c) $x < -1$ or $x > 1$

8.   $r = -3, a = 2; r = -1\frac{1}{2}, a = 8$

9.   $S_n = \frac{3}{4}(5^n - 1); 14$

10.   $S = \sum_{r=1}^{\infty}\left(\frac{x-1}{2}\right)^r$ ; GP, $r = \frac{1}{2}(x-1); -1 < x < 3$

11.   $2^{n+2}$; $a = 8, r = 2$; no

12.   $4\frac{1}{2}$

13.   $\frac{2}{3}$

**Exercise 19C  (page 122)**

1.   (a) $1 + \frac{x}{1!} + \frac{x^2}{2!} + \frac{x^3}{3!} + \frac{x^4}{4!} + \cdots$

    (b) $1 - \frac{x^2}{2!} + \frac{x^4}{4!} - \frac{x^6}{6!} + \cdots$

    (c) $1 - \frac{1}{2}x + \frac{1}{3}x^3 - \frac{1}{4}x^4 + \cdots$

    (d) $2x - \frac{4}{3}x^3 + \frac{4}{15}x^5$

    (e) $1 - \frac{x}{1!} + \frac{x^2}{2!} - \frac{x^3}{3!} + \frac{x^4}{4!}$

    (f) $1 + x + x^2 + x^3 + x^4$

2.   If $f(x) = \ln x$ then $f'(0)$ is undefined

3.   (a) 4.46
    (b) 0.0953

(c) −0.356
(d) 0.866
(e) 1.02
(f) 0.833

*   radians, because Maclaurin's theorem involves differentiation

**Exercise 19D  (page 125)**

Proofs

**Exercise 19E  (page 127)**

1.   $\frac{1}{2r-1} - \frac{1}{2r+1}; \frac{2n}{2n+1}$

2.   $\frac{1}{r} - \frac{1}{r+2}$;

    $\frac{3}{4} - \frac{2n+3}{2(n+1)(n+2)}$ or $\frac{n(3n+5)}{4(n+1)(n+2)}$

3.   $\frac{1}{3}\left(\frac{1}{r} - \frac{1}{r+3}\right)$;

    $\frac{11}{18} - \frac{3n^2 + 12n + 11}{3(n+1)(n+2)(n+3)}$; yes, $\frac{11}{18}$

4.   $n(n+2)$; no.

**Exercise 19F  (page 129)**

1.   (a) Darren
    (b) Ricardo

2.   3216

3.   $\frac{3}{10} + \frac{3}{100} + \frac{3}{1000} +$ , $\frac{1}{3}$

4.   (a) 1.08
    (b) 59,970
    (d) \$ 3000

5.   $r = 1.12$, \$ 6375

6.   (a) 0.675 m
    (b) 7.86 m
    (c) 8.4 m

7.   (b) \$ 10,800
    (c) 26 years

**8.** **(a)** **(i)** $5000 \times 1.06^n$
  **(ii)** 24 years
 **(b)** **(i)** 20%
  **(ii)** 23 years
 **(c)** in (a) $\to \infty$,
  in (b) $\to$ 30,000

## Exercise 19G   (page 131)

**1.** $6\frac{1}{4}$

**2.** **(a)** $-28500$
 **(b)** proof

**3.** **(a)** proof
 **(b)** $n^2$; $\frac{1}{6}N(N+1)(2N+1)$

**4.** $3(2^n-1)$

**5.** **(a)** 750
 **(b)** $\dfrac{2187}{16}$

**6.** 63

**7.** $2n-\frac{1}{2}$

**8.** $n(2n+3)$

**9.** **(a)** 12
 **(b)** $\pm 8$

**10.** **(i)** **(a)** $3n$
  **(b)** $\dfrac{-0.9}{1.9}(1-(-0.9)^n)$
  **(c)** $n-1$ when $n$ is odd,
   $n$ when $n$ is even
  **(d)** $\frac{1}{2}n(3n+11)$
 **(ii)** b is convergent, c is
  periodic

**11.** $-750$

**12.** **(i)** **(a)** $2^{r-1}$
  **(b)** $2^{r-1}$
  **(c)** $2^r-1$
 **(ii)** 20

**13.** $\frac{4}{5}$; 18

**14.** **(a)** 247050
 **(b)** 16

**15.** $r=0.917$, $a=40$

**16.** proof

**17.** **(a)** $k(3k+1)$
 **(b)** proof
 **(c)** $\frac{1}{12}n(n+1)(3n^2+11n+10)$

**18.** $x<-5,\ -1<x<1, x>5$

**19.** **(a)** $d=28$; $p=40, q=68$
 **(b)** $r=2$; $p=24, q=48$

**20.** $x=4, y=2$

## Exercise 19X   (page 133)

**1.** **(a)** $1-4+9-16$
 **(b)** $\displaystyle\sum_{n=1}^{\infty}(2n+1)$

**2.** $2\frac{2}{3}$

**3.** **(a)** 1
 **(b)** $\pm\sqrt{3}$

**4.** proof

**5.** $\frac{1}{2}$

**6.** **(a)** 2
 **(b)** 12

**7.** **(a)** proof
 **(b)** $2n^2(n+1)^2$

**8.** **(a)** proof
 **(b)** $k=1225$

**9.** **(a)** proof
 **(b)** $\frac{1}{6}n(n+1)(n+2)$

**10.** $x<-4$ or $-1<x<1$ or $x>4$

**11.** proofs

**12.** **(a)** $18(\frac{1}{3})^n$ or $6(\frac{1}{3})^{n-1}$ or $2\cdot 3^{2-n}$
 **(b)** $r=\frac{1}{3}$
 **(c)** 9

**13.** proof

**14.** **(a)** $\dfrac{1}{(3n-2)(3n+1)}$
 **(b)** $\dfrac{n}{3n+1}$
 **(c)** $\frac{1}{3}$

**15.** 0.9

**16.** **(a)** $(d=\log_a c)$
 **(b)** 21

**17.** **(a)** $r=\frac{1}{8}$
 **(b)** $\dfrac{4}{7}\left(1-\left(\dfrac{1}{8}\right)^n\right)$
 **(c)** proof

**18.** **(a)** $\dfrac{n}{n+1}$
 **(b)** yes, to 1

**19.** $3n^2+8n$

**20.** **(a)** 11, 28, 51, 80, 115
 **(b)** proof

**21.** proof

**22.** $x=1\frac{3}{4}, y=-2\frac{1}{2}$

**23.** **(a)** $8(1-(\frac{1}{2})^n)$
 **(b)** 8

**24.** proof

**25.** 765

**26.** **(a)** $4n-6$
 **(b)** proof
 **(c)** **(i)** $-2$
  **(ii)** 4

**27.** proofs

**28.** proof

**29.** **(a)** 7
 **(b)** 2 ($-2$ is not valid)

**30.** (a) $a = \ln 3, d = \ln 3$
(b) $210 \ln 3$
(c) proof

**31.** $443

**32.** (a) $20 \times 1.1^{n-1}$
(b) 18th

**33.** (a) (i) $\frac{2}{5}Q$ (ii) $\frac{14}{25}Q$
(ii) $\frac{14}{25}Q$
(b) 100 units

**34.** (a) $ 29.13
(b) $ 507.39

**35.** (c) $ 6392

**36.** (a) $1200(0.96)^{n-1}$
(b) $30000(1 - 0.96^n)$
(c) 10055
(d) 30000

# CHAPTER 20

<u>**Exercise 20A**</u> **(page 144)**

**1.** (a) $1 + 13x + 78x^2 + 286x^3 + \ldots$
(b) $1 - 15x + 105x^2 - 455x^3 + \ldots$
(c) $1 + 30x + 405x^2 + 3240x^3 + \ldots$
(d) $128 - 2240x + 16800x^2 - 70000x^3 + \ldots$

**2.** (a) 84
(b) $-1512$
(c) $-30618a^4$
(d) 20412

**3.** 2

**4.** (i) 3
(ii) $-7722$

**5.** 5

**6.** $a = 4, n = 9$

**7.** (i) $-48384$
(ii) 218.75

**8.** 48

**9.** (i) 20000
(ii) 4.375

**10.** $-6$

**11.** 4.375

**12.** $-280$

**13.** 672

**14.** 6561

**15.** 9405

**16.** 4

**17.** 7

**18.** $-3$

**19.** $a = 2, b = 11$

**20.** $a : b = 8 : 7$

**21.** 5

**22.** 11

**23.** $\frac{1}{4}$

**24.** $-3.36$

**25.** $8x + 8x^3$

**26.** $217 + 88\sqrt{6}$

<u>**Exercise 20B**</u> **(page 147)**

**1.** 1.005010010005001

**2.** (a) $32 + 80y + 80y^2 + 40y^3 + 10y^4 + y^5$
(b) (i) 32.8080401001
(ii) 24.76099

**3.** $1 - 11x + 55x^2 - 165x^3 + 330x^4 + \ldots, \quad 0.801$

**4.** (a) 1.0721
(b) 0.913517
(c) 67.937

**5.** (a) $1 - x + x^2 - x^3 + \ldots$
(b) $1 + x - \frac{1}{2}x + \frac{1}{2}x^2$
(c) $2 - \frac{x}{12} - \frac{x^2}{288} - \frac{x^3}{21236} - \ldots$

6.  $1 - \dfrac{1}{2}x + \dfrac{3}{8}x^2 - \dfrac{5}{16}x^3 + \cdots$ , 0.193

7.  $2 + \dfrac{1}{4}x - \dfrac{1}{64}x^2 + \dfrac{1}{512}x^3 + \cdots$ , 2.236

8.  $f(x) = \dfrac{3}{1 + 3x} + \dfrac{2}{1 - 2x}$,
    $5 - 5x + 35x^2 - 65x^3 + \ldots \; |x| < \frac{1}{3}$

9.  [write as $(1 - x)^{\frac{1}{2}}(1 + x)^{-\frac{1}{2}}$, then expand and multiply]
    $1 - x + \frac{1}{2}x^2 - \frac{1}{2}x^3 + \ldots \quad |x| < 1$

## Exercise 20C   (page 149)

proofs

## Exercise 20X   (page 149)

1.  $\frac{1}{2}$

2.  $-6720$

3.  112

4.  (a)  7/18
    (b)  189/16

5.  2

6.  (a)  $16 + 16(2 - u)x + 8(3 - 4u)x^2 + \ldots$
    (b)  (i)  $\frac{3}{4}$
         (ii)  20

7.  $-7654\frac{1}{2}$

8.  $a = \frac{2}{3}, n = 9$

9.  $a = 4; -80$

10. (a)  $\dfrac{n!}{k!(n - k)!}$ ;  $\dfrac{n!}{(k - 1)!(n - k + 1)!}$
    (b)  proof

11. $\pm 2$

12. proofs

13. (a)  proof
    (b)  $x = 7$
         ( $x$ cannot be $-2$ since it must be a positive integer)

14. (a)  $8 + x(8u - 12) - x^2(6 - 12u)$
    (b)  $\frac{1}{2}$

# CHAPTER 21

## Exercise 21A   (page 154)

1.  12%

2.  0.023

3.  (a)  2m
    (b)  42.75 m²

4.  3.74 m; 0.023

5.  (a)  $3.25 \le l < 3.35$
    (b)  8

6.  (a)  0.27%
    (b)  0.008

7.  1 cm

8.  (a)  0.9224
    (b)  4
    (c)  0.003%

9.  0.00011

10. (a)  $1 + 24x + 264x^2 + 1760x^3$
    (b)  $x = 0.01$;  1.26816
    (c)  1.268241795
    (d)  0.00645%

## Exercise 21X   (page 155)

1.  2.0%

2.  (a)  35.65 m²
    (b)  2.0 %

3.  (a)  0.001
    (b)  0.3 %

PURE MATHEMATICS FOR CAPE VOL. 2

4. 20.75 cm²

5. 1.73 %

6. (a) $24.4 \le l < 25.5$
   (b) (i) 0 cm² (if the tile were measured correctly there would be no absolute error, so minimum error = 0)
       (ii) 4.04%

# CHAPTER 22

## Exercise 22A (page 159)

1–4. proof

5. (i) 3
   (ii) $2\frac{1}{2}$

6, 7. proof

## Exercise 22B (page 165)

1. (a) 0.2
   (b) 0.8
   (c) 0.7

2. (a) 0.23
   (b) 0.79

3. (a) 2.646
   (b) 0.850
   (c) 1.7632
   (d) 1.4973
   (e) 0.79206

4. (c) 2.28

5. 13.80

6. (b) 2.91

7. (c) 2.93

8. (b) 2.13

9. (a) 3
   (b) 4.721

10. (b) 1.373

11. (c) $x_1 = 0.88629$, $x_2 = 0.88798$, $x \approx 0.89$
    (d) $x_1 = 0.88629$, $x_2 = 0.88804$
    (e) easier to use (but it takes more iterations to reach an accurate approximation)

## Exercise 22X (page 166)

1 – 3. proofs

4. (a) proof
   (b) 0.2

5. (a) proof
   (b) 0.2

6. (a) proof
   (b) [1, 1.5]

7. (a) proof
   (b) 0.2

8. (a) proof
   (b) proof
   (c) $x_{n+1} = \dfrac{2x_n^3 + 3}{3x_n^2 + 1}$

9. (a) – (c) proof
   (d) 1.13

10. (b) (iii) [0.75, 1]
    (iv) $\beta = \alpha \dfrac{\sin \alpha - \alpha^2}{\cos \alpha - 2\alpha}$
    (v) 0.877

11. proof

12. (a) proofs
    (b) 0.818 (further application of N-R gives 0.771 as a better approximation)

13. (a) – (b) proofs

14. (a) $x + e^x = 0$

15. (a)

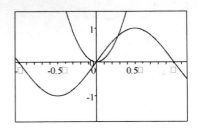

(c) $[\frac{1}{4}\pi, \frac{1}{3}\pi]$ or [0.5, 1.0] etc
(d) 0.943

# CHAPTER 23

## Exercise 23A (page 173)

1. 1320

2. 15

3. 24

4. (a) 12
   (b) 7

5. (a) 120
   (b) $5^5 = 3125$

6. (a) 12
   (b) 14

## Exercise 23B (page 175)

1. Combination $^7C_4 = 35$

2. Permutation $3! = 6$

3. Permutation $^{25}P_{11} = 1.8 \times 10^{14}$

4. Combination $^8C_5 = 56$

5. Permutation $10! = 3.6 \times 10^6$

6. Combination $^{20}C_{12} = 125,970$

## Exercise 23C (page 176)

1. $7! = 5040$

**2.** 7920

**3.** 816

**4.** 604,800

**5.** 20

**6.** 3003

**7.** 120

**8.** (a) $2.69 \times 10^7$
    (b) $1.84 \times 10^7$

**9.** (a) 286
    (b) 6

**10.** (a) 56
     (b) 10
     (c) 1
     (d) 30

**11.** (a) 1680
     (b) 1120

**12.** (a) 8568
     (b) 252
     (c) 3360

**13.** $5.9 \times 10^{14}$

**Exercise 23D** (page 179)

**1.** 840

**2.** 10,080

**3.** 3360

**4.** 10,080

**5.** 240

**6.** 360

**7.** 86,400

**8.** (a) 24
    (b) 50

**9.** 3,628,800; 45

**10.** (a) 5005
     (b) 720

**11.** 1287

**12.** 150

**13.** 480

**14.** 38,760;
     (a) 12,320
     (b) 5320

**15.** 715

**16.** 5040;
     (a) 144
     (b) 1440

**17.** (a) 40,320
     (b) 1152
     (c) 5040

**18.** (a) 1440
     (b) 2880

**19.** (a) 462
     (b) 210
     (c) 200
     (d) 281

**20.** (a) 840
     (b) 60
     (c) 120

**21.** (a) 30,240;
     (b) (i) 840
         (ii) 2520
         (iii) 240

**22.** (a) (i) 16
         (ii) 90
     (b) 112

**23.** 840; 120

**24.** 10,080

**25.** 1260; 120

**26.** (a) 144
     (b) 2160

**27.** (a) 360
     (b) 120

**28.** (a) 40,320
     (b) 10,080

**29.** (a) 9
     (b) 6

**Exercise 23E** (page 188)

**1.** {2, 4, 6, 8, 12, 16, 18, 24}

**2.** the 36 temperatures, 12 for each fluid

**3.** the 6 permutations of the 3 crops as they are allocated to the 3 different fields

**4.** $\frac{2}{3}$

**5.** $\frac{1}{3}$

**6.** (a) 4/11
    (b) 9/11

**7.** (a) $\frac{1}{4}$
    (b) $\frac{1}{6}$

**8.** (a) $\frac{1}{32}$
    (b) $\frac{1}{8}$
    (c) $\frac{1}{2}$
    (d) $\frac{3}{4}$
    (e) $\frac{3}{4}$

**9.** (a)

| | 0 | 1 | 2 | 3 | 4 | 5 |
|---|---|---|---|---|---|---|
| H | 0 | 1 | 2 | 3 | 4 | 5 |
| T | 0 | 2 | 4 | 6 | 8 | 10 |

     (i) $\frac{1}{6}$
     (ii) $\frac{1}{12}$
     (iii) 0
     (b) $\frac{1}{3}$

**10.** (a) 3/16
     (b) $\frac{1}{2}$
     (c) $\frac{3}{8}$
     (d) $\frac{1}{8}$

**11.** (a) 1/25
     (b) 1/13

**12.** 6/25

**13.** $\frac{1}{12}$

**14.** (a) 3/5
(b) 11/25
(c) 13/50
(d) 19/20

**15.** (a) 13/50
(b) 7/50
(c) 13/25
(d) 17/26
(e) 37/50

**16.** (a) 1/20
(b) 3/10
(c) 17/61
(d) 69/100

**17.** (a) no
(b) yes

**18.** (a) 0; 13/15
(b) 2/15

**19.** (a) 5/8
(b) 3/32

**20.** (a) $\frac{1}{6}$
(b) 4/9

**21.** $\frac{1}{6}$, yes

**22.** (a) $\frac{1}{5}$

**Exercise 23F**   (page 194)

**1.** (b) 9/100
(c) 16/25

**2.** (a) 73/100
(b) 0.2376

**3.** (b) 0.6775

**4.** 13/18

**5.** 7/12

**Exercise 23 G**   (page 196)

**1.** 15/32

**2.** $\frac{3}{8}$

**3.** (a) $\frac{2}{3}$
(b) $\frac{1}{6}$

**4.** (a) $\frac{1}{3}$
(b) $\frac{1}{6}$

**5.** A and B are independent, A and C are independent, but B and C are not independent

**6.** (a) 11/24
(b) 5/8

**7.** (a)

|  | Fruit | Other | Total |
|---|---|---|---|
| Birds nest | 2 | 4 | 6 |
| No nest | 5 | 9 | 14 |
| Total | 7 | 13 | 20 |

(b) 9/20
(c) 2/7

**8.** (a) 3/20
(b) $\frac{2}{3}$
(c) yes

**9.** (a) 12/25
(b) $\frac{1}{5}$
(c) 16/25
(d) 5/9

**10.** 4/15

**11.** 2/5

**12.** (a) $\frac{1}{2}$
(b) 7/15

**13.** 16/27

**14.** (a) 5/18
(b) 6

**15.** 9/20 (45%)

## Exercise 23X (page 198)

1. 2520

2. (a) 72
   (b) 375

3. 420

4. (a) 720
   (b) 1 235 520

5. (a) 120
   (b) 20

6. 220

7. the integers 0 to 300

8. the ordered pairs $(x, n)$ where $x$ represents any one of the 4 ticket offices, and $n$ is an integer between 0 and 75

9. 65/318

10. (a) 2/9
    (b) 5/9

11. (a) 65/236
    (b) 12/59
    (c) 79/118
    (d) 26/37

12. (a) 22/67
    (b) 7/67
    (c) 60/67
    (d) 27/67
    (e) 2/7
    (f) 1/9
    (g) 1/11

13. (a) 2/5
    (b) 4/5

14. (a) any of the pairs represented by xs in the diagram

|       | $R_1$ | $R_2$ | G | $B_1$ | $B_2$ |
|-------|-------|-------|---|-------|-------|
| $R_1$ |       | x     | x | x     | x     |
| $R_2$ | x     |       | x | x     | x     |
| G     | x     | x     |   | x     | x     |
| $B_1$ | x     | x     | x |       | x     |
| $B_2$ | x     | x     | x | x     |       |

   (b) (i) $\frac{1}{5}$
       (ii) 7/10

15. 2/7

16. (a) 3600
    (b) 2400
    (c) 5/7

17. A 4/7  B 2/7  C 1/7

18. (a) 625
    (b) (i) 250
        (ii) 3/5

19. (a) {HHH, HHT, HTH, HTT, THH, THT, TTH, TTT}
    (b) (i) 3/8
        (ii) $\frac{1}{2}$

20. (b) (i) 4/5
        (ii) 34/37
        (iii) 17/21

21. (a) 12144
    (b) 48/253
    (c) 35/253 (0.138)

22. (a) 180
    (b) 8/15

23. (a) 0.543
    (b) 0.457

# CHAPTER 24

## Exercise 24A (page 207)

1. (a) (c) (f) (g) (i) (k) (n) have no solutions.

   (b) $\begin{pmatrix} -1 & 4 \\ 5 & 5 \end{pmatrix}$

   (d) $\begin{pmatrix} 14 & 4 & 16 \\ 6 & 10 & -2 \end{pmatrix}$

   (e) $\begin{pmatrix} -5 & 11 \\ 4 & 1 \end{pmatrix}$

   (h) $\begin{pmatrix} 2 \\ 3 \end{pmatrix}$

   (j) $(13)$

   (l) $(-9 \ 13 \ 20)$

   (m) $\begin{pmatrix} 19 & 0 \\ 0 & 19 \end{pmatrix}$

   (o) $\begin{pmatrix} 14 & -6 & -16 \\ -9 & 4 & 9 \\ 2 & 15 & 17 \end{pmatrix}$

   (p) $\begin{pmatrix} 8 & 17 \\ 7 & 3 \end{pmatrix}$

   (q) $\begin{pmatrix} -7 & 11 \\ 1 & 12 \end{pmatrix}$

   (r) $\begin{pmatrix} 5 & 5 \\ 5 & 10 \end{pmatrix}$

   (s) $\begin{pmatrix} -7 \\ 27 \end{pmatrix}$

   (t) $(-7 \ 27)$

2. O + A = A + O = A and O + D = D + O = D   so O is the identity element for matrix addition.

3. IH = HI = H and ID = DI = D so I is the identity element for matrix multiplication.

## Exercise 24B (page 209)

1. (a) $\begin{pmatrix} -5 & 3 \\ 2 & -1 \end{pmatrix}$

**(b)** $\dfrac{1}{14}\begin{pmatrix} 3 & -2 \\ -5 & 8 \end{pmatrix}$

**(c)** no inverse

**(d)** $\dfrac{1}{13}\begin{pmatrix} -2 & 7 \\ 1 & 3 \end{pmatrix}$

**(e)** $\dfrac{1}{16}\begin{pmatrix} 2 & 5 \\ -2 & 3 \end{pmatrix}$

**(f)** $\begin{pmatrix} -1 & \frac{1}{2} \\ -2 & \frac{1}{2} \end{pmatrix}$

**(g)** no inverse

**(h)** $\dfrac{1}{12}\begin{pmatrix} -2 & 2 \\ 1 & 5 \end{pmatrix}$

**(i)** $\begin{pmatrix} -1 & 4 \\ 2 & -7 \end{pmatrix}$

**(j)** $\begin{pmatrix} -\frac{5}{4} & \frac{3}{4} \\ 2 & -1 \end{pmatrix}$

**2.** $X = \begin{pmatrix} 5 & 9 \\ -8 & -14 \end{pmatrix}$,

$Y = \begin{pmatrix} -14 & 6 \\ -12 & 5 \end{pmatrix}$

**3.** $P = \begin{pmatrix} \frac{6}{7} & -1 \\ -\frac{8}{7} & 0 \end{pmatrix}$,

$Q = \begin{pmatrix} -\frac{5}{7} & \frac{1}{7} \\ \frac{1}{7} & \frac{11}{7} \end{pmatrix}$

## Exercise 24C  (page 210)

**1.**
(a) $x = 4,\ y = 2$
(b) $x = 4,\ y = 5$
(c) $x = 2,\ y = 3$
(d) $x = 1,\ y = 4$
(e) $x = 2.5,\ y = 4$
(f) $x = 1,\ y = 7$
(g) $x = 3,\ y = 0$
(h) $x = 2,\ y = 3$

## Exercise 24D  (page 212)

**1.**
(a) 12
(b) −8
(c) 0

**(d)** 12

**2.**
(a) $1, -1\frac{1}{2}$
(b) $-1, 2\frac{1}{2}$
(c) $-2, 1, 4$
(d) $2, -4$
(e) $1, 2, -3$
(f) $1, -5 \pm \sqrt{34}$

**3.**
(a) $2(x^3 + a^3)$
(b) $p(q-p)(r-q)$

## Exercise 24E  (page 216)

**1.** $P^{-1} = \begin{pmatrix} 1 & 0 & 0 \\ 1 & 1 & 0 \\ -5 & -2 & 1 \end{pmatrix}$

**2.** no inverse

**3.** $R^{-1} = \begin{pmatrix} -3 & 1 & 1 \\ 2 & -1 & 0 \\ 7 & -2 & -2 \end{pmatrix}$

**4.** $S^{-1} = \dfrac{1}{4}\begin{pmatrix} 30 & -16 & -10 \\ -6 & 4 & 2 \\ -19 & 10 & 7 \end{pmatrix}$

**5.** no inverse

**6.** $U^{-1} = \dfrac{1}{6}\begin{pmatrix} 10 & -2 & -3 \\ 6 & 0 & -3 \\ -4 & 2 & 3 \end{pmatrix}$

## Exercise 24F  (page 217)

**1.** $x = 2, y = 4, z = -1$

**2.** $x = -2, y = 3, z = 1$

**3.** $x = 5, y = 1, z = 1$

## Exercise 24G  (page 219)

**1.** $x = 3, y = -2, z = 1$

**2.** $x = -1, y = 5, z = 2$

**3.** $x = 2, y = 3, z = -1$

## Exercise 24H  (page 225)

**1.**
(a) one, $x = 1, y = -1, z = 1$
(b) infinite, all points satisfying
$x = t, y = t - 1, z = 1$
(c) none
(d) infinite, all points satisfying
$x = 28t, y = 12t, z = -11t$
(e) one, $x = -3, y = 1, z = 4$
(f) one, $x = 1, y = 3, z = -1$
(g) infinite, all points satisfying
$x = t,\ y = -2 - 3t, z = 5t + 11$
(h) none
(i) none
(j) infinite, all points satisfying
$x = \frac{1}{2}(1 - t), y = t, z = \frac{1}{2}(5t + 7)$
(k) $x = 2, y = 1, z = 3$
(l) none

**2.**
(a) (i) $p \neq -10$
  (ii) $p = -10, q = 2$
  (iii) $p = -10, q \neq 2$
(b) (i) $p \neq -8$
  (ii) $p = -8, q = -2\frac{1}{2}$
  (iii) $p = -8, q \neq -2\frac{1}{2}$
(c) (i) $q \neq 1$
  (ii) $q = 1, p = 8$
  (iii) $q = 1\ p \neq 8$
(d) (i) $p \neq 4$
  (ii) $p = 4, q = 8$
  (iii) $p = 4, q \neq 8$
(e) (i) $p \neq 2.8$
  (ii) $p = 2.8,\ q = 4.8$
  (iii) $p = 2.8, q \neq 4.8$
(f) (i) $p \neq -10$
  (ii) $p = -10, q = 0$
  (iii) $p = -10, q \neq 0$
(g) (i) $q \neq -44$
  (ii) $q = -44, p = -\frac{1}{2}$
  (iii) $q = -44\ p \neq -\frac{1}{2}$
(h) (i) $p \neq -\frac{3}{2}$
  (ii) $p = -\frac{3}{2}, q = 14$
  (iii) $p = -\frac{3}{2}, q \neq 14$

**3.**
(i) $p = 3$
(ii) impossible

**4.**
(i) $p = -6, q = -12$; all values
of $x$ and $y$ satisfying the
equation $5x - 3y - 6$
(ii) same equation – same line

**Exercise 24I  (page 228)**

1.  (a) (13, 14, 20)
    (b) (1, 2, 3)

2.  (a) (27, 18, 29)
    (b) (2, 2, 3)

3.  **ABC** = (7650), this represents the total cost in $ of labour and lamination for all rackets per week

4.  (a) $\begin{pmatrix} 6 & 4 & 3 \\ 8 & 5 & 2 \end{pmatrix}$

    (b) $\begin{pmatrix} 0.20 \\ 0.25 \\ 0.35 \end{pmatrix}$

    (c) $\begin{pmatrix} 3.25 \\ 3.55 \end{pmatrix}$

    (d) (6.80) which represents the total cost of the shopping for the two students.

5.  (a) (372 224 368 156)
    (b) total 13,356
    (c) 51, 27, 48, 18
    (d) yes: total laden weight is 144T

6.  Delphians with 26 points

7.  (a) (86 96 102)
    (b) $5294

**Exercise 24X  (page 230)**

1.  (a) (i)
    $\begin{pmatrix} 3 & 2 & -1 \\ 1 & -1 & 2 \\ -2 & -3 & 1 \end{pmatrix}\begin{pmatrix} x \\ y \\ z \end{pmatrix} = \begin{pmatrix} 2 \\ 14 \\ 2 \end{pmatrix}$

    (b) $x = 3, y = -1, z = 5$

2.  $p = 12, q = 4$

3.  (a) $p = 4$

(b) all $x, y$ for which $4x - y = 2$

4.  P (2, 1, 4)

5.  (a) $\begin{pmatrix} 11 & -6 \\ 26 & -12 \end{pmatrix}$

6.  (a) $\left(\begin{array}{ccc|c} 1 & 2 & 3 & 7 \\ 2 & 2 & -1 & 0 \\ 3 & -4 & 2 & 7 \end{array}\right)$

    (b) $\left(\begin{array}{ccc|c} 1 & 2 & 3 & 7 \\ 0 & 2 & 7 & 14 \\ 0 & 0 & 28 & 56 \end{array}\right)$

    (c) (1, 0 , 2)

7.  $x = 1, y = 0, z = 2$

8.  (i) $p^2 \neq 64 \ (p \neq \pm 8)$
    (ii) $p = 8$
    (iii) $p = -8$

9.  proof

10. (a) $\begin{pmatrix} 1 & 1 & -1 \\ 2 & -1 & 1 \\ 3 & 0 & 2 \end{pmatrix}\begin{pmatrix} x \\ y \\ z \end{pmatrix} = \begin{pmatrix} 2 \\ 1 \\ 1 \end{pmatrix}$

    (b) (i) $\begin{pmatrix} -2 & -1 & 3 \\ -2 & 5 & 3 \\ 0 & -3 & -3 \end{pmatrix}$

    (ii) $\begin{pmatrix} -6 & 0 & 0 \\ 0 & -6 & 0 \\ 0 & 0 & -6 \end{pmatrix}$
    (iii)

11. (a) $\begin{pmatrix} 4 & 0 & 0 \\ 0 & 4 & 0 \\ 0 & 0 & 4 \end{pmatrix} = 4\mathbf{I}$

    (b) $\frac{1}{4}\mathbf{B}$

12. (a) $2p + 2q + 6r$
    (b) $c = 2p + 4q$
    $z = 2p + 2q + 6r$
    $b = 6p + 4q + 4r$

    (c) $\begin{pmatrix} 2 & 4 & 0 \\ 2 & 2 & 6 \\ 6 & 4 & 4 \end{pmatrix}\begin{pmatrix} p \\ q \\ r \end{pmatrix} = \begin{pmatrix} c \\ z \\ b \end{pmatrix}$

    (d) $\mathbf{X} = \mathbf{M}^{-1}\mathbf{D}$
    (e) $p = 3, q = 6, r = 2$

# CHAPTER 25

**Exercise 25A**

1.  (a) $2 - 3i$
    (b) $4 + 2i$
    (c) $-3 - 5i$
    (d) $-1 + 8i$
    (e) $1 - 7i$
    (f) $8i$
    (g) $14 + 13i$
    (h) $26 - 20i$
    (i) $14 + 8i$
    (j) $-2 + 26i$
    (k) $9 - 19i$
    (l) $-42 + 2i$
    (m) 4
    (n) 20
    (o) 34

2.  (a) $5 - 2i$
    (b) $-7 - 4i$
    (c) $-2 + 6i$
    (d) $-2 - 6i$
    (e) $-18 + 20i$
    (f) $-50 - 26i$
    (g) $-43 + 6i$
    (h) $-43 - 6i$
    (i) $-27 + 34i$
    (j) $-27 - 34i$
    (k) 10
    (l) $4i$
    (m) $-14$
    (n) 29
    (o) 65

3.  (a) 3
    (b) 1
    (c) $-17$
    (d) 23
    (e) 0

## Exercise 25B

1. (a) $\frac{4}{5} - \frac{7}{5}i$

   (b) $-\frac{7}{13} + \frac{17}{13}i$

   (c) $\frac{1}{17} + \frac{4}{17}i$

   (d) $2 - 3i$

   (e) $\frac{15}{17} + \frac{8}{17}i$

   (f) $\frac{32}{65} + \frac{4}{65}i$

   (g) $\frac{1}{29} + \frac{41}{29}i$

   (h) $\frac{4}{5} - \frac{17}{5}i$

   (i) $-\frac{41}{73} - \frac{12}{73}i$

   (j) $-\frac{4}{17} + \frac{18}{17}i$

2. (a) $\pm(3 - 2i)$

   (b) $\pm(6 + i)$

   (c) $\pm(5 - 3i)$

   (d) $\pm(5 - 2i)$

   (e) $\pm(1 + 2i)$

   (f) $\pm(4 - 3i)$

   (g) $\pm(2 + 3i)$

   (h) $\pm\left(\frac{1}{\sqrt{2}} + \frac{1}{\sqrt{2}}i\right)$

3. (a) $4 \pm 2i$

   (b) $\frac{1}{2} \pm i$

   (c) $-1 \pm 2i$

4. (a) $x^2 - 6x + 10 = 0$

   (b) $x^2 - 2x + 5 = 0$

   (c) $x^2 + 2x + 10 = 0$

   (d) $16x^2 - 16x + 5 = 0$

5. $3 + 2i$, $x^2 - 6x + 13$

6. (a) $z^2 - (6 - i)z + 11 - 3i$

   (b) $z^2 - (2 + 3i)z - 1 + 3i$

   (c) $z^2 + (3 + 2i)z - 7 + 17i$

7. $z^2 - 3(1 + i)z + 2 + 3i$

## Exercise 25C

1. $z_1 = 4 + 2i$, $z_2 = -1 + 3i$, $z_3 = 2 - i$, $z_4 = -1 - 2i$

2.

3. (a) $\sqrt{2}, \frac{1}{4}\pi$

   (b) $5, -0.93$

   (c) $2\sqrt{2}, \frac{3}{4}\pi$

   (d) $2, -\frac{1}{3}\pi$

   (e) $13, -1.97$

   (f) $3, \frac{1}{2}\pi$

   (g) $4, \pi$

   (h) $2, -\frac{1}{2}\pi$

   (i) $5, 0$

   (j) $4, -\frac{5}{6}\pi$

4. (a) $5\sqrt{2}, -0.14$

   (b) $1, -\frac{1}{2}\pi$

   (c) $1, -1.43$

   (d) $\sqrt{2}, -\frac{3}{4}\pi$

   (e) $\sqrt{10}, 0.32$

5. (a) $4, \frac{1}{3}\pi$

   (b) $\frac{1}{2}, -\frac{1}{6}\pi$

6. (a) $2\sqrt{10}$

   (b) $10$

   (c) $\frac{1}{5}\sqrt{10}$

7. (a) $\sqrt{97}$

   (b) $\frac{1}{2}\sqrt{2}$

8. (a) $-\frac{1}{2} + \frac{1}{2}i$

   (b) $\frac{1}{2}\sqrt{2}, \frac{3}{4}\pi$

9. (a) $1 - 3i$

   (b) $\frac{1}{2} - \frac{1}{2}i$

   (c) $\frac{1}{2}\sqrt{2}, -\frac{1}{4}\pi$

10. (a) $2$

    (b) $\frac{5}{6}\pi$

    (c) $-\frac{1}{3}\pi$

11. (a) $-2 + 4i$

    (b) $\frac{1}{15} - \frac{1}{30}i$

12. (b) $\frac{1}{2}\sqrt{2}, -\frac{3}{4}\pi$

13. (a) $6$

    (b) $x + \dfrac{x}{x^2 + y^2}$,

    $y - \dfrac{y}{x^2 + y^2}$

## Exercise 25D

1. (a) $(x - 2)^2 + y^2 = 9$; circle centre $(2, 0)$, radius 3

   (b) straight line $2x = 5$

   (c) $(x - 1)^2 + (y - 1)^2 = 1$; circle, centre $(1, 1)$, radius 1

   (d) straight line $y = 2$

   (e) line $y = x$ in first quadrant

   (f) line $\sqrt{3}\, y = x - 2$ in first quadrant

2. (a) circle $3x^2 + 3y^2 - 2x - 4y = 0$;

   (b) $x^2 + y^2 + 2x + 8y + 16 = 0$; circle, centre $(-1, -4)$, radius 1

   (c) $x^2 + y^2 = 1$; circle, centre $(0, 0)$, radius 1

   (d) straight line (that part below the $x$-axis) $y + \sqrt{3}\, x + 3\sqrt{3} = 0$

   (e) straight line $x = y$

   (f) straight line $4y = 1$

3. circle $x^2 + y^2 + 4x = 0$

## Exercise 25E

1. $\tan 3\theta = \dfrac{3\tan\theta - \tan^3\theta}{1 - 3\tan^2\theta}$

2. proof

3. (a) $\cos 5\theta + i\sin 5\theta$

   (b) $\cos 3\theta - i\sin 3\theta$

   (c) $\cos 4\theta - i\sin 4\theta$

   (d) $-\sin 3\theta - i\cos 3\theta$

4.  **(a)** $-4(1 + i)$
    **(b)** $8(i + i)$
    **(c)** $-8(1 - i\sqrt{3})$
    **(d)** $\frac{1}{8}(1 - i\sqrt{3})$

5.  $-64i$

6.  $\frac{1}{2}(1 \pm i\sqrt{3})$

7.  **(a)** $\cos 2\theta + i \sin 2\theta$
    **(b)** $2\cos 4\theta + 2i \sin 4\theta$
    **(c)** $5\cos\theta - 5i \sin\theta$

8.  **(a)** $-1$
    **(b)** $\frac{1}{2}\sqrt{2} - \frac{1}{2}\sqrt{2}i$
    **(c)** $ei$

9.  **(a)** $e^{i3\theta}$
    **(b)** $\sqrt{2}e^{-i\frac{1}{4}\pi}$
    **(c)** $e^{i\frac{1}{2}\pi}$
    **(d)** $4e^{i\frac{5}{3}\delta}$

10. 11. proofs

## Exercise 25X

1.  **(i)** $10 + 10i$
    **(ii)** $10\sqrt{2}, \frac{1}{4}\pi$
    **(iii)** $iw = -10 + 10i, \overline{w} = 10 - 10i$
    **(iv)** $20\sqrt{2}$

2.  $\dfrac{x^2 + y^2 - 1}{(x+1)^2 + y^2} + \dfrac{2iy}{(x+1)^2 + y^2}$

    **(i)** $\dfrac{2y}{x^2 + y^2 - 1} = 1$

    **(ii)** $x^2 + y^2 - 2y - 1 = 0$,
    $x^2 + (y-1)^2 = 2$ which is a circle

    **(iii)** circle, centre $(0, 1)$,
    radius $\sqrt{2}$

3.  **(a)** $\pm(3 - 2i)$
    **(b)** **(i)**

    $\text{Re } w = x\left(1 + \dfrac{1}{x^2 + y^2}\right)$

    $\text{Im } w = y\left(1 - \dfrac{1}{x^2 + y^2}\right)$

**(ii)** $x^2 + y^2 = 1$ (circle, centre origin, radius 1)

4.  **(a)** $\dfrac{x^2 + y^2 + x - 2}{(x+2)^2 + y^2} + \dfrac{3iy}{(x+2)^2 + y^2}$

    **(b)** **(i)** $x^2 + y^2 + x - 3y - 2 = 0$
    **(ii)** $(x + \frac{1}{2})^2 + (y - \frac{3}{2})^2 = 4\frac{1}{2}$
    **(iii)** centre $(-\frac{1}{2}, \frac{3}{2})$, radius $\frac{3}{2}\sqrt{2}$

5.  $\pm(1 + 2i)$

6.  **(a)** $13$
    **(b)** $u = 11, v = 2$

7.  $x = \frac{4}{5}, y = \frac{8}{5}$

8.  $|k| < 2 \ (-2 < k < 2)$

9.  $|k| < 3 \ (-3 < k < 3)$

10  $\frac{18}{25} + \frac{1}{25}i$

11. $\sqrt{13}$

12. **(a)** **(i)** $5 + i$
    **(ii)** $18 - i$
    **(iii)** $-\frac{6}{25} - \frac{17}{25}i$
    **(b)** $x^2 - (5 + i)x + 18 - i = 0$

# APPENDIX 3

| Question | Answer | Question | Answer | Question | Answer | Question | Answer | Question | Answer |
|---|---|---|---|---|---|---|---|---|---|
| 1 | C | 21 | C | 41 | A | 61 | C | 81 | D |
| 2 | C | 22 | A | 42 | B | 62 | A | 82 | D |
| 3 | D | 23 | B | 43 | B | 63 | D | 83 | B |
| 4 | B | 24 | D | 44 | C | 64 | B | 84 | A |
| 5 | A | 25 | B | 45 | C | 65 | C | 85 | B |
| 6 | C | 26 | D | 46 | A | 66 | A | 86 | A |
| 7 | C | 27 | C | 47 | C | 67 | A | 87 | D |
| 8 | A | 28 | C | 48 | C | 68 | D | 88 | B |
| 9 | B | 29 | D | 49 | D | 69 | B | 89 | B |
| 10 | C | 30 | B | 50 | B | 70 | C | 90 | C |
| 11 | B | 31 | B | 51 | C | 71 | C | 91 | A |
| 12 | D | 32 | D | 52 | C | 72 | D | 92 | C |
| 13 | D | 33 | A | 53 | B | 73 | C | 93 | B |
| 14 | B | 34 | C | 54 | D | 74 | A | 94 | D |
| 15 | A | 35 | D | 55 | B | 75 | C | 95 | A |
| 16 | C | 36 | C | 56 | A | 76 | A | 96 | C |
| 17 | B | 37 | B | 57 | A | 77 | B | 97 | A |
| 18 | A | 38 | C | 58 | C | 78 | B | 98 | C |
| 19 | B | 39 | B | 59 | A | 79 | C | 99 | B |
| 20 | C | 40 | B | 60 | D | 80 | A | 100 | D |

# APPENDIX 4
## Paper 1

| Item | Key | Item | Key | Item | Key |
|---|---|---|---|---|---|
| 1 | D | 16 | D | 31 | D |
| 2 | A | 17 | A | 32 | C |
| 3 | C | 18 | A | 33 | B |
| 4 | D | 19 | C | 34 | B |
| 5 | A | 20 | C | 35 | D |
| 6 | B | 21 | C | 36 | A |
| 7 | B | 22 | B | 37 | D |
| 8 | C | 23 | B | 38 | B |
| 9 | D | 24 | A | 39 | C |
| 10 | A | 25 | B | 40 | A |
| 11 | B | 26 | B | 41 | A |
| 12 | C | 27 | C | 42 | C |
| 13 | D | 28 | C | 43 | C |
| 14 | A | 29 | D | 44 | B |
| 15 | C | 30 | C | 45 | D |

**Paper 2**

1. (a) (ii) $-e^{-x}[x^2+2x+2]+c$

(b) (i) a) $\dfrac{3}{1+9x^2}$

b) $\frac{2}{3}\tan^{-1}(3x)+\frac{1}{18}\ln(1+9x^2)+k$

(c) $\frac{1}{2}\ln(4+y^2)=\ln\tan x+c$

2. (a) (i) $k=\frac{1}{37}\ln 2$

(ii) $2.5(\ln 2)^{\frac{53}{37}}$

(iii) 2024

(b) (i) $u=2,\ v=1$

(ii) $y=Ae^{-3x}+Be^{-x}+\sin 3x+2\cos 3x$

3. (a) (i) $A=\frac{1}{2},\ B=-\frac{1}{2}$

(ii) $\dfrac{1}{2}\left(1-\dfrac{1}{2n+1}\right)$

(iii) $\frac{1}{2}$

(b) (i) $n(3n-1)$

4. (a) (i) $r=\frac{1}{8}$

(ii) $\dfrac{4}{7}\left[1-\dfrac{1}{2^{3n}}\right]$

(b) $\cos 2x=1-2x^2+\frac{2}{3}x^4$

(c) (i) $1+x+\frac{1}{2}x^2+\frac{1}{2}x^3$

(ii) $7.14141,\ |x|<1$

5. (a) (i) $\frac{2}{9}$

(ii) $\frac{5}{9}$

(b) (i) a) 0.547

b) 0.76

c) 0.267

d) 0.329

(ii) 0.297

6. (a) $-4,\ 1\pm\sqrt6$

(b) $x_1=3,\ x_2=-5,\ x_3=1$

(c) (i) $1+i$

# Index

diagram

ed in the United States
V00001B/71-252/P